SUPPLY CHAIN RISK

A Handbook of Assessment,
Management, and Performance

INT. SERIES IN OPERATIONS RESEARCH & MANAGEMENT SCIENCE

Series Editor: Frederick S. Hillier, Stanford University
Special Editorial Consultant: Camille C. Price, Stephen F. Austin State University
Titles with an asterisk (*) were recommended by Dr. Price

Bouyssou et al/ *EVALUATION AND DECISION MODELS WITH MULTIPLE CRITERIA: Stepping stones for the analyst*
Blecker & Friedrich/ *MASS CUSTOMIZATION: Challenges and Solutions*
Appa, Pitsoulis & Williams/ *HANDBOOK ON MODELLING FOR DISCRETE OPTIMIZATION*
Herrmann/ *HANDBOOK OF PRODUCTION SCHEDULING*
Axsäter/ *INVENTORY CONTROL, 2nd Ed.*
Hall/ *PATIENT FLOW: Reducing Delay in Healthcare Delivery*
Józefowska & Węglarz/ *PERSPECTIVES IN MODERN PROJECT SCHEDULING*
Tian & Zhang/ *VACATION QUEUEING MODELS: Theory and Applications*
Yan, Yin & Zhang/ *STOCHASTIC PROCESSES, OPTIMIZATION, AND CONTROL THEORY APPLICATIONS IN FINANCIAL ENGINEERING, QUEUEING NETWORKS, AND MANUFACTURING SYSTEMS*
Saaty & Vargas/ *DECISION MAKING WITH THE ANALYTIC NETWORK PROCESS: Economic, Political, Social & Technological Applications w. Benefits, Opportunities, Costs & Risks*
Yu/ *TECHNOLOGY PORTFOLIO PLANNING AND MANAGEMENT: Practical Concepts and Tools*
Kandiller/ *PRINCIPLES OF MATHEMATICS IN OPERATIONS RESEARCH*
Lee & Lee/ *BUILDING SUPPLY CHAIN EXCELLENCE IN EMERGING ECONOMIES*
Weintraub/ *MANAGEMENT OF NATURAL RESOURCES: A Handbook of Operations Research Models, Algorithms, and Implementations*
Hooker/ *INTEGRATED METHODS FOR OPTIMIZATION*
Dawande et al/ *THROUGHPUT OPTIMIZATION IN ROBOTIC CELLS*
Friesz/ *NETWORK SCIENCE, NONLINEAR SCIENCE and INFRASTRUCTURE SYSTEMS*
Cai, Sha & Wong/ *TIME-VARYING NETWORK OPTIMIZATION*
Mamon & Elliott/ *HIDDEN MARKOV MODELS IN FINANCE*
del Castillo/ *PROCESS OPTIMIZATION: A Statistical Approach*
Józefowska/*JUST-IN-TIME SCHEDULING: Models & Algorithms for Computer & Manufacturing Systems*
Yu, Wang & Lai/ *FOREIGN-EXCHANGE-RATE FORECASTING WITH ARTIFICIAL NEURAL NETWORKS*
Beyer et al/ *MARKOVIAN DEMAND INVENTORY MODELS*
Shi & Olafsson/ *NESTED PARTITIONS OPTIMIZATION: Methodology And Applications*
Samaniego/ *SYSTEM SIGNATURES AND THEIR APPLICATIONS IN ENGINEERING RELIABILITY*
Kleijnen/ *DESIGN AND ANALYSIS OF SIMULATION EXPERIMENTS*
Førsund/ *HYDROPOWER ECONOMICS*
Kogan & Tapiero/ *SUPPLY CHAIN GAMES: Operations Management and Risk Valuation*
Vanderbei/ *LINEAR PROGRAMMING: Foundations & Extensions, 3rd Edition*
Chhajed & Lowe/ *BUILDING INTUITION: Insights from Basic Operations Mgmt. Models and Principles*
Luenberger & Ye/ *LINEAR AND NONLINEAR PROGRAMMING, 3rd Edition*
Drew et al/ *COMPUTATIONAL PROBABILITY: Algorithms and Applications in the Mathematical Sciences*
Chinneck/ *FEASIBILITY AND INFEASIBILITY IN OPTIMIZATION: Algorithms and Computation Methods*
Tang, Teo & Wei/ *SUPPLY CHAIN ANALYSIS: A Handbook on the Interaction of Information, System and Optimization*
Ozcan/ *HEALTH CARE BENCHMARKING AND PERFORMANCE EVALUATION: An Assessment using Data Envelopment Analysis (DEA)*
Wierenga/ *HANDBOOK OF MARKETING DECISION MODELS*
Agrawal & Smith/ *RETAIL SUPPLY CHAIN MANAGEMENT: Quantitative Models and Empirical Studies*
Brill/ *LEVEL CROSSING METHODS IN STOCHASTIC MODELS*

~A list of the early publications in the series is found at the end of the book~

SUPPLY CHAIN RISK

A Handbook of Assessment,
Management, and Performance

edited by

George A. Zsidisin
Bob Ritchie

 Springer

Editors
George A. Zsidisin
Bowling Green State University
Bowling Green, OH
USA

Bob Ritchie
Lancashire Business School
University of Central Lancashire
United Kingdom

ISSN: 0884-8289
ISBN: 978-0-387-79933-9   e-ISBN: 978-0-387-79934-6
DOI: 10.1007/978-0-387-79933-9

Library of Congress Control Number: 2008930187

© 2008 by Springer Science+Business Media, LLC
All rights reserved. This work may not be translated or copied in whole or in part without the written permission of the publisher (Springer Science+Business Media, LLC, 233 Spring Street, New York, NY 10013, USA), except for brief excerpts in connection with reviews or scholarly analysis. Use in connection with any form of information storage and retrieval, electronic adaptation, computer software, or by similar or dissimilar methodology now know or hereafter developed is forbidden.
The use in this publication of trade names, trademarks, service marks and similar terms, even if the are not identified as such, is not to be taken as an expression of opinion as to whether or not they are subject to proprietary rights.

Printed on acid-free paper

springer.com

To my beautiful wife Laural, for her love and patience with me, and my three wonderful sons, Nick, Lucas and Blaise, who teach me every day the importance of life.

George A. Zsidisin

To my wonderful wife and soul-mate Celia whose patience, support and love make each day worth living.

Bob Ritchie

Contents

Acknowledgements xvii

Chapter 1: Supply Chain Risk Management – Developments, Issues and Challenges 1
George A. Zsidisin and Bob Ritchie
1.1 Introduction 1
1.2 Background to SCRM 2
1.3 Supply Chain Risk 3
1.4 Structure of the Handbook 5
   1.4.1 Risk Analysis, Assessment and Tools 5
   1.4.2 Supply Chain Design and Risk 7
   1.4.3 Supply Chain Risk Management 8
   1.4.4 Supply Chain Security 10
1.5 Background of ISCRiM and Handbook 11
1.6 Future Developments in SCRM 11

SECTION ONE - RISK ANALYSIS, ASSESSMENT AND TOOLS 13

Chapter 2: Assessing the Vulnerability of Supply Chains 15
Bjørn Egil Asbjørnslett
2.1 Introduction 15
   2.1.1 Background 16
   2.1.2 Objective 16
   2.1.3 Approach – Mission-Oriented 17
2.2 Concepts and Definitions 18
   2.2.1 Search for Effectiveness and Efficiency 19
   2.2.2 Vulnerability Analysis versus Risk Analysis 21
2.3 Vulnerability of Supply Chains 21
   2.3.1 The mission of Logistics and SCM 21
   2.3.2 Resilience in Supply Chain Systems and SCM 22
   2.3.3 Factors Contributing to Vulnerability and Resilience 23
2.4 Vulnerability Analysis 23
   2.4.1 Flow Sheet of the Analysis 23
      2.4.1.1 Step One: Definition of Scope of Work 25
      2.4.1.2 Step Two: Description of Context 25
      2.4.1.3 Step Three: Taxonomy Development 26
      2.4.1.4 Step Four: Scenario Development 27

2.4.2 Documenting Scenarios in the Analysis 28
　　　　2.4.2.1 Step Five: Criticality Ranking 29
　　　　2.4.2.2 Step Six: Scenarios of Importance 29
　　　　2.4.2.3 Step Seven: Potential for Reducing Likelihood and
　　　　　　　　Consequence 30
　　2.4.3 The Second Round 32
　2.5 Summary 32
　References 32

Chapter 3:　Risk Management in Value Networks 35
　　　　　　Jukka Hallikas and Jari Varis
　3.1 Introduction 35
　3.2 Value Networks 36
　　3.2.1 Business Ecosystem Health and Risks 37
　　3.2.2 Value Creation and Risk Management in Networks 39
　3.3 Connecting Industry Change with Risk Analysis 43
　　3.3.1 Trend Analysis on Network Level: Case ICT-Industry 44
　　3.3.2 Future Drivers and Scenario Creation 45
　3.4 Risk Profiles of Interconnected Actors 47
　3.5 Discussion 51
　References 52

Chapter 4:　Predicting and Managing Supply Chain Risks 53
　　　　　　Samir Dani
　4.1 Introduction 53
　4.2 Uncertainty and Risk 54
　4.3 Risks in the Supply Chain 55
　4.4 Supply Chain Risk Management 56
　4.5 Proactive Supply Chain Risk Management 58
　4.6 Predicting Supply Chain Risks 59
　　4.6.1 Tools for Risk Prediction 59
　　　　4.6.1.1　Data Mining 59
　　　　4.6.1.2　Failure Mode Effect Analysis (FMEA) 61
　4.7 The Predictive – Proactive Methodology 62
　4.8 Conclusion 64
　References 64

Chapter 5:　Assessing Risks in Projects and Processes 67
　　　　　　Barbara Gaudenzi
　5.1 Introduction 67
　5.2 Managing Uncertainty and Risks in Projects and Processes 70
　　5.2.1 Project and Process Objectives 71
　　5.2.2 Project and Process Management 71
　　5.2.3 Stakeholder Interests 72
　　5.2.4 Performance Measurement 72

5.3 Risk Assessment 74
    5.3.1 Step 1 76
    5.3.2 Step 2 76
    5.3.3 Step 3 77
    5.3.4 Step 4 78
    5.3.5 Step 5 78
5.4 Some Risk Factors in Processes and Projects 79
5.5 Conclusions 80
References 81

Chapter 6:   Risk Management System – A Conceptual Model 83
             Arben Mullai
6.1 Introduction 83
6.2 Variations in Terms and Definitions 84
6.3 A Unified Concept of the Risk Management System 85
6.4 Main Elements of the Risk Management System 86
6.5 Risk Analysis 88
    6.5.1 Preparing/Setting Up 88
    6.5.2 Analysis Process 89
    6.5.3 Conclusions and Recommendations 91
6.6 Risk Evaluation 91
    6.6.1 Select Criteria 91
    6.6.2 Compare Risks 92
    6.6.3 Rank Risks 93
    6.6.4 Develop Strategies and Measures 93
6.7 Risk Management 94
    6.7.1 Identify Options 94
    6.7.2 Decision-Making 95
    6.7.3 Planning 95
    6.7.4 Implement 96
    6.7.5 Follow-Up and Monitoring 96
6.8 Risk Communication 96
6.9 Re-Assessment – A Continuous Cyclic Process 98
6.10 Summary 98
References 98

Chapter 7:   Using Simulation to Investigate Supply Chain Disruptions ... 103
             Steven A. Melnyk, Alexander Rodrigues, and Gary L. Ragatz
7.1 Introduction 103
7.2 Understanding Supply Chain Disruptions: Background 104
7.3 Modeling Supply Chain Disruptions: A Framework 106
    7.3.1 The Triggering Event 106
    7.3.2 The Simulation Model 110
7.4 Parameters and Experimental Factors 111
    7.4.1 Experimental Factors 111
    7.4.2 Performance Measures 112

7.5 Generating and Analyzing Simulation Data 113
7.6 Classical Statistical Analysis of Simulation-Based Experimental Data ... 114
7.7 Intervention Analysis: Evaluating the Transient Response 116
7.8 Outlier Detection 117
7.9 Dealing with Replications: Time Series Considerations 117
7.10 Other Applications 118
7.11 Methodological Triangulation 118
7.12 Concluding Comments 119
References 120

SECTION TWO - SUPPLY CHAIN DESIGN AND RISK 123

Chapter 8: Single Versus Multiple Sourcing: A Supply Risk Management Perspective 125
Constantin Blome and Michael Henke
8.1 Introduction 125
8.2 The Decision of Single and Multiple Sourcing 126
8.3 Evaluation of the Single and Multiple Sourcing Decision
    from a Risk Perspective 128
8.4 Integrating Supply Risk Management into an Enterprise
    Risk Management System 133
References 134

Chapter 9: The Role of Product Design in Global Supply Chain Risk Management 137
Omera Khan, Martin Christopher, and Bernard Burnes
9.1 Introduction 137
9.2 Trends in the Textile and Clothing Industry 138
9.3 The Trend to Global/Off Shore Sourcing 139
9.4 Research Method 141
9.5 Case Study of Marks and Spencer: Part One 143
   9.5.1 Company Introduction 143
   9.5.2 A Shift to Global Sourcing 144
   9.5.3 Central Procurement 145
   9.5.4 The Impact of Changed Supplier Relationships 145
9.6 Case Study: Part Two 146
   9.6.1 Intelligent Risk Taking 146
   9.6.2 Consolidation of the M&S Supply Chain 147
   9.6.3 Internalising the Design Process to Manage Supply Chain Risk 148
9.7 The Hidden Risks of Off Shore Sourcing 149
9.8 Strategies for Managing Risk in a Global Market 150
9.9 Conclusions 151
References 152

Chapter 10: How Much Flexibility Does It Take to Mitigate Supply
	Chain Risks? 155
	Christopher Tang and Brian Tomlin
	10.1 Introduction.. 155
	10.2 Supply Chain Risks and the Role of Flexibility 157
		10.2.1 Supply Risks.. 157
		10.2.2 Process Risks... 158
		10.2.3 Demand Risks 159
		10.2.4 Rare-but-Severe Disruption Risks 160
		10.2.5 Other Risks.. 161
		10.2.6 The Role of Flexibility in Mitigating Risks.............. 163
	10.3 The Power of Flexibility: How Much Flexibility Do You Need?......... 164
		10.3.1 Supply-Cost Risk: The Benefit of Flexibility via Multiple
			Suppliers.. 166
		10.3.2 Supply-Commitment Risk: The Benefit of Flexibility
			via a Flexible Supply Contract................................... 166
		10.3.3 Process Risk: The Benefit of Flexibility via Flexible
			Manufacturing Processes 167
		10.3.4 Demand Risk: The Power of Flexibility via Postponement 168
		10.3.5 Demand Risk: The Power of Flexibility via Responsive Pricing.. 169
	10.4. Concluding Remarks 170
	References... 172

SECTION THREE - SUPPLY CHAIN RISK MANAGEMENT 175

Chapter 11: Enterprise and Supply Risk Management...................................... 177
	Michael Henke
	11.1 Introduction: Enterprise Risk Management
		for Supply Management 177
	11.2 ERM Starting Points 179
	11.3 ERM Monitoring and Managing of Supply Risks................. 181
	11.4 Advances in Operationalizing Tasks for Supply Controlling
		and Risk Management 183
	11.5 Conclusions... 184
	References... 185

Chapter 12: Pre-Contract Risk in International PFI Projects 187
	Simon A. Burtonshaw-Gunn
	12.1 Introduction.. 187
	12.2 Construction Prime Contracting and the Management of Risk.............. 189
	12.3 Risks in International PFI Projects 192
	12.4 Conclusion 195
	References... 197

Chapter 13: Supply Chain Risk Management for Small
and Medium-Sized Businesses 199
Uta Jüttner and Arne Ziegenbein
13.1 Introduction.. 199
13.2 The Research Project 200
   13.2.1 Researcher-Driven Objective 201
   13.2.2 Project Plan 201
13.3 Requirements for Managing Supply Chain Risks in SMEs 202
   13.3.1 The Four Partnering Companies 202
   13.3.2 Findings from the Initial Supply Chain Risk Analyses ... 204
   13.3.3 Requirements for the Practical SCRM Methodology 206
   13.3.4 Evaluating Existing Supply Chain Risk Management
         Approaches.. 206
       13.3.4.1. Scope of Analysis 206
       13.3.4.2 Objectives 207
       13.3.4.3 Implementation 208
13.4 A Practical SCRM Approach for SME 209
   13.4.1 Phase 1 – Identification of Supply Chain Risks............... 211
       13.4.1.1 Step 1.1 – Defining the Supply Chain for Analysis 211
       13.4.1.2 Step 1.2 – Supply Chain Mapping........................ 212
       13.4.1.3 Step 1.3 – Identification of Relevant Supply Chain Risks... 212
   13.4.2 Phase 2 – Assessment of Supply Chain Risks 213
       13.4.2.1 Step 2.1 – Assessment of Supply Chain Risks
          and Mitigation Measures 213
       13.4.2.2 Step 2.2 – Analysis of the Supply Chain Portfolio............... 213
   13.4.3 Phase 3 – Supply Chain Risk Mitigation 214
       13.4.3.1 Step 3.1 – Identification of Mitigation Actions................... 214
       13.4.3.2 Step 3.2 – Assessment of Mitigation Actions 214
       13.4.3.3 Step 3.3 – Decision on Mitigation Actions
          and Action Plans... 215
13.5 Conclusion 216
References.. 216

Chapter 14: Psychological Foundations of Supply Chain Risk Management..... 219
Michael E. Smith
14.1 Introduction.. 219
14.2 Perception in Risk... 219
14.3 The Challenged Decision Maker 220
14.4 The Shortcuts We Take 221
14.5 The Number of Suppliers Issue – Reconsidered 225
14.6 Practical Steps for Supply Management Professionals 230
References.. 231

Chapter 15: Behavioural Risks in Supply Networks 235
    M. Seiter
  15.1 Introduction 235
  15.2 Conceptual Model 236
  15.3 Research Methodology 240
    15.3.1 Instrument 240
    15.3.2 Sample and Data Collection 240
    15.3.3 Scale Development 240
  15.4 Results 241
    15.4.1 Analysis of the Measures and Model Fit 241
    15.4.2 Structural Model Results 242
  15.5 Discussion 243
  15.6 Managerial Implications 244
  15.7 Conclusions 244
  References 245

Chapter 16: SCRM and Performance – Issues and Challenges 249
    Bob Ritchie and Clare Brindley
  16.1 Introduction 249
  16.2 Risk and Performance 251
  16.3 Risk – Uncertainty and Risk Management 252
  16.4 Performance 254
    16.4.1 Timeframe – Risk and Performance 254
  16.5 Performance and Risk Metrics 256
  16.6 Risk-Performance: Sources, Profiles and Drivers 259
  16.7 Risk Management Responses 261
  16.8 Risk and Performance Outcomes 262
  16.9 Empirical Application – Illustrative Case 263
    16.9.1 Case Study 264
    16.9.2 Supply Chain Members' Awareness of SCRM 264
    16.9.3 Risk Drivers – Identification, Evaluation and Prioritization 265
    16.9.4 Perception of the Risk/Performance Interaction 265
    16.9.5 Risk Management Responses 266
  16.10 Conclusions and Future Developments 267
  References 269

Chapter 17: Dominant Risks and Risk Management Practices in Supply Chains 271
    Stephan M. Wagner and Christoph Bode
  17.1 Introduction 271
  17.2 Nomenclature and Conceptual Framework 273
    17.2.1 Supply Chain Risk 273
    17.2.2 Supply Chain Disruption and Supply Chain Risk Sources 274
      17.2.2.1 Demand Side Risks 275
      17.2.2.2 Supply Side Risks 276

   17.2.2.3 Regulatory, Legal, and Bureaucratic Risk...........................276
   17.2.2.4 Infrastructure Risks 277
   17.2.2.5 Catastrophic Risks 277
  17.2.3 Supply Chain Vulnerability and Its Drivers 278
  17.2.4 Supply Chain Risk Management..279
   17.2.4.1 Cause-Oriented Supply Chain Risk Management Practices 279
   17.2.4.2 Effect-Oriented Supply Chain Risk Management Practices 280
 17.3 Questionnaire Development and Data Collection 281
 17.4 Results and Discussion 283
  17.4.1 Supply Chain Risks..283
  17.4.2 Supply Chain Risk Management..285
 17.5 Managerial Implications and Conclusions 286
 References..287

SECTION FOUR - SUPPLY CHAIN SECURITY 291

Chapter 18: Food Supply Chain Security: Issues and Implications 293
   Douglas Voss and Judith Whipple
 18.1 Introduction..293
 18.2 Importance and Challenges of Food Supply Chain Security 295
 18.3 Why Firms Focus on Supply Chain Security 295
 18.4 Security Best Practices in the Food Industry...297
 18.5 Encouraging Security Enhanced Supply Chains 301
 18.6 The Impact of Security of Supplier Selection in the Food Industry.......302
 18.7 Conclusion 303
 References..304

Chapter 19: Supply Chain Security: A Dynamic Capabilities Approach 307
   Chad Autry and Nada Sanders
 19.1 Introduction..307
 19.2 Supply Chain Security Concepts 309
  19.2.1 Supply Chain Security and Security Management 309
  19.2.2 Supply Chain Security Orientation 309
 19.3 Securing Supply Chains Through Capability Development 312
 19.4 Building Supply Chain Continuity Through Security
   Management Capabilities 314
  19.4.1 Processes 315
   19.4.1.1 Build Security into the Process..315
   19.4.1.2 Supply Chain Visibility 316
   19.4.1.3 Sourcing Strategy 317
   19.4.1.4 Inventory Management..318
   19.4.1.5 Product and Process Redesign..318
   19.4.1.6 Demand Based Management..319
   19.4.1.7 Forecasting...320

    19.4.2 Technology 321
        19.4.2.1 Radio Frequency Identification 322
        19.4.2.2 Specific Security Applications of Technology 323
    19.4.3 Human Resources 325
        19.4.3.1 Top Management Support 326
        19.4.3.2 Employee 326
        19.4.3.3 Suppliers 326
        19.4.3.4 Governmental Security Initiatives 327
  19.5 Conclusion 328
  References 328

Chapter 20: Securing Global Food Distribution Networks 331
    Nicole Mau and Markus Mau
  20.1 Challenges in Providing Food Safety and Quality 331
  20.2 Problems in the Food Supply Chain 332
  20.3 Requirement Shift Through Globalized Procurement
       and BtoB-Solutions 334
  20.4 Risk Management 335
  20.5 Vertical Integration Versus Risk Management 336
  20.6 Efforts to Solve Supply Chain Security Problems 337
  20.7 A Potential Solution 338
  20.8 Setup of Supply Chain Security 341
  References 344

Index 345

ACKNOWLEDGEMENTS

This book consists of the collective work of many individuals and organizations. It would have been impossible to write and edit this book on supply chain risk without the contributions and efforts of numerous scholars and practitioners. There are many individuals to thank, and we apologise in advance for potentially omitting some contributors. First, we would like to thank the members and associates affiliated with the International Supply Chain Risk Management (ISCRiM) network (described in greater detail in Chap. 1), as well as the Universities that have sponsored its annual meetings during the past 8 years. These schools include Manchester Metropolitan University, Lund Institute of Technology, Cranfield University, the European Business School in Germany, Lappeenranta University, and the Norwegian University of Science and Technology. Next, we would like to thank all of the authors who have contributed chapters to this book: Bjørn Asbjørnslett, Jukka Hallikas, Jari Varis, Samir Dani, Barbara Gaudenzi, Arben Mullai, Steven A. Melnyk, Alexander Rodrigues, Gary L. Ragatz, Constantin Blome, Michael Henke, Omera Khan, Martin Christopher, Bernard Burnes, Christopher Tang, Brian Tomlin, Simon A. Burtonshaw-Gunn, Uta Jüttner, Arne Ziegenbein, Michael Smith, Mischa Seiter, Clare Brindley, Stephan M. Wagner, Christoph Bode, Douglas Voss, Judith Whipple, Chad Autry, Nada Sanders, Nicole Mau, and Markus Mau. If it was not for their diligent work and patience, we would never have been able to have this book in the first place. We would like to thank the numerous companies and individuals who participated in the various research projects on supply chain risk. Their insights and generosity of time allowed the authors of these chapters to gather and analyze data that formed the genesis of this book. We would also like to thank Springer publishers for their willingness to work with us on this project, specifically Fred Hillier, Gary Folven, Carolyn Ford, and Concetta Seminara-Kennedy for their guidance and patience in publishing this book. In addition, we would like to thank Jon Petro for his tireless work in formatting this book. Last but never least we would like to thank our families for all of their support during the writing of this book. Without them and their support this project would never have been completed.

Chapter 1: Supply Chain Risk Management – Developments, Issues and Challenges

George A. Zsidisin* and Bob Ritchie

*Corresponding author: Department of Management, Bowling Green State University, USA

1.1 Introduction

The management of risk in supply chains has now become an established, albeit fairly recently, element in the fields of Supply Chain Management (SCM), corporate strategic management and Enterprise Risk Management (ERM). In addition to such cross-functional contributions, Supply Chain Risk Management (SCRM) contributes to the decision making processes in most functional areas within a business (e.g. marketing decisions concerning product delivery lead times; health and safety management within production operations). The essence of this Handbook is the capture, interpretation and dissemination of the latest developments in research, practice and policy in what is proving to be a very rapidly developing field. The text is designed to appeal to researchers, scholars, policy makers and practitioners alike, whilst seeking to ensure that what is presented is well grounded in robust empirical methodologies and evidence or accurately represents the structures, practices and processes employed in industry. Like all developing fields of study, SCRM draws on other disciplines and fields of study. Hence, many of the contributing authors approach the issues from distinct and differing perspectives.

This opening chapter of the Handbook will seek to summarize the evolution in the field to date, providing introductory explanations of concepts and definitions employed and identifying the key issues and concerns that provide the essence of SCRM. An explanation of the overall structure of the Handbook is provided together with a brief resume of the focus and outcomes of each of the remaining chapters. The Chapter concludes with a review of the current trends in the field and suggestions about the next phase in its development.

1.2 Background to SCRM

Defining the terms in a multi-disciplinary and still-developing field is not without its problems since many authors choose to highlight particular dimensions or perspectives appropriate to their focus of attention. Although this may arguably produce an almost infinite variety of definitions, most of these are generally considered to be consistent and complementary. As editors, we have not sought to impose particular definitions of constructs, concepts and terms on our co-authors as this may constrain our colleagues and negate one of the purposes of this Handbook, the development of insightful and rich research which pushes forward our knowledge and understanding of this practice-oriented field. However, it may be of value to the reader to have some appreciation of the scope and terminology employed before embarking on the remainder of the chapters.

A preliminary definition of the term *supply chain* would encompass the linkage of stages in a process from the initial raw material or commodity sourcing through various stages of manufacture, processing, storage, transportation to the eventual delivery and consumption by the end consumer. This might suggest that supply chains are concerned primarily with logistics. However, the conceptualization of the term *supply chain* in the context of the *Supply Chain Management* (SCM) field is significantly more diverse. Previously, SCM was typically a reactive mode of management seeking to insulate the business from the risks of supply chain disruptions, especially from major suppliers immediately upstream, primarily engaged with the assessment of buffer stocks to minimize the undesirable consequences of such disruptions. SCM today demands a much more proactive, strategic and corporate approach, engaging with the other organizations throughout the supply chain in seeking to gain sustainable competitive advantage and profitability through leaner, more agile, efficient, resilient, comprehensive and customer-focused strategies. Developments of this nature may not automatically reduce the risks and indeed may certainly change the profile of risks encountered if not increasing them. SCRM is a necessary partner to the rapidly developing SCM field.

Effective SCM is concerned with the interchange of information, communications and relationship development, potentially throughout the entire supply chain, upstream to the raw material supply sources and downstream to the end consumer of the goods and services. Increasingly, recognition needs to be given to the comprehensive package of services that accompanies most products to the marketplace. Technical services and support are the more evident type of service, although financial services (e.g. loans, insurance), marketing and sales support (e.g. point-of-sale promotions) and staff training are now established as important elements in the supply chain mix. Equally, recognition should be given to the fact that every intermediary or organization in the supply chain is a customer, entitled to similar considerations as the final consumer in the chain. The expectations of supply chain or channel members may extend beyond the quality of the supplied resources to those of dependability, reliability, security and responsiveness of the supply chain to mitigate any dislocations wherever they may happen in the chain.

A schematic vision of a supply chain may conjure up a vision of a linear relationship between a supplier and a focal manufacturing company and possibly the linkage downstream to the retailer. The reality is much more complex. The consideration may encompass several stages upstream in the chain, often termed first, second and third tier suppliers or alternatively first, second and third tier customers or distributors. At each of the stages going upstream or downstream there are likely to exist a multiple set of organizations interacting in a commercial sense. The term *network* is often used to refer to this multitude of interacting organizations with each organization being located as a *node* in the network. The majority of these nodes whilst making an important contribution are less likely to be critical to the success of the supply chain. The focus of SCM is inevitably focused on the smaller subset of organizations whose contribution in terms of product or service supplies is likely to be critical. However, organizations would be wise to monitor all of the remaining suppliers and distributors to identify those that may become critical in the future, since disruptions to the supply of even low value components have the capacity to dislocate an entire supply chain.

SCM may be viewed at both the operational and strategic levels. Operationally, the focus would be towards the effective and efficient functioning of Purchasing, Goods Inward, Warehousing, Stock Control and Distribution. SCM whilst not divorced from the operational level involvement is more concerned with managing the functioning of the supply chain at the strategic level. Operational issues and concerns often demand strategic planning and management to resolve and equally strategic developments will impact at the operational level at some time in the future.

In summary, SCM is no longer a purely reactive activity seeking to improve the capacity of the organization to absorb potential external shock waves, primarily directed along a linear supply route whilst seeking to minimize the disruption. It is now a more proactive activity engaging with a complex network of upstream and downstream partners seeking collectively to enhance competitive advantage, added value, lean operations, agility and profitability at the same time as managing a more complex interaction of risks. Issues are now about the benefits and risks associated with multiple sourcing, sharing the consequences of risks across the supply network, sharing information, building relationships and establishing trust. All of these developments are illustrated and examined in detail in subsequent chapters. The next section provides an overview of the key concerns and components of SCM and SCRM, as a prelude to summarizing the structure of the Handbook and its constituent chapters.

1.3 Supply Chain Risk

An initial definition of supply chain risk is encompassed by Zsidisin (2005 p. 3) as "the potential occurrence of an incident or failure to seize opportunities with inbound supply in which its outcomes result in a financial loss for the [purchasing] firm." This generic definition is articulated and developed by many of the authors in the Handbook, often focusing on its application in particular industrial sectors or business contexts. We have deliberately chosen not to promote a definition for the term *risk* for two reasons. Firstly, seeking agreement on a definition has proved problematic for most fields of study across significant time periods. Secondly, prescribing a particular definition is likely to prove counter-productive in generating and encouraging the different perspectives and approaches adopted by our co-authors. However, risk itself within the context of the supply chain may be categorized in a number of dimensions:

1. Disruptions to the supply of goods or services, including poor quality, which cause downtime and consequent failure to satisfy the customer's requirements on time.
2. Volatility in terms of price may result in difficulties in passing on price changes to the customer and potentially have consequences in lost profit.
3. Poor quality products or service, either upstream or downstream, may impact on the level of satisfaction of the customer with consequences for future revenues and possibly more immediate claims for financial compensation.
4. The reputation of the firm, often generated by issues not directly related to the supply chain itself, may pose risks. Inadvertant comments by senior executives or the failure to endorse certain protocols may damage the reputation of the organization.

The remaining 19 chapters in the Handbook all provide different examples of supply chain risks including those listed above and others more specific to different supply chain contexts (e.g. risks associated with the transportation of dangerous goods). The authors also outline the nature of the impact these risks have on the focal organization and others in the supply chain. In each case the chapters conclude with potential solutions, presented and evaluated as part of the SCRM process.

The components of SCRM may again be differentially defined in terms of the subdivision of the particular strands, although most definitions would encompass the following:

1. *Risk identification and modeling* – incorporating the sources and characterization of risks, what may trigger them and the relationship to the supply chain functioning effectively and efficiently.
2. *Risk Analysis, Assessment and Impact Measurement* in terms of likelihood of occurrence and potential consequences.
3. *Risk Management* – generating and considering alternative scenarios and solutions, judging their respective merits, selecting solutions and undertaking the implementation.
4. *Risk Monitoring and Evaluation* – monitoring, controlling and managing solutions and assessing their impact on business performance outcomes.

5. *Organizational and Personal Learning including Knowledge Transfer* – seeking to capture, extract, distill and disseminate lessons and experiences to others within the organization and its associated supply chain members.

The SCRM approach comprises a set of interacting considerations or activities in relation to the supply chain. The essence of SCRM is taking a more pro-active approach to managing risks in the supply chain in advance of their occurrence. This does not necessarily ensure that all such potential risks can be identified in advance or if identified, sufficiently well resolved to prevent some or all of the negative consequences. In summary, SCRM like other management systems is dependent on good quality management in terms of knowledge, abilities, experiences and skills. The concepts, tools and technologies provide support but are unable, as yet, to replace the judgments required in most risky decision situations.

1.4 Structure of the Handbook

The Handbook has been structured into four main sections, reflecting the composition of the SCRM field discussed earlier:

1. Risk Analysis, Assessment and Tools
2. Supply Chain Design and Risk
3. Supply Chain Risk Management
4. Supply Chain Security

Although we have classified each of the chapter contributions into one of these four sections, the categorization of many of these chapters within its respective section only partly reflects its focus. The absence of any widely accepted framework for categorizing research in this field reflects the novel and evolving nature of SCRM as well as the SCM field itself. The interpretation of the term *risk* as a construct is still being debated, similar to that of supply chain management, where there are many different perspectives that exist. This is due to the fact that risk is a deceptively complex concept, as alluded to in the beginning of this chapter. As mentioned previously, we have not sought to prescribe particular conceptual definitions but rather to encourage authors to explore such concepts, frameworks and models in the context of their particular research perspective. Therefore, many of the chapters in this book simultaneously touch upon several topics or facets of supply chain risk and its management. The 30 authors contributing to this text are drawn from 11 different countries and are recognized international authorities in research, practice and policy making associated with SCRM and the wider domain of SCM.

1.4.1 Risk Analysis, Assessment and Tools

The focus of the six chapters in this first section concern the stages in the SCRM approach incorporating differing perspectives or models associated with the conception of the supply chain, approaches to identifying and classifying risks, how these might be measured and broad strategies designed to manage the risks.

Dr. Bjorn Asbjornslett tackles the issues associated with analyzing the potential vulnerability of supply chains, enabling a more proactive approach to SCRM. Having introduced the concepts of vulnerability and resilience in the context of the supply chain, the author examines the contributory factors influencing both vulnerability and resilience. The chapter concludes with the exposition of a detailed approach to assessing the vulnerability of the supply chain, providing practical illustrations, guidance and advice on associated tools that may be employed to support the analysis and assessment.

The consideration of supply chains more as value networks, adding value throughout the process to the ultimate consumer, is a central feature of the chapter provided by *Dr. Jukka Hallikas* and *Dr. Jari Varis*. This approach assists in recognizing both the multi-dimensional and complex interfaces between the various organizations (i.e. nodes) in the supply network as well as highlighting the importance of tangible (e.g. products) and intangible (e.g. customer satisfaction) value added within the network. Utilizing the Information and Communications Technology (ICT) sector, the authors demonstrate an approach to modeling the value flows between the key players. This in turn enables the organization to highlight the critical risk sources, their causes and potential consequences. The approach provides an important tool for the management team involved in SCRM.

Dr. Samir Dami develops a holistic perspective on SCRM recommending a proactive approach to managing supply chain risk. Following an exploration of the key definitions relating to uncertainty and risk within the supply chain context, the chapter considers potential approaches to the categorization of supply chain risks in terms of sources and potential impact. Having addressed some of the potential solutions to managing commonly occurring risks, two analytical tools, Data Mining and Failure Mode Effect Analysis are examined as possible approaches to uncover potential risk sources and to effect some scale of measuring these. The author concludes that irrespective of the tools employed, the emphasis should be towards a more proactive approach to SCRM.

Dr. Barbara Gaudenzi contends that managers need to address and manage risks at the project and process level if they are to respond effectively to the competitive and dynamic environment. Linking projects and processes, the author examines the key drivers for success and imaginatively demonstrates the parallels between these and the link to risk assessment. The chapter evolves around a series of four propositions which are used to identify the appropriate risk assessment method. The chapter concludes with an approach to risk assessment for projects and processes that spans both internal and external stakeholders.

Chapter 6, produced by *Arben Mullai*, features risk management relating to the transportation of dangerous goods. Using this sector as the contextual backcloth, the author develops a conceptual model of the risk management system. The key dimensions of risk analysis, risk evaluation and risk management are explained, justified and demonstrated in the context of dangerous goods transportation. A very practical approach is employed and the reader is guided through each of the stages in the process by means of conceptual models and more importantly illustrative examples of the decision parameters and requirements.

The concluding chapter in this section, Chap. 7, presented by *Professors Steven Melnyk, Alexandre Rodrigues and Gary Ragatz*, focuses on supply chain disruptions. Eschewing previous approaches featuring case study and anecdotal evidence, the authors examine the contribution of discrete event computer simulation. The benefits associated with simulating the impact of different forms of disruption may be examined in terms of the impact on performance and permit the testing and evaluation of different risk management strategies and policies. The contribution of two specific statistical approaches are presented and analyzed in terms of different forms of supply chain disruption. The authors conclude that simulation may provide an important contribution to analyzing the management of supply chains but recognize that this route poses important challenges for future research.

1.4.2 Supply Chain Design and Risk

The three chapters in the second section, *Supply Chain Design and Risk*, focus on how organizations can shape their supply chains to address the risks inherent in all supply chains or those risks more commonly associated with their specific supply chain.

Chapter 8, authored by *Professors Constantin Blome and Michael Henke*, examines a traditional and commonly applied mechanism that firms have employed to manage the risk inherent with single sourcing, the use of multiple sourcing. The chapter examines the pros and cons of the two alternatives against the various characteristics of supply chain risk. The authors contend that generic solutions to such sourcing decisions are inappropriate as the context of the decision will strongly influence the nature of the risks involved. The authors indicate when organizations need to consider a multiple sourcing policy and the various environmental conditions that may warrant the use of more than one supplier.

Chapter 9, written by *Dr. Omera Khan, Professor Martin Christopher and Professor Bernard Burnes*, recognizes the strategic role of product design in global supply chain risk management and explores the influence of product design on the structure and management of the supply chain and its risks. Utilizing the fashion and clothing manufacturing sector in the UK as the context, the authors demonstrate that the rapidly changing tastes in the fashion market dictate the exposure to risks and their management. They demonstrate using the longitudinal case study associated with Marks and Spencers, one of the market leaders in fashion retailing in the UK, that strategies involving "off-shoring" design and manufacture have proved to be counter-productive and that multiple sourcing may also prove problematic in seeking sustained competitive advantage. This design-led risk management approach offers a novel approach to mitigating supply chain risk.

The third chapter (Chap. 10) in this section, authored by *Professors Christopher Tang and Brian Tomlin*, examines the contention that strategies designed to enhance supply chain flexibility will mitigate supply chain risks. The authors provide a comprehensive analysis of different categories of risk and the impact that flexibility might have on these in terms of likelihood and consequences.

A mathematical model is developed and analyzed to demonstrate that the optimization of risk may be best achieved through more limited degrees of flexibility rather than seeking maximum flexibility. The authors challenge what might be the accepted conventional wisdom in relation to flexibility and security providing appropriate guidance for the way forward in practice.

1.4.3 Supply Chain Risk Management

The third section addresses a number of different perspectives on the role and more general approaches to SCRM. The drive to develop and agree industry-wide standards on SCRM and SCM within wider frameworks such as Enterprise Risk Management is addressed as well as how best to approach the adoption from the perspective of the highly significant Small and Medium-Sized Enterprise (SME) sector. Particular issues are addressed that arise as a consequence of the globalization of many supply chains with consequences for the SCRM approach to developing relationships and trust. The attitudes and behaviors of individuals and groups of practicing decision makers faced with supply chain risks remains a fundamental concern of those seeking to develop SCRM solutions and approaches. Four of the chapters are directly related to this theme whilst another articulates the consequences of such behaviors in relation to both quantitative and qualitative performance metrics for the business.

Professor Michael Henke examines the potential positioning and the relationship between the emerging standards such as Enterprise Risk Management (ERM) and Supply Chain Risk Management in Chap. 11. The ERM process is examined in the context of the three dimensions: management processes, risk management components and the organizational entities. The author contends that the ERM approach offers a logical conceptual framework for the development of Supply Chain Risk Management, although recognizing that much work is still required to make this operational.

Private Financing Initiatives (PFI) and Public Private Partnerships (PPP) have become an established approach to public infrastructure investment and management internationally, albeit utilizing differing terminology. *Professor Simon Burtonshaw-Gunn* explains the background to these developments in Chap. 12, highlighting the challenges posed in terms of risk management. The focal point of the discussion is the pre-contractual stages in the context of the international construction industry, with the author providing a comprehensive schedule of pre-contract risk considerations. The practical experience of the author in construction project management has resulted in the provision of insightful, well-informed and practical advice for those managing risks in this area.

The development, research and discussion of supply chain risk management has to date predominantly targeted the larger organization as opposed to the Small and Medium Sized Enterprise. Yet in most developed and developing economies, the SME sector is responsible for the majority of employment, sales and added value. *Professors Uta Juttner* and *Arne Ziegenbein* help to redress this imbalance in Chap. 13, recognizing that most SMEs are more exposed to supply chain risks whilst simultaneously disadvantaged by lack of management resources and expertise. The authors develop a structured approach to the identification, assessment and mitigation of supply chain risks on the basis of four in-depth case studies in Switzerland. The approach is well explained and illustrated and is clearly delineated in terms of the three phases and the sequence of actions and decisions within each phase.

Professor Michael Smith reminds us in Chap. 14 that underlying all of the frameworks, organizational and inter-organizational structures, procedures and processes designed to support supply chain risk management, is the psychological behavior of key decision makers or influencers. Risk perception by individuals and groups are the under-pinning to risk behavior since identification, assessment and management requires thinking, judgment and decision making in situations with imperfect information. Evidence suggests that most supply chain risk managers employ heuristics in their decision processes, although the author observes that statistical training can improve the quality of risk perceptions and management. The chapter concludes with well reasoned practical advice and guidance for the risk management practitioner.

The author of Chap. 15, *Mischa Seiter*, focuses attention on an often underrepresented risk source in supply chains, behaviors by individuals and organizations which may often disrupt supply chains and cause deterioration in quality, absence of inventory and price escalation. The presence of opportunistic behavior suggests a willingness to break both explicit and implicit contracts and agreements in the pursuit of competitive advantage and profit. A conceptual three-stage model is developed, generating a series of independent variables and their potential to encourage or detract from opportunistic behavior throughout the supply chain. The resulting hypotheses are tested using a sample of 104 large scale German organizations. Interesting and supportable results relating to partner selection, communications quality and sharing cost accounting information all yielded an impact directly or indirectly on opportunistic behavior.

As more attention, resources and time are devoted to supply chain risk management, the question increasingly posed is that of value for money. *Professors Bob Ritchie* and *Clare Brindley* address this and associated issues in Chap. 16. The authors investigate the anticipated conceptual link between risk and performance and examine the range of performance metrics or measurement systems, from single parameter measures (e.g. Return on Investment) or multi-dimensional (e.g. Balanced Scorecard). Further complexities are examined including the time frame and the question of enterprises seeking to achieve a balanced portfolio of investments and supply chain risks. Case study evidence drawn from a manufacturing organization and focused on its downstream supply chain activities provide supporting evidence for the framework as well as insights to, at times, the more idiosyncratic performance measures. The framework developed and the identified components provide valuable guidance and advice for practitioners and future researchers.

Professor Stephan Wagner and *Christoph Bode* investigate the attitudes and behavior of senior executives in Germany towards supply chain disruption and supply chain risk management. The results of their large-scale empirical study are summarized in Chap. 17, following a useful presentation of the development and clarification of the nomenclature in the SCRM field. They conclude that German executives are less concerned in preparing for catastrophic failure or disruptions of the supply chain and devote more attention to the more routine supply and demand side issues. Their findings also suggest that there is an absence of a systematic approach and practices of supply chain risk management, which corresponds with other studies in other countries. This highlights a need for more concerted efforts in terms of promoting and disseminating supply chain risk management and its associated concepts, methodologies and practices.

1.4.4 Supply Chain Security

This fourth section deals predominately with a specific source or driver of supply chain risk, which involves the intentional actions of parties to disrupt the flows of supply chains or to interfere with the product or service, resulting in loss of quality, delivery time and customer satisfaction.

The maintenance of high levels of supply chain security in the food supply chain is claimed to be critical to the reassurance of the consumer. *Professors Douglas Voss and Judith Whipple* address the issues associated with food supply chains in Chap. 18. The authors recognize that security is not solely concerned with preventing theft and product damage, intentional or unintentional, but also about disruptions to supply which may prove life-threatening or expensive in terms of recovery. The authors derive and define ten key competencies associated with supply chain security, explaining how these operate in practice and interrelate with one another. The results of an empirical study involving experts in the food supply sector identify the priority afforded to security. The authors draw insightful conclusions from their research in terms of the impact of security on business performance and the importance of supply chain security in relation to competitive advantage.

The authors of Chap. 19, *Professors Chad Autry* and *Nada Sanders*, address the issues associated with business continuity planning in the face of significant threats from the external environment. A framework encompassing the technology, processes and human resource dimensions is derived from the literature and developed to articulate the nature of the firm-level capabilities required. The authors advocate a dynamic capabilities approach which incorporates these three dimensions and harnesses both internal and external supply chain resources to address future disaster planning. The chapter concludes that the approach to the effective management of supply chain security requires a change in attitudes throughout the firm to embrace supply chain security as a common goal on par with profitability.

The concluding Chap. 20 examines the implications for security as food supply chains become increasingly global. Long distances, more intermediaries and greater complexity of the resulting distribution networks pose new challenges for effective supply chain risk management. The authors, *Dr. Nicole Mau* and *Professor Markus Mau*, articulate the key issues and illustrate these with practical examples. The conclusions drawn from their analysis highlights the importance of information flows upstream and downstream within the supply chain, the importance of transparency and the need to manage and control the supply chain throughout its entirety. Recognition is also given to the need for improved co-ordination and transparency between the national and regional agencies responsible for policy, oversight and monitoring of the global food supply network. A comprehensive control platform is required for the decision makers to manage supply chain security effectively.

1.5 Background of ISCRiM and Handbook

The primary purpose of this research Handbook is to collect and share various streams of research and trends in supply chain risk and its management. This Handbook is the second collection of such manuscripts predominately from the members of the International Supply Chain Risk Management (ISCRiM) network. ISCRiM was formed on October 2001 at the Manchester Metropolitan University in Crewe, UK. At this time, the study of risk in the supply chain was at its infancy. Seven active researchers in this emergent field, Bob Ritchie, Clare Brindley, Ulf Paulsson, Robert Lindroth, Andreas Norrman, Simon Burtonshaw-Gunn, and George Zsidisin, were the founding members. It was at this initial meeting that these individuals decided to form ISCRiM and meet on an annual basis to share and update each other on the research projects that they were conducting on supply chain risk. During this meeting the group also jointly decided to write a book to collect their thoughts and research findings associated with the topic of supply chain risk. This first book, titled "Supply Chain Risk" and edited by Clare Brindley, consisted of 12 chapters dichotomized into research frameworks and techniques and applications associated with supply chain risk.

Since our initial work and meeting on supply chain risk in 2001, this subject has emerged to being on the forefront of study in academe as well as discussed in the board rooms of many corporations. Today, the ISCRiM network consists of ~30 scholars drawn from Europe and the United States, with the majority of members in European countries including the United Kingdom, Germany, Sweden, Finland, Denmark, Switzerland, Italy, and Norway.

During the annual meeting in 2006 hosted by the European Business School in Oestrich-Winkel, Germany, the ISCRiM membership recognized the incredible growth of research on supply chain risk, in conjunction with the parallel growth of the ISCRiM network itself, the group jointly decided to write this second book.

1.6 Future Developments in SCRM

Attempting to forecast future developments in any field is a risky business in itself. However, most colleagues in presenting their research and practical experiences in their chapters have either directly or indirectly alluded to the developments that they anticipate or see as essential in their specialist areas. The distillation of these points suggests the following:

1. Developing new frameworks to capture the more complex, dynamic, continually evolving and multi-faceted nature of today's supply chains.
2. Increasing attention towards the application of SCRM in the smaller business (SME) context.
3. Identifying contingency factors and the generation of appropriate models and decision heuristics.
4. Increasing attention towards developing appropriate performance measurements and methods to assess the impact of risk and the associated SCRM actions in both quantitative and qualitative metrics.
5. Applying network theory and concepts to understanding the behavior of networks and the organizations operating within them.
6. Developing greater insights into the approaches to engender improved relationships and trust between supply chain partners and how these might be sustained in global supply chains.
7. Recognizing the inherent attitudes and behavors of individuals and groups involved in the SCRM process.
8. Emphasizing the development and training of individuals to competently manage supply chain risks, in terms of undergraduate, postgraduate and professional development programs.
9. Providing more in-depth research and development in specialist sub-fields.
10. Developing and applying performance standards to ensure management of risks where supply chain security is significant.
11. Developing more robust analytical tools and frameworks to support decision makers.

Above all, the need remains to provide platforms and vehicles for the interaction, exchange and engagement between practitioners, policy makers and researchers in the SCM and SCRM fields. ISCRiM will continue to provide one such vehicle, seeking to ensure excellence and promoting leading-edge research amongst its members and the wider SCRM community.

SECTION ONE - RISK ANALYSIS, ASSESSMENT AND TOOLS

Chapter 2: Assessing the Vulnerability of Supply Chains

Bjørn Egil Asbjørnslett

Senior Consultant, MARINTEK Solutions

2.1 Introduction

Supply chain systems are becoming increasingly lengthy and complex, reflecting the dynamic and global marketplace. Adopting a more proactive approach to dealing with new and changing risks and vulnerabilities emerging within or influencing the system may be a wise action to secure the mission of the supply chain system.

In this chapter, an approach to analysing vulnerability in a supply chain system is presented as a means to reduce risk, to become better prepared to manage the system's vulnerabilities and to improve the system's resilience. The analysis establishes the relationship between relevant threats and risks, and the potential scenarios and consequences that determine the vulnerability of the supply chain system. This is designed to generate a deliberate and conscious management process, seeking to establish an acceptable degree of vulnerability and risk within the supply chain system.

The approach proposed comprises two basic preliminary steps. Firstly, a choice has to be made with respect to the focus and magnitude of the study, or the scope of work the analysis shall cover. Secondly, a description of the supply chain context to be analysed is required, including the specific supply chain management roles, functions and activities currently present or prospectively required. Both steps are important in establishing a sound basis for further analysis. The approach focuses on understanding the context from another angle, a vulnerability angle. This facilitates a more scalable approach, permitting a more conscious decision to be taken on which part of the demand and supply chain(s) and SCM is most appropriate to address.

2.1.1 Background

The industrial world as we know it today has become a global network of demand and supply nodes, interlinked through interacting logistics systems. The Internet and related 'e-services' have opened up the demand and supply markets of the world, so that the 'next-door' marketplace could as well be the 'next-continent' marketplace. The supply chain systems are complex entities with multiple physical and virtual relationships, and multiple internal and external interfaces. High demands are put on both the quality of the products and services, and on the supply chain regularity and dependability. Whether the product is to be a part of a more complex product, or the final product for consumption or use by consumers or professional users, the product is expected to be available when needed, and as promised. In the search for improved effectiveness and efficiency the supply chain systems are reengineered according to modern concepts, and made more specialized and dependent with less tolerance for failure.

Supply chain systems are subject to changes originating both from within their market place, as a result of changes in inter-organisational constructions, and from changes in demand from external sources. Such changes are a natural part of business and should be planned and prepared for, and dealt with proactively. Addressing such changes is the primary theme of vulnerability analysis presented in this chapter. As supply chains become longer and parts of larger networks of demand and supply nodes and interacting logistics nodes and modes, they become more prone to the negative attributes of systems; indeterminacy, complexity, flexibility, sensitivity, reliability and vulnerability (Meister 1991). A question then arises: How can we develop or revise supply chain systems so that we realise the inherent opportunities without taking on unacceptable vulnerability?

2.1.2 Objective

The main objective of vulnerability analysis presented in this chapter is to bring to light scenarios, and thereby threats and risks that make the supply chain system vulnerable, together with actions that may make the system more resilient in dealing with these risks and vulnerabilities. The analysis should be regarded as a means to becoming more proactive, dealing with relevant vulnerability issues prior to the occurrence of critical events, incidents or accidents. This may be seen as part of change and the change management procedures engaged in supply chain management. This essentially proactive approach may enable decision makers to uncover new areas or factors of risk and vulnerability, before implementing and operating a revised supply chain system.

The approach and structure of vulnerability analysis is designed to reach these overall objectives through achievement of the following sub-objectives;

1. Understand the nature and types of factors that may pose threats and risks to the achievement of the supply chain system's short and long term mission.

2. Understand the scenarios (processes and mechanisms) through which these threats, risks and vulnerabilities may evolve.
3. Understand how through the use of vulnerability scenarios, the likelihood and consequences of such threats may be reduced and managed in a cost- and service effective manner, whilst achieving an acceptable vulnerability level.

Using the vulnerability analysis in such a proactive way should help in balancing out the risk and the vulnerability side of opportunity development to improve effectiveness and efficiency in supply chain systems.

2.1.3 Approach – Mission-Oriented

The focus of vulnerability analysis relates to the supply chain system's mission as the supplier of time and place utility in a larger dependent network, e.g., an operational mission, as well as the supply chain system's ability to regain its position within its marketplace, e.g., a strategic mission.

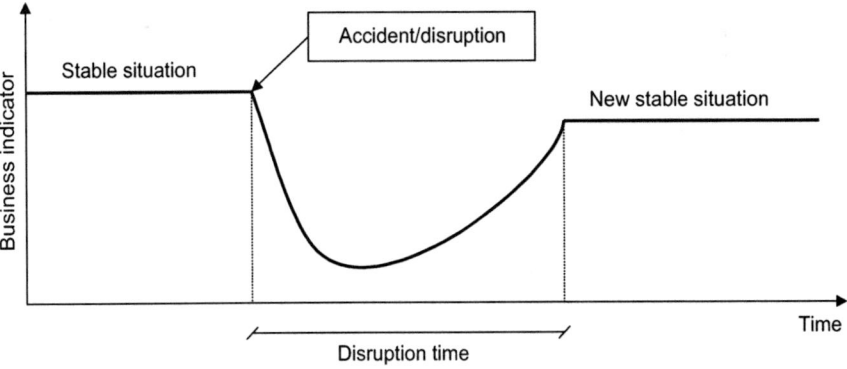

Fig. 2.1 Regaining stability after an accidental event or disruption

Logistics and supply chain systems should be resilient, meaning that they should be able to fulfil their mission of logistics service even if the initial attempt fails to succeed. This is directly related to the mission of logistics. For example, a slogan for postal services has been: 'The mail shall be delivered. If the mail may not be delivered, it shall be returned. Then the mail shall be delivered!' e.g., the postal delivery (logistics) mission must be accomplished! Focusing on the mission of logistics and supply chain management the resilient capabilities should contribute to proactive preparation for and 'optimal' treatment of situations that stress the mission of the logistics system.

In a supply chain network or a tightly coupled supply chain, critical events like strikes, fires or bankruptcy in one node in a supply chain may lead to interruptions or reduced capacity in the supply network. Common for all these examples is that various threats affect the mission of the system, and lead to a costly operation to 'sort things out' and regain stability. This is illustrated in Fig. 2.1.

Figure 2.1 shows that given a stable situation, an accident or disruption will lead to a reduction in the mission of the system, as measured by critical performance indicators for the company or supply chain. The generic term 'business indicator' is used, that should be seen as a direct measure for the supply chain's mission, both short term operationally and long term strategically. Another important part in Fig. 2.1 is the disruption time, e.g., the time from the accident or disruption having an effect on the business indicator until the supply chain system has reached a new stable situation, again delivering its mission. The new stable situation for a resilient system may be lower or higher than the former stable situation, as measured by the business indicator. With respect to two system definitions, a robust system will retain its system structure intact and continue with the previous situation, whilst a resilient system will adapt to regain a new stable position.

2.2 Concepts and Definitions

Vulnerability is a concept that may be used to characterize a supply chain system's lack of robustness or resilience with respect to various threats that originate both within and outside its system boundaries. The vulnerability of a supply chain system may be manifested both in its infrastructures – both nodal and modal, its processes, as well as the operation and management of the supply chain. The nodal part of the supply chain infrastructure is the nodes in the supply chain, e.g., ports, terminals, warehouses and transhipment points, check or security gates, while the modal parts of the supply chain is the transportation modes, e.g., road, rail, air or waterborne transport modes, including deep-sea, short-sea and inland waterways. A variety of demands, both from the society in general, and from related businesses, will also add to the supply chain system's vulnerability. A supply chain system may be vulnerable with respect to as diverse threats as; technical failures, human errors, criminal acts, environmental impacts, accidents, loss of key personnel, strikes, variation in energy prices, etc. An example of a contemporary aspect that could have tremendous impact on the long term vulnerability of supply chain systems are the environmental impact and related sustainability of the systems, e.g., as denoted through new carbon footprint measures. The global environmental concerns and sustainability focus could lead to radical changes in the supply chain system, and expose strategic and tactical vulnerability in present supply chain constructions. Another example could be the dependence on single, large port sites in many deep-sea maritime logistics chains.

The vulnerability concept has been used in several contexts with rather different meanings. The definition of the vulnerability concept as it is used here is;

> the properties of a supply chain system; its premises, facilities, and equipment, including its human resources, human organization and all its software, hardware, and net-ware, that may weaken or limit its ability to endure threats and survive accidental events that originate both within and outside the system boundaries.

Several other concepts are related to vulnerability. Among these are; threats and the risk concept, as well as some concepts that are 'opposite' to vulnerability; robustness, resilience, and damage tolerance. A supply chain system is said to be robust, or resilient, with respect to a threat, if the threat is not able to produce any 'lethal' effects on the system. We define robustness as a system's ability to resist an accidental event and return to do its intended mission and retain the same stable situation as it had before the accidental event. Resilience may be defined as a system's ability to return to a new stable situation after an accidental event. As such, robust systems have the ability to resist, while resilient systems have the ability to adapt.

The analysis is a scenario-based analysis, e.g., it is based on scenarios that commence with a potential accidental event. We define an accidental event as the event which makes the disruption or accident visual. For an acute accident this may be the accident itself, or in a gradual deterioration of a system's performance this may be when a given threshold is crossed.

A threat is a stable, latent, adverse factor that may manifest itself in an accidental event. It is not the threat itself, but the accidental event and the possible events that follow from it, the scenario, that is important in an analysis, as well as finding the often hidden interactions between factors that by themselves would not lead to an accident, but where complex interactions, among factors and events, may lead to a system accident.

It should be remarked that when we use the term 'accident', it is used for all events that may cause a considerable reduction in the mission of the supply chain system. That means that the event may be an accidental event that contributes either direct mission-reducing consequences, or consequences that may materialize over time.

2.2.1 Search for Effectiveness and Efficiency

A saying states that 'if you don't take risk, you will not drink champagne'. The message is that there is always a relationship between the risk one is willing to take in one's choices and the benefits that may be reaped from these choices. How much risk one is willing to take to enhance effectiveness and efficiency, in proportion to increased vulnerability is the underlying dilemma one, directly or indirectly, faces when designing new and improved supply chain systems. Hollnagel (2004) raises some concerns around the search for effectiveness and efficiency, and has given it the acronym ETTO, for the efficiency-thoroughness trade-off;

> Thoroughness means that they try as best they can to do the right thing and do it in the right way, e.g., to choose the correct action and to carry it out as well as possible. Efficiency means that they try to do this without spending too much effort in order to meet the demands of the situation, regardless of whether these demands are imposed by

an external source or of their own making. This efficiency-thoroughness trade-off (ETTO) is a common feature of human performance that seems to play a role at the level of individuals and at the level of organizations alike. (Hollnagel 2004, p. 153)

Hollnagel's ETTO principle can be related to Perrow's (1984, 1999) definition of systems according to their degree of coupling and interaction. According to Perrow, interactions in a system may range from linear to complex. Linearity means 'interactions in an expected sequence', while complexity means 'interactions in an unexpected sequence'. From Perrow's discussion of complex and linear systems one might believe that a linear system is an optimal system, but Perrow claims that there are two reasons why we (want to) have complex (vulnerable) systems:

1. Complex systems are more efficient than linear systems (in the narrow terms of supply chain efficiency, neglecting accident and disruption hazards).
2. Often, we have complex systems because we do not know how to produce the output through linear systems.

Then according to Perrow we have complex systems both because we want to take advantage of the possibilities the complex systems give, in our search for efficiency and effectiveness, but also because we do not sufficiently understand how a (given) complex system may be converted into a linear system, e.g., we have limitations in our learning capability. The learning capability may only be improved by using the system, and therefore a complex system may only be converted to a linear system through a learning process. It is during this learning process that complex systems are especially vulnerable due to our limited knowledge of all interactions. To support this learning process a pro-active vulnerability analysis, as the one we present in this chapter, may increase the rate by which knowledge about the system is gained, before and while one is using the system. The analysis may as such be an aid in building a more effective and efficient system that becomes more robust or resilient, and thereby less vulnerable. This may be seen as a parallel to the proactive approach to analysing changes in aviation that are part of ICAO's new safety management system concept (ICAO 2005).

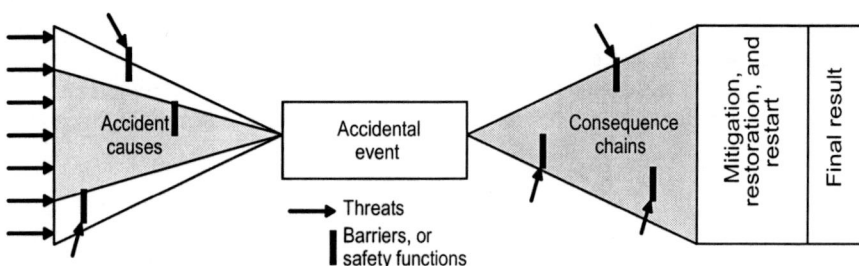

Fig. 2.2 Vulnerability analysis versus risk analysis

2.2.2 Vulnerability Analysis versus Risk Analysis

The main difference between a risk analysis and a vulnerability analysis is the focus of the analysis. Risk analysis is focused towards the human, environmental and property impacts of an accidental event (IEC 300-3-9, 1995), while a vulnerability analysis is focused towards the system mission and the survivability of the system.

In a risk analysis three questions make up the basis of the analysis: (i) what can go wrong, (ii) how likely is it to happen, and (iii) what are the consequences (see e.g., Kaplan 1997). A vulnerability analysis, on the other hand, focuses upon (a) an extended set of threats and consequences, (b) adequate resources to mitigate and bring the system back to new stability, and (c) the disruption time before new stability is established, as shown in Fig. 2.1.

Figure 2.2 illustrates the difference between a risk analysis and a vulnerability analysis. The shaded triangles show the scope of a risk analysis, while the white parts show the additional elements in a vulnerability analysis. The threats may also affect the barriers and safety functions, e.g., by intentional harm against a barrier. The last, but not the least important part of a vulnerability analysis, is the efficiency and adequacy of the available resources to mitigate, restore or rebuild the system. As stated above, the focus of a vulnerability analysis is on the mission and survivability of the system, and as such the analysis is not finished before elements that shall bring the system back to new stability are analysed.

2.3 Vulnerability of Supply Chains

2.3.1 The Mission of Logistics and SCM

A supply chain system is made up of a number of elements. The elements may be grouped in several ways depending on what one would like to address. Several authors within the safety theory domain have addressed the functions of supply chain systems, and how these functions are related to the vulnerability of the system, see e.g., Meister (1990), Perrow (1984, 1999) and Reason (1990). However, safety science and SCM/logistics differs somewhat in scope. Safety science conceives safety as a goal per se. The mission of logistics and SCM however, is to provide place and time utility. This means providing 'place and time' services that are reliable and flexible, e.g., that can be trusted and that can meet and adapt to required changes. In this context, a logistics system is defined in a broad sense, covering transport of personnel, components, products, materials, and energy, along with the transfer of information, services and funds. A reliable and efficient logistics system is an important prerequisite for most aspects of contemporary business systems. Malfunctions and delays in a logistics system may lead to significant economic losses and may also cause safety and security problems. All logistics systems are exposed to hazards and threats of different types that may cause negative or adverse consequences, such as delays, material losses, extra costs, pollution, and/or injuries and fatalities. The hazards/threats may be random or deliberate, and may be related to technology, operations, and environmental, and/or societal aspects. It is important to take precautions during planning/design of logistics systems, and also during daily operations and control, to address mission threatening vulnerabilities and to make systems resilient or robust – on a cost-effective basis.

In summary, one can say that risks lead to mission threatening vulnerabilities in a logistics or supply chain system. To cope with these vulnerabilities one must make the logistics and supply chain system resilient. Then one should find good theory and knowledge that could lead to resilient planning and management of the logistics mission. Safety theory and one of its contemporary approaches, *Resilience Engineering*, is a good source of knowledge to develop from.

2.3.2 Resilience in Supply Chain Systems and SCM

Paradoxically, the more efficient a logistics chain is, the more vulnerable it may be, as suggested by the ETTO principle (Hollnagel 2004) referred to earlier. The need for resilience reflects a need for strategies to maintain high efficiency in a dynamic and complex environment. Also within the safety science domain, the increasingly complex world has enforced a shift of assumptions from those of a stable, controllable setting enabling functional closure, to those of a more dynamic and complex world.

Within the safety domain, an increasing unease with conventional safety practices have developed from the experience of fundamental surprises that seemingly escape the attention of 'normal' safety practices. On the theoretical side, the challenge of Perrow's (1984, 1999) normal accident theory contributed to a marked change. Perrow's argument is that we are experiencing the consequences of our own constructions, thus producing interactive complexity and tight couplings which facilitate the uncontrollable propagation of unintended side-effects. Perrow attempts to draw a normative cross line, inspired by historical experience: 'man's reach has always extended his grasp' (Perrow 1999:11). His premise is based on a fundamental distinction between centralized vs. de-centralized principles of organization, arguing that high interactive complexity and tight coupling demand different organizational principles in that respect, thus normatively concluding that we should avoid the combination of high interactive complexity and tight coupling. From a pure safety standpoint, it could be argued that the damage potential of logistics does not fit the scope of Perrow's discussion, e.g., compared to nuclear technology. From a SCM point of view however, the damage potential of malfunctioning logistics is severe enough!

A number of related perspectives have been developed with a similar mission as *high reliability organisations* (HRO), that is, conceptualising *resilient organizations*. Recently, *Resilience Engineering* (Hollnagel et al. 2006) has been presented as a more comprehensive and elaborate approach to the same goal, that is, overcoming the *normal accident theory* (NAT) challenge, encompassing the notion of complexity. The journey from classical risk analysis approaches based on functional closure to HRO and Resilience Engineering, sparked by the NAT perspective, signifies a need for safety theory to be reinterpreted in a much more dynamic setting. This line of development decreases the distance between the safety domain and the resilient logistics and supply chain domain, opening up the possibility that the former can (theoretically) inform the latter.

2.3.3 Factors Contributing to Vulnerability and Resilience

Understanding the threats and risks that are the underlying factors affecting the vulnerability of a supply chain system is an important part of a vulnerability analysis. We will here use the term 'factor' instead of threat to show that the 'level' of the factor may decide whether it is a threat or a barrier. A company's safety climate may, e.g., be a threat (contributing to vulnerability) if it is poor, or a barrier against accidents if it is strong. The factors contributing to vulnerability address the elements and relations where vulnerability may materialize. The factors depend on the type of system. These factors are present in the internal organization of physical and human capital, within the system's relationships with its surrounding elements, or external to the system. The factors may be used to create check-lists for a vulnerability analysis of a supply chain system. By relating the factors to the dimensions of the system model, further structured insight may be gained, and the factors' impact may be easier to perceive. These factors contributing to vulnerability and resilience provide an essential basis for the vulnerability analysis we present in the next section.

2.4 Vulnerability Analysis

In this section we present a pro-active approach to vulnerability analysis of supply chain systems. By pro-active we mean that the analysis shall be conducted prior to any accidental event, and not as a consequence of an accidental event, as well as focusing on the potential mission-oriented consequences of an accidental event and the resources that may enable the system to regain stability. The approach is based on the set of factors (threats) that are present in the system, and the accidental events and scenarios that may materialize from these threats.

2.4.1 Flow Sheet of the Analysis

The vulnerability analysis follows from and develops further a generic approach to risk assessment for socio-technical systems:

1. Characterize context – workplace, organization, target/purpose, demands, resources

2. Identify risks (accident types) – scenarios, tasks, activities, personnel, disturbances
3. Analyse risks (accident potential) – probability, consequences, failure modes, failure types
4. Decide on countermeasures – policies, defences, monitoring, procedures, communication

Figure 2.3 shows a flow-sheet of the vulnerability analysis procedure. In the first two steps, planning and preparation of the analysis should be conducted. Then, in step three, a structured overview of the factors (threats) that may affect the vulnerability of the system are developed. Based on the set of vulnerability factors, a rough list of accident scenarios are developed in step four. In step five, the rough list is evaluated, and the most likely scenarios are given quantitative weights and ranked in order of importance. The most critical (vulnerable) scenarios are then, in step six, evaluated with respect to likelihood and consequence, both without and taking mitigating resources into account. Step seven evaluates measures for reducing likelihood and consequence of the scenarios. Concluding the analysis requires verification of the actions proposed based on the results of the analysis. This may require either a second round of analysis or specific functional analyses for areas with unacceptable vulnerability.

Three worksheets are used in the vulnerability analysis. In step four, identifying and analysing scenarios based on the proposed set of factors are done in work-sheet no.1. Then, in step five, work-sheet no.2 structures a quantitative analysis of

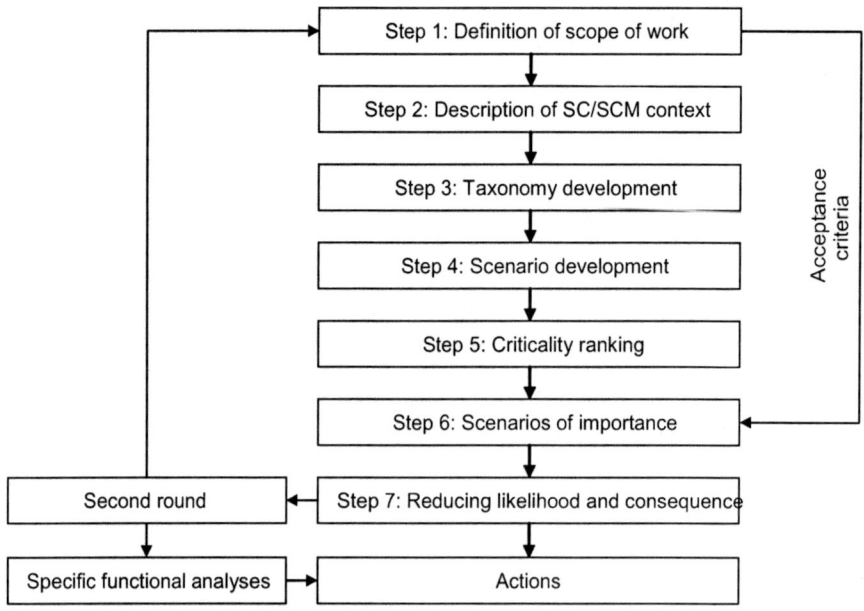

Fig. 2.3 Flow-sheet of the vulnerability analysis

the scenarios providing a basis for a criticality ranking of the scenarios. In step seven, work-sheet no.3 is used to identify measures to reduce the vulnerability by focusing on the likelihood and the consequences of the scenario.

In sum, the seven steps of the analysis may be grouped into three;

1. Understanding the context-specific threat and risk picture of the given supply chain and SCM context, and structure this into a taxonomy of the vulnerability factors.
2. Analyse and rank the vulnerability scenarios, resulting in a criticality ranking of the scenarios.
3. Handling of the vulnerability through cost- or service-effective likelihood or consequence reducing measures, bringing the vulnerability down to an acceptable level.

This leads to (i) increased and structured awareness of the vulnerability in the given supply chain context and the SCM thereof; (ii) ability to analyse the vulnerability of one's SC and SCM context in a structured way, and finally (iii) a conscious treatment of acceptable vulnerability.

2.4.1.1 Step One: Definition of Scope of Work

The first part step of the analysis defines the frames and targets for the analysis. This includes setting objectives for the analysis, the level of the analysis (see e.g., Norrmann and Lindroth 2004), as well as establishing acceptance criteria. It is also important that time and resources are allocated to cover the scope of the analysis, and that the resources are committed. The analysis should be carried out by a multi-disciplinary group, able to cover all relevant aspects of the supply chain, in close cooperation with the relevant functional departments and the management group of the focal company. Establishing effective information and communication channels to relevant stakeholders within the inter-organisational supply chain, external to the focal company is an important requirement.

One aspect that should be given due attention at this stage is the acceptance criteria. The acceptance criteria will come into use when evaluating whether a vulnerability scenario is acceptable or not (step six), and will also be part of establishing and evaluating the measures of consequence and mitigation aspects that are relevant to the focus (step five). Any differences in acceptance criteria for un-mitigated versus mitigated consequences should also be considered.

2.4.1.2 Step Two: Description of Context

Understanding the specific context is important in all logistics and supply chain improvement work. Due to this, process mapping and development has become an important knowledge area within supply chain management. The specific supply chain context should be described, with emphasis given to the areas that one wants to focus on, as defined in step one. The context may be described in many ways, but it is important to cover all relevant areas, as well as the relation between them.

One structure to use could be to consider separately the infrastructure, processes, management, relationships and interfaces.

Infrastructure should cover all nodes and modes used along the supply chain. Processes should cover all processes from the demand side, to delivery of goods. The processes could further be separated into the logistical flows of information, goods, services and financial funds. However, at a first run-through of the analysis, it is recommended not to make the context description too fine-grained, but rather make notes of how the context could further be detailed if required. Management should cover the management roles and responsibilities, including information and communication technology that are used for SCM. Finally, interfaces and relationships should be documented, e.g., through the use of stakeholder mapping. Contractual matters are an important part of the relationship mapping, and where areas of mutual vulnerability are expected, these should be given due attention in contact with the inter-organisational supply chain actors.

2.4.1.3 Step Three: Taxonomy Development

The objective of this step is to establish a structured set of vulnerability originating factors that should be used as a basis for developing the vulnerability scenarios. As such, the set of factors may be used as checklists both in conducting the vulnerability analysis, as well as to develop a structured consciousness with respect to the diverse set of factors that may affect the vulnerability of supply chains and SCM.

Taxonomies may be helpful in making constructive approaches to structuring different contexts, as demonstrated by the many two-by-two taxonomies used, e.g., Kraljic's classification taxonomy for supply (Kraljic 1983). A drawback of taxonomies may be that the user gets stuck in a particular way of approaching and thinking. However, we believe that the advantage of having a structured starting point outweighs that, especially for the first rounds. However, as there exist several ways in structuring a taxonomy the decision maker may for their own purposes develop a set of different taxonomies as they develop better knowledge about how to use them, and for which purpose. Such taxonomies will in themselves collect significant knowledge that may have considerable value in its own right, e.g., as in the insurance underwriting process, where taxonomies like the ones described here are used both as formal written documents, and as tacit knowledge of the individual underwriter or group of underwriters as part of the risk assessment process.

The taxonomies may be structured in many ways, and in Fig. 2.4 a structure based on internal versus external factors is presented. To develop a taxonomy several other starting-points may be used. For example:

1. Internal/external versus long term/short term.
2. System attributes; Infrastructure factors; Human and Organizational factors; External factors.
3. DEPOSE (Perrow 1984, 1999) – Design, Equipment, Procedures, Operators, Environment, and Supplies and Material.

4. Seven dimensions of resilience (Foster 1993); Social, System characteristics, Economic, Environmental, Time and timing, Operational, and Physical dimensions.
5. Supply chain and SCM specific – see e.g., Norrmann and Lindroth (2004).

Presenting taxonomies through a fishbone diagram, as in Fig. 2.4, is a good way to communicate the areas to be 'aware of' and to identify that either individually or interacting they may be the originating factor leading to an accidental event or disruption.

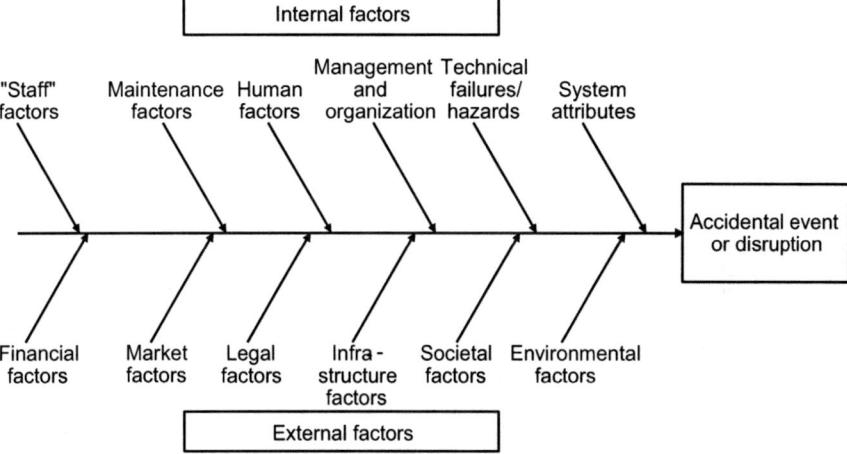

Fig. 2.4 Fishbone diagram of internal and external factors contributing to vulnerability

2.4.1.4 Step Four: Scenario Development

We will use the following definition of a scenario: A scenario is a sequence of possible events, originating from an accidental event, where the events may be separated in time and space, and where barriers to prevent the sequence are part of the scenario. The scenario is as such the chain of events that must be analysed, while the accidental event is the first significant deviation or event that the analysis commences from. In our interpretation, a scenario ends when the system is temporarily stabilized, but before post-accident actions to mitigate, restore or rebuild are initiated. An author within safety and risk theory (Hollnagel 2004) states that: 'One of the important ingredients in understanding that a risk exists, and also in being able to understand the details, is to have imagination. This does not just mean wild speculations but rather the ability systematically to explore the set of possible events' (Hollnagel 2004, pp 181).

Hollnagel uses the term requisite imagination, based on the concept requisite variety from cybernetics and control theory, about 'the ability to speculate constructively about the possible ways in which something can go wrong' (Hollnagel 2004, pp 182). Hazard and operability analysis, HAZOP, is a technique based on the use of process descriptions (originally used on PID's for process facilities) and checklists. As such, HAZOP studies combine steps two and three above, e.g., a good understanding about the supply chains and SCM context combined with good taxonomies acting as checklists for the analysis. Another approach to being imaginative in a structured way is TRIZ, a theory of problem solving originating from Russia (Kaplan 1996). Both Hollnagel (2004) and Kaplan (1997) state the necessity of being innovative and imaginative in a structured way to be able to explore and develop the potential risk and vulnerability space, hence both TRIZ and HAZOP may contribute in such processes, TRIZ as a structured technique for guiding invention and HAZOP to proactively hinder problems in (technical) systems based on well documented processes.

2.4.2 Documenting Scenarios in the Analysis

The first step of the vulnerability analysis is performed based on the work-sheet presented in Table 2.1. In the first column (a), factors that may contribute to vulnerability are listed. The set of factors developed in step three may act as a basis for a checklist over which areas and factors to remember when preparing the analysis. Then in the next column (b), scenarios starting with accidental events, which materialize from the threat, are identified and described. This may help to identify topics that may be improved by small means as well as contributing to increase the focus on everyday factors, and thereby lead to organizational learning. When a scenario is described it is important to establish whether the scenario is likely to threaten the survivability of the supply chain mission or not, column (c). This is done to limit the number of scenarios for further attention, so that scenarios that are not likely to create problems may be kept out of further analyses. The term 'likely' should be understood so that not only the likelihood of an event, but also the consequences and vulnerability aspects related to the system's mission should be given due attention. The next step then is to locate the potential immediate effects in the scenario (d). Establishing an oversight over immediate effects is necessary due to the need to analyse resources, systems and plans for mitigation, restoration, or rebuilding. These resources, systems and plans may both be internal (e) as well as external (f). Detailed checklists of barriers and mitigating, restoring or rebuilding resources should be used to establish the status quo of these resources, systems and plans. Finally, remarks may be made about each scenario (g). To summarize, the objective of the first step of the analysis is to find and describe scenarios that have consequences above a certain level (scenarios of relevance), and how the system is prepared to handle these given scenarios.

Table 2.1 Worksheet no.1; documenting vulnerability scenarios

Threat	Scenario	Likely? (yes/no)	Potential immediate effects?	Resources/systems/plans for mitigation, restoration, rebuilding, etc.		Remarks
				Internal	External	
(a)	(b)	(c)	(d)	(e)	(f)	(g)

2.4.2.1 Step Five: Criticality Ranking

Step five is a quantitative analysis to establish a ranking of the scenarios according to their criticality (i.e. degree of emergency of attention required). The analysis is carried out based on work-sheet no.2 presented in Table 2.2. The work-sheet is rather similar to a work-sheet developed by the US Federal Emergency Management Agency (FEMA). The input to each column in worksheet no.2 is given a weight from one to four, where 'one' means least critical while 'four' means most critical.

Table 2.2 Worksheet no.2; ranking criticality of scenarios

Scenario description	Likelihood of scenario	Consequences of scenario			Resources to mitigate, rebuild, restore, etc.		Total
		Service	Cost	'Other' ...	Internal	External	
	(4-1)	(4-1)	(4-1)	(4-1)	(4-1)	(4-1)	
1	(1)	(2)	(3)	(4)	(5)	(6)	(7)
2							
3							
4							
5							
...							

Work-sheet no. 2 has three main parts. In the first part, the likelihood of the scenario (1) is registered on a scale from one to four. The second part, (2)–(4), addresses the consequences of the scenario with reference to service, cost, and 'other' impacts on the scale from one to four. These consequences are of different kinds, so one should establish an understanding or rating among them as to which is the more important. Service and cost are important aspects of logistics, and we have therefore used them specifically. 'Other' is used to highlight that consequence categories should be chosen and stated specifically to fit the aim of the analysis. For a vulnerability analysis it is especially important to understand how the various consequences affect the mission or survivability of the supply chain. It is also possible to use the analysis specifically towards selected areas of interest by focusing only on these or by giving them higher weighting. The third part, (5)–(6), addresses the presence of mitigating, restoring or rebuilding resources, both internal and external, and gives them weights on the scale from one to four. The resources, systems or plans to mitigate, restore or rebuild, are critical in a vulnerability analysis, as they constitute the tools that shall bring the system back to new stability.

2.4.2.2 Step Six: Scenarios of Importance

Which scenarios that are most important to attend to (most critical), may, for example, be shown in a consequence-likelihood matrix as in Fig. 2.5. The consequence-likelihood matrix shows the likelihood and consequence ranking of the accidental event, as well as the consequence ranking after the effect of resources to mitigate, restore or rebuild have been accounted for. The direct consequence ranking of the accidental event is shown by the circle with the scenario's number inside. The leftward black line with the downward black triangle at the end, illustrates the effect that resources to mitigate, restore and rebuild have on the consequence ranking of the scenario, with regard to the systems survivability due to the accidental event. The different shading in the matrix visualizes the importance to attend to

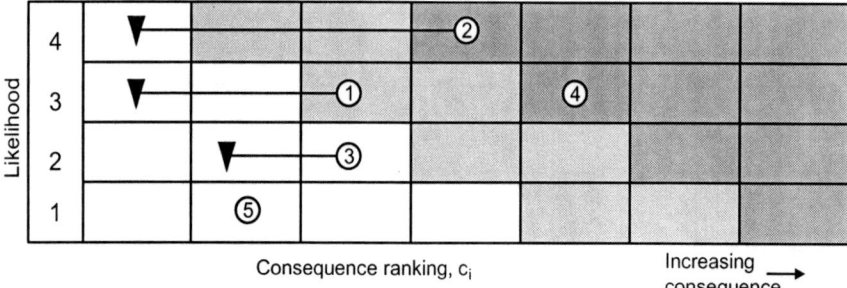

Fig. 2.5 Presenting scenarios of importance in a likelihood/consequence diagram

among the scenarios. The more to the right, and the higher in the matrix, the more important is a scenario. As such we have 'low-criticality' scenarios in the lower left corner (white), and 'high-criticality' scenarios in the upper right corner (dark grey shading). The 'criticality areas' should be based on the acceptance criteria developed in step one, both for the un-mitigated and the mitigated consequences.

Of the scenarios presented in Fig. 2.5 we see that scenario no.1 is of medium criticality, due to the potential to reduce this to 'low' criticality through available resources to mitigate, restore or rebuild,. Scenario no. 4 on the other hand, which is of 'high' criticality, has no potential reduction in consequence ranking due to barriers or the absence of mitigating, restoring and rebuilding resources. Hence, this should be addressed well before scenario one.

The result of the analysis is a list of critical scenarios that may be used 'backwards' in the second part of the analysis, to guide actions to prevent causes or interactions that may lead to an accidental event, or improve barriers and mitigating, restoring or rebuilding resources, systems and plans that may limit the consequences of an accidental event.

2.4.2.3 Step Seven: Potential for Reducing Likelihood and Consequence

After establishing the most critical scenarios, two topics must be addressed. First, one has to analyse the potential causal chain from the root causes leading up to the accidental event, to establish how actions can be made to reduce the likelihood of the individual causal chains. Secondly, one has to understand how the consequences of the accidental event can be reduced through addressing design, operational and contingency aspects. Reduction of the likelihood of causes is considered more important than reduction of consequences (Norwegian Petroleum Directorate 1990). This consideration is reflected in work-sheet no. 3 in Table 2.3, which is based on the same idea as adopted in the risk analysis regulations issued by the Norwegian Petroleum Directorate (1990).

Table 2.3 Worksheet no.3; evaluating measures with potential to reduce likelihood and consequence

Scenario Description	Reduction of likelihood		Reduction of consequences		
	Measures to avoid or reduce a threat	Measures to reduce the probability of an accidental event	Measures related to design and passive barriers	Measures related to operations and active barriers	Measures related to mitigation
1	(i)	(ii)	(iii)	(iv)	(v)
2					
3					
4					
5					
...					

This final part of the analysis, as presented in Table 2.3, addresses both the likelihood and consequence side of the vulnerabilities. Likelihood reducing measures take precedence over consequence reducing measures, so core aspects, e.g., threats and design, take precedence over events and actions. The likelihood reducing part of the analysis is divided into two. First, measures that may reduce a threat should be addressed (i), then measures that may reduce the probability of a threat to develop into an accidental event (ii). Here we see that the analysis focuses primarily at the core (the threat), then the manifestation of the threat. After addressing reduction of likelihood, the focus is turned to reduction of consequences, which is divided into three. First, measures related to the design of the system, and its passive barriers should be addressed (iii), then measures related to operations, procedures and active barriers (iv), and finally measures related to mitigation, restoring and rebuilding to bring the system back to a new stable situation (v).

The scenario may be uniquely determined by a given threat, but may also be dependent on a combination of more than one threat. This should be remembered when addressing the likelihood side of the final part of the analysis. The threats are ever present factors in the system, while the scenario start with the materialization of the threat, the accidental event we want to avoid. Therefore a thorough and creative identification and description of scenarios is important, as described in step four. This will help to identify topics that may be improved by small means as well as contribute to increase the focus on everyday factors, and thereby lead to organizational learning.

2.4.3 The Second Round

Vulnerability analysis, as a risk analysis tool, is a top-down analysis aimed at finding and providing focus to the most important of the key or critical elements. Hence, after having completed the first round of the analysis, one is left with a set of critical vulnerable elements to handle. This will often require additional analysis to be able to reduce the vulnerability in the most cost or other resource effective way. Such further analyses will often require specific analyses that will be related to the functional area in which the vulnerable elements were discovered.

2.5 Summary

Many supply chains are particularly vulnerable because management is not fully aware of the threats that the system is exposed to and the vulnerable situation these threats impose on the supply chain. This chapter has outlined an approach to structure, understand and analyse vulnerability in supply chains. The approach could be a part of SCM, especially to test changes in vulnerability after changes in the supply chain network. The approach is based on seven steps, with three main issues;

1. Understand the types of threats and risk that may threaten the supply chain system's short and long term mission and comprehend the underlying factors driving such risks.
2. Analyse and rank the scenarios (processes and mechanisms) through which these risks and vulnerabilities may mature.
3. Understand how the vulnerability scenarios, their likelihood and consequences, may be reduced in a cost- and service effective manner, reaching an acceptable vulnerability level.

We believe that through this approach, vulnerability and resilience management may become a practical part of SCM, establishing a formal and conscious process of understanding, analysing and handling risks and vulnerabilities in supply chains.

References

FEMA, US Federal Emergency Management Agency, www.fema.gov.
Foster H.D. (1993) Resilience theory and system evaluation. In J.A. Wise, V.D. Hopkin and P. Stager (eds.) *Verification and Validation of Complex Systems: Human Factors Issues*. NATO Advances Science Institute, Serie F: Computers and System Sciences, Vol. 100, Springer, Berlin.
Hollnagel E. (2004) *Barriers and Accident Prevention*. Ashgate Publishing Ltd., Aldershot.
Hollnagel E. Woods D. and Leveson N. (eds.) (2006) *Resilience Engineering*. Ashgate Publishing Group.

ICAO (2005) *ICAO Safety Management Manual*. Doc 9859, AN/460, 2005.
IEC 300-3-9 (1995) Dependability management – Part 3: application guide – Section 9: risk analysis of technological systems. International Electrotechnical Commission, Geneva.
Kaplan S. (1996) *An Introduction to TRIZ The Russian Theory of Inventive Problem Solving*. Ideation International Inc.
Kaplan S. (1997) The Words of Risk Analysis. *Risk Analysis*, Vol. 17, No. 4, pp. 407–417.
Kraljic P. (1983) Purchasing must become supply management. In *Harvard Business Review*, September/October, pp. 109–117.
Meister D. (1991) *Psychology of System Design*. Elsevier, New York.
Norrmann, A. and Lindroth R. (2004) Categorization of Supply Chain Risk and Risk Management. In C. Brindley (ed.), *Supply Chain Risk*, Ashgate Publishing Ltd., Aldershot.
Norwegian Petroleum Directorate (1990) *Regulations Concerning Implementation and Use of Risk Analyses in thePpetroleum Activities withGguidelines*. Norwegian Petroleum Directorate, Stavanger, Norway.
Perrow, C. (1984/1999) *Normal Accidents: Living with High-risk Technologies*. Basic Books, New York.
Reason, J. (1990) *Human Error*. Cambridge University Press, Cambridge.

Chapter 3: Risk Management in Value Networks

Jukka Hallikas and Jari Varis

Technology Business Research Center and School of Business, Lappeenranta University of Technology, Finland

3.1 Introduction

One of the most challenging issues in the anticipation of risks in the business environment is to understand the dynamics of industry change and to recognize the relevant indicators. In large companies, different kinds of business and market intelligence systems and departments may collect great amounts of data and indicators related to the possible changes, but the difficulty lies in the interpretation of this data, and making sound decisions based on these indicators. It has been noted that the ability to anticipate where lucrative opportunities are likely to arise distinguishes top-performing firms from ordinary companies (Fine 1996). Marketing and technological forecasting capabilities are thus critical, especially for companies which are active in turbulent, high-technology markets. On the other side of the coin, risk and uncertainty are inherent in the hoped-for windows of opportunity, and forecasting ability is a critical capability in companies (Fine 1996). "Fortune favours the prepared firm" as Cohen and Levinthal (1994 p 1) have pointed out.

Business intelligence, as well as the business strategy literature, has historically concentrated on the individual corporation as the appropriate unit of analysis (Fine 1996). Lately, the focus has been shifting towards the extended organization and network level, which may cause additional difficulties in forecasting the future risks and opportunities. Recent literature also takes into account the competitor network when analyzing risks in value networks (Gilad 2004). As companies are increasingly relying on different kinds of collaborative arrangements in their business models (Draulans et al. 2003; Contractor and Lorange 2002), they naturally become also more and more dependent on other companies capabilities and resources, which makes their situation more unpredictable regarding possible changes in the business environment.

Especially in knowledge-based high-technology markets, such as the ICT-industry (Information and Communications Technology), the pace of change is so fast that it is a very challenging task to keep track of all possible indicators of change, and to make reliable decisions about future actions. The features of the ICT industry – convergence, technology and knowledge intensity, turbulence, short life-cycles of products, high levels of technological and market uncertainty etc. – make the management of a company a demanding task. Furthermore, the intangible nature of the products increases the complexity of exchanges in the value network.

In these kind of turbulent industries the boundaries of firms become fuzzier, and lead to a symbiotic life between companies. At the same time, the knowledge-based economy also favours the generation of collaborative networks as a way to respond to the fragmented technological knowledge demands (Contractor and Lorange 2002; Varis 2004). For instance, operators are unlikely to develop by themselves all the services and applications that the market requires. It is probable that the content and services are provided by a myriad of third party organizations (Peppard and Rylander 2006), which can be seen in the increased alliance activity in these industries.

The primary contribution of this chapter is to illustrate different approaches to risk management in value networks. We present different theoretical backgrounds for analyzing networks and illustrate some examples of the analyses that can be applied to support risk management at the network level.

3.2 Value Networks

The fundamental decision problem in all businesses is how value is created, shared and captured in the complex relationships between customers, competitors, suppliers, and other network actors. The broad concept of value can be regarded as a trade-off between benefits and sacrifices (Parolini 1999), and the value of an offering is formed through a set of relationships with customers and other stakeholders (Ramirez and Wallin 2000). Value in business relationships constitutes the intangible and tangible elements of customer and supplier perceptions of the value of an offer. The risks associated with value creation, on the other hand, pose a threat of value escalation in these relationships. This may be caused by the unwanted transfer of knowledge when engaging in close cooperation with another company.

The term "value chain," first introduced by Porter (1985), demonstrates how value creating activities have been organized within a single firm. In addition to the importance of studying single value chains in industries, the value network approach provides a holistic approach for exploring the relationships and business logic of several, combined value chains. This approach is derived from the literature of inter-firm relationships, where the underlying value creation assumptions are based on the sharing of complementary capabilities, creating new knowledge through these capabilities, and learning from several partners in the networks of industrial relationships (Doz and Hamel 1998). The value network approach takes into account the tangible and intangible dynamic interchange between various actors in industry-wide supply chains and networks. Although the risk management in supply networks and relationships has become more important in recent years (Harland et al. 2003; Agrell et al. 2004; Hallikas et al. 2004), there is still a growing need to consider risk management in terms of the value networks of interrelated businesses and to take industry dynamics into account in the analysis.

What has changed in many industries, such as ICT, is that single firms have become more dependent on each other due to the outsourcing and sharing of complementary resources and capabilities between partners in industry networks. This kind of network economy has emphasized the importance of external linkages with other companies. Thus, the original value chain level framework for a single organization no longer fits the analysis of value creation activities. Next, we will briefly outline the concept of the value network and its connection to risk management.

3.2.1 Business Ecosystem Health and Risks

Business ecosystem theory is closely related to the value network risk management consideration. A business ecosystem can be outlined as the environment beyond the core business and the value network. Sometimes the term business ecosystem has been used more as a conceptualization or analogy to value networks. As defined by Iansiti and Levien (2004), the business ecosystem can be seen as interdependencies between several business actors and the business environment in which they operate.

According to Iansity and Levien (2004), there are strong parallels between business networks and biological ecosystems. They recommend that companies should systematically identify the organizations with which their future is most closely intertwined and determine the dependencies that are most critical to their business. When a company can be dependent on hundreds or even thousands of other businesses, it is obvious that the members of a network can rise and fall together. The rise and fall of the Internet business can be seen as an example of this kind of a common fate in an ecosystem.

A business ecosystem should survive unforeseen technological change, just like a biological ecosystem must tolerate environmental changes. The health of the ecosystem is an important indicator of how the ecosystem is able to produce value and share it between the members of the system. From the risk management standpoint, companies should position themselves within the business ecosystem that is most able to generate value now and also into the future, and which is committed to sharing wealth with its members.

The evolution of the business ecosystem provides a framework for considering the strategies and relationships in the value networks in terms of the business life-cycle. Here the business ecosystem characterizes the opportunity and risk environment which allow different value networks and core businesses to grow and co-evolve. According to Moore (1996), this co-evolution is defined as a complex interplay between competitive and cooperative business strategies. Furthermore, the analysis of a business ecosystem reveals that the industry will continuously redefine its boundaries, breaking those traditionally associated with the industry. Moore (1996) defines the effectiveness of an ecosystem as one that must add value, provides economies of scale, generates opportunities for continuing innovation, and is continuously investing in and expanding its community of allies.

One way to analyze risks in networks is to evaluate the health of the ecosystem with suitable indicators. Figure 3.1 illustrates an example of possible ecosystem health indicators, the sales growth (%) and average return on invested capital (ROIC %), related to the IT subcontracting industry and selected original equipment manufacturing companies (OEM) during the years 2000-2005.

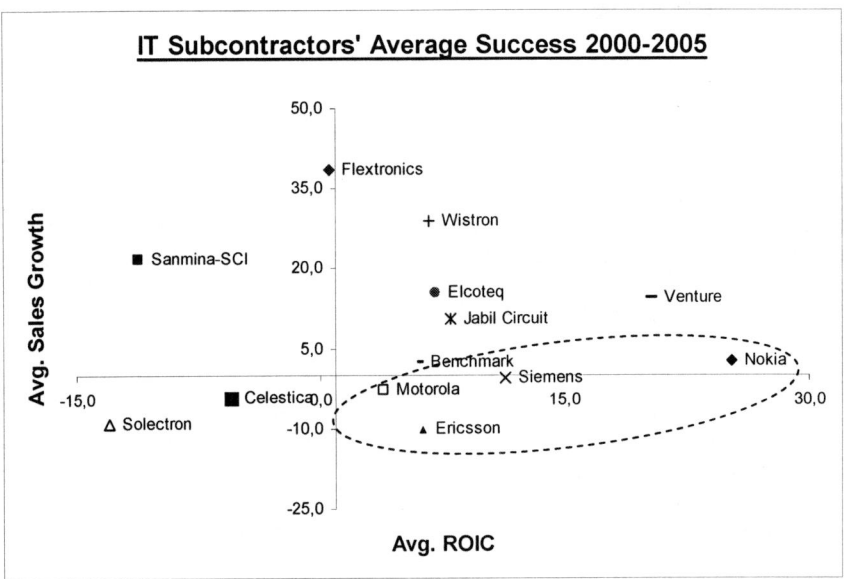

Fig. 3.1 Health of an IT subcontracting ecosystem

As shown in the circled area in Fig. 3.1, all the selected OEM companies are situated in the middle right side of the quadrant, implying relatively steady sales growth and positive return on invested capital. The subcontractors' position, on the other hand, varies to a greater extent. Some of the subcontractors have employed a growth strategy, while others are situated close to their OEM partners. Some companies' performance indicators are rather poor, however, indicating uncertainty in the total health of the ecosystem. Our data indicates, as seen in the figure, that the risks associated with the OEMs seem to be transferred to some extent to the subcontractors in the case sample. This hypothesis might be interesting to test with a larger sample.

There are various strategies for an organization to position itself in the business ecosystem, each of which also presents certain risks. The suitability of different strategies depends on several factors, including the complexity of the business environment and the complexity of asset-sharing relationships in the ecosystem. We will discuss two basic strategies, keystone strategy and niche player strategy. The description of these two strategies is based on the ideas of Iansiti and Levien (2004).

The keystone strategy is based on the analogy with biological keystones, which have an ability and motivation to create and sustain health and act as a health regulator in an ecosystem. Keystones are powerful leaders in an ecosystem creating platforms that link other members in the system and create niches and innovation opportunities for the ecosystem members. The major risk for the keystone player is the inability to sustain health in the ecosystem. Health is ensured for example by sharing information, and sharing critical assets in the ecosystem, as well as by creating opportunities for the members and assuring that efficiencies and innovations take place. Finally, keystones aim at sharing benefits and reducing risk with the ecosystem members and increasing the robustness and sustainability of the ecosystem as whole. Keystone strategies are mostly suited for large and innovative companies.

The focus of the niche strategy is specializing in capabilities needed in the ecosystem domain. Niche player advantage in the ecosystem is created by leveraging and developing the critical assets and capabilities needed in the ecosystem. One example of a niche player is the application developer for a standard platform product. The niche players are dependent on the existing solutions and develop complementary components to support the effectiveness of these solutions. The risk of the niche strategy is that the value of complementary resources and the capabilities of the niche actors depend on the success of the ecosystem platform.

3.2.2 Value Creation and Risk Management in Networks

The concept of the value creating system has been added to the strategic management literature to express the entire set of activities and companies linked to produce value for both end-customers and actors in the system. According to Normann and Ramirez (1993 pp. 65–66) the focus of strategic analysis is not the company or even the industry but the value-creating system itself, within which different economic actors – suppliers, business partners, allies, customers – work together to produce value. Much of the prior understanding of value creation is related to the value chain analysis of Porter (1985). Value chain analysis describes how the value increases in the value chain/system. According to Parolini (1999), the net value obtained by the actors is related to the profits made by the value creating system as a whole. This value is distributed among the members in a system. The actors may achieve a larger share of the profits according to their ability to manage and control activities:

1. that are more critical to the final result
2. whose output has some unique elements in comparison with other goods
3. whose contribution can be perceived by purchasers
4. whose performance requires skills and resources that cannot be easily imitated
5. which correspond to the system bottlenecks
6. whose control makes it possible to influence the behaviour of the entire value creating system

Parolini (1999) has developed a framework for the structural analysis of the nodes in a value-creating system from the value network perspective. It is based on the exploration of the attractiveness of value adding nodes by considering elementary units, e.g., the nodes that indicate sets of economically inseparable activities in a value creating system or a value network. The framework for the analysis of the structural attractiveness of the nodes corresponds with Porter's (1985) analysis of the competitive positioning of an organization.

The objective of value network analysis is to explore the structure for its interaction with several actors in a network of relationships. Instead of on organizations' internal perspective on value creation, value network analysis provides a powerful way of exploring external and internal linkages of an organization to generate value for the customers and participants in a network. The fundamental addition to the traditional intra-organizational strategy and risk analysis is that in value network analysis, the relationships and structure of the value network influence the created value and risk to the organization.

There are several approaches and levels for mapping inter-firm value flows. The traditional input-output logic of modelling the performance of the system lies behind many approaches to the mapping of business chains and networks. Here, each stage (organization) in a chain can be mapped as flows of inputs and outputs in the value chain of activities. In general, by using value network mapping it is possible to:

1. define and receive common understanding on the systemic structure of the value network and value creating roles between the actors
2. understand the value flows between the actors in a network (both tangible and intangible)
3. identify and assess the risks associated with the value flows

From the risk management perspective, the business-to-business structure of industries determines how the value is created and shared among the members in the network. According to Allee (2000), a value network generates economic value through complex dynamic exchanges between firms. These exchanges go beyond the transaction of goods and services among firms, covering the knowledge transfer and sharing of intangible benefits. Furthermore, these attributes are important in mapping the relationships in a value network and can be briefly described as:

1. Goods, services, and revenue involving for example contracts and invoices from transactions.
2. Knowledge referring to for example, the exchange of strategic information, planning knowledge, process knowledge, technical knowledge, collaborative design policy development.
3. Intangible benefits involving the exchange of value and benefits that go beyond the actual services and that are not accounted via traditional financial measures, for example the sense of community and customer loyalty.

Figure 3.2 depicts an example value network relating to some relevant actors in the main business model of Google. The map presents both direct and indirect linkages between the actors in the network. However, only direct linkages are included in the example map. Table 3.1 presents a summary of the flows between the focal company (e.g., Google) and selected actors. Analysis of the indirect linkages to the business model has been omitted.

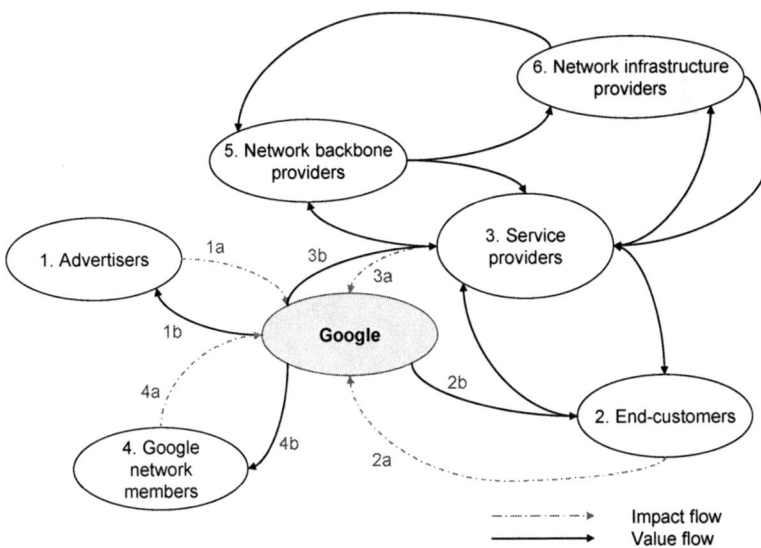

Fig. 3.2 Example value network of the business model for Google

The example demonstrates the connection of value network analysis with modern business model thinking. Here, the business model of a single firm is outlined via tangible and intangible flows between the actors in the network, addressing the essential role of systemic business logic as well as knowledge and information sharing. Another important characteristic of the value network method is to analyze the value network as an open system of interconnected elements. This systemic and holistic approach is likely to provide a more comprehensive way of understanding the dynamics of value network development. Applying the method of Allee (2000), we have separated the value interchange into value flows and impact flows, both describing the divergent tangible and intangible exchange between the actors in the network. In addition, the risks of each value network relationship have been mapped according to the value and impact flows.

Table 3.1 Value and impact flows between actors in the value network of Google

	Impact (a)	Value proposition (b)	Risks
1. Advertiser	Fees from advertisement

Advertising information

Customer requirements and needs for the R&D | Global/Local visibility in the Internet
Targeted advertisement based on customer search behaviour
Delivery of relevant ads targeted to search results or web content (Google AdWords program)

Customer information | Brand reputation
Price escalation
User base size
Effectiveness of Internet advertisement
Trust of advertisers |
| 2. End-customer | User information | Effective search technology
User friendly interface
E-mail with mobile client
Instant communication
File sharing | Competitiveness of search technology

Trust of end-customers |
| 3. Service provider | Customer access
Customer identity
Location information
Billing services | Volume
Global information platform for service delivery
Search technology | Restricted access to the services
Price escalation of access |
| 4. Google network member | Displaying of Google's advertisements on web sites | Sharing of fees with network members | Brand reputation
Trust between network members |

The method of understanding the characteristics of value network activities requires the analysis of relationships between the actors in a network. As shown in the example value network map (Figure 3.2), the analysis of direct linkages between one focal actor and its direct relationships provide a systematic approach for analyzing the value flows between the actors in a value network. However, more comprehensive mapping is needed to understand the dynamics of the relationships from the value network perspective. This means that instead of just mapping the direct input and output flows, it is also necessary to address the indirect relationships, because this makes it possible to outline the patterns of change from both the direct and indirect perspective. The Google case is a good example of this kind of systemic nature of a value network business model, where most of the direct value flows are free to the end-customer, but these flows generate wealth indirectly to the whole ecosystem, which finally feeds back to Google as advertisement financial flow. In this kind of value creating network, the indirect linkages in the network may pose a high risk for the success of the business model.

3.3 Connecting Industry Change with Risk Analysis

When considering risk management at a value network level, the risk management action is derived from the strategic positioning of the company in its value network. Here, the main concern is related to the robustness of the competitive positioning of the firm's business model to create value to its end-customers and partners. As pointed out by Gilad (2004) shifts and dissonance related to industries and markets cause the most serious risks to the companies in industry value networks. A dynamic interpretation of industry and value networks is therefore needed to anticipate the changing competitive structure of industries, and to reduce the strategic failures of companies. We will outline the connection and positioning of the industry change drivers and uncertainties in relation to the value network by using the ICT sector as an example.

To address the risk of dynamic behaviour of the industry value network, it is necessary to identify and analyze the factors that drive the change. According to Grant (2002), there are two factors that drive the industry evolution: demand growth and the production and diffusion of knowledge. Demand growth illustrates the pattern of life-cycle evolution at a particular industry level, commonly characterized by an s-shaped curve. The creation and diffusion of knowledge, on the other hand, illustrates the dynamics of technological change, product and process innovations and, in general, the indicators and drivers of change in the industry.

Traditional industry analysis literature has divided the change attributes into factors arising from the internal (transactional) business environment, and factors emerging from the external (contextual) business environment (van der Heijden 1996). The identification of external change forces is essential in order to address the significant and uncertain factors that drive the change at the industry and value network level. Here, the external factors are also likely to provide the essential weak indication of changing conditions in the business environment of the value network.

Conceivably the most commonly used taxonomy for the identification and clustering of external factors is the so called PEST – an analysis, which divides the macro environmental forces into Political (and regulatory) forces, Economic forces, Societal forces, and Technological forces. Examples of these factors are presented briefly in the following:

1. Political/Regulatory Factors
    (a) Development of EU-level rules for Voice over Internet (Voip) regulation
    (b) Delayed regulation of Internet-based communication services
    (c) End-user welfare and increasing competition as an objective of regulation
2. Economic Factors
    (a) Emergence of new market areas with growth potential (China, India, Africa)
    (b) The value of mobile communication business is growing
    (c) The cost of service infrastructure is decreasing

3. Societal Factors
    (a) Aging of the population
    (b) Global growth of mobile communication
    (c) Globalization of services
4. Technology Factors
    (a) Communication services are mostly based on the IP-protocol
    (b) Increasing importance of security as a value driver
    (c) Shift towards standardized interfaces and modular system designs

We argue that it is possible to understand many of the value network changes and risks by analyzing the effect of external change factors. This would be facilitated by analyzing the cause-and-effect relationships between the various external and internal factors of the value networks.

3.3.1 Trend Analysis on Network Level: Case ICT-Industry

Trend analysis of the value network level is one way to anticipate future business risks in a company's environment. The purpose of the analysis is to detect the driving factors and uncertainties that companies in the ICT industry find important. When considering these forces and uncertainties either as opportunities or threats, they expose important value creation opportunities for the companies in the industry.

We studied the financial reports of 28 companies for the year 2005 to analyze the trends in the ICT sector. The researchers went through these reports of ICT companies and collected references relating to issues that may have importance in the future development of this industry. Cause and effect relationships of driving factors and uncertainties were addressed to demonstrate the interaction between attributes in the analysis. However, all causes and effects could not be traced from the company reports. We employed the PEST categorization of the elements in order to aggregate similar attributes into separate tables. This facilitated the compilation of data from several company and actor sources. The process, which can be applied iteratively in the analysis, is the following:

1. Collect the data of driving factors and uncertainties from selected company reports
2. Formulate a cause-implication table, which connects the driving factors and uncertainties into the larger context
3. Create actor-specific tables (e.g., Internet, Operator)
4. Classify the drivers and uncertainties according to PEST (political, economic, social and technological) classes to combine the change attributes.

The researchers also evaluated which sectors the trends may affect inside the ICT industry. The most important (most often mentioned) trends were classified in different PEST classes. As a conclusion, the researchers provided a list of the most important trends and their estimated effect to different industry players.

This kind of analysis is related to the need to understand the change in industry, which knowledge is not always easy to come by, as McGahan notes in her article in Harvard Business Review (McGahan 2004). She points out that companies misread clues and arrive at false conclusions all the time. Furthermore, she claims that there are two types of threat of obsolescence. The first is the threat to core activities, and the second is the threat to core assets. These threats may lead to problems for the activities to generate profits or for assets to generate value as in the past. Changes in the business environment may require new ways of dealing with competitors and consideration of the use of alliances to defend new competition from outsiders.

Collaboration can also make the risks lower in knowledge-intensive sectors. First, the costs and risks can be shared with collaborative partners. Second, alliance partners can provide increasing network returns to scale and economies of scope. Third, alliances can be used as a device to win the "learning races" in these markets, where speed is of the essence, especially in situations where multiple technologies are applied through collaboration (Contractor and Lorange 2002). It seems that the presented trend analysis can give interesting reflections on the future development paths of an industry. The capacity to anticipate future development from the "weak signals" in the particular industry may result in better performance in the race for sustainable competitive advantage.

3.3.2 Future Drivers and Scenario Creation

One of the most commonly used frameworks for analyzing the risks and opportunities in an industry consists of 5 competitive forces (buyer power, supplier power, substitute threat, rivalry, threat of new entrants), which in a particular sense determine the organizations' competitive power position in the industry. When connecting the specific industry change drivers/scenarios into the analysis, the 5-force model can be used as a powerful tool to explore the dynamic tensions in the industry value network positions. In general, Porter's framework can be used as a platform for analyzing the impact of selected change drivers on the future value network position of a firm. This will provide a basis for anticipating the risks that affect the company's future structural position in the network.

In order to explore the change impact of selected industry driving forces on the risks associated with competitive positioning, we use changes in the traditional telecommunication services as a case example. Table 3.2 illustrates the example list of "what if" type driving variables that are considered to be important in the future of the ICT sector in general. Each variable presents a plausible but hypothetical argument about the future industry condition, and the total set of variables constitute a hypothetical scenario for the future of the industry. The scenario is based on a larger survey of expert opinions associated with the exploration of future industry development. Example driving forces are plausible industry driving variables that are connected to the future of a single actor group within the ICT value network. In the example, these driving forces are connected to evaluate the changes and risk of the competitive position of an actor group. These can be outlined as single scenarios providing a different context and impact on the single actors within the industry value network. The variables as a whole constitute a single scenario for the future of the industry.

Table 3.2 Example of the connection of driving force attributes into the competitive positioning of an actor group within the value network

Assessment (-1, 0, +1) (can be a composite effect)	Buyer power	Supplier power	Substitute threat	Rivalry	Entrant threat
Demand & Societal					
Internet services become mobility-aware	0	0	-1	0	1
Aging of end-users create new types of services	0	0	1	0	1
End users participate actively in service development and content creation	0	-1	-1	0	1
Access to mobile services becomes independent of access and operator	1	-1	1	1	1
Quality of new wireless access channels becomes sufficient for POTS-quality telecom	1	-1	1	0	1
Majority of end users do not need to trust the operator for need the security	1	-1	1	0	1
Opportunities for new profitable telecommunication services arise	0	-1	-1	1	1
Video, TV & other high-bandwidth service usage grows	0	-1	-1	0	1
Majority of customers prefer converged bundles to segregated services	0	0	-1	0	0
Industry structure					
Revenue sharing and valuation models promote joint production of mobile services	0	0	-1	0	0
Bandwidth of mobile networks becomes sufficient to provide all demanded services	0	0	-1	0	0
A service can be easily offered via different access channels	1	0	1	0	1
Consolidation of incumbent service operators continues	-1	-1	0	-1	-1
New mobile service business models arise extensively outside industry incumbents	1	-1	1	0	1
Political & Regulatory					
Regulators' cooperation increases globally	1	0	1	1	1
Regulation is influenced by industry incumbents and it affects free competition	-1	-1	-1	0	-1
Entry for service operators is liberalized	1	-1	1	0	1
Economical					
Income level in developing countries increases	1	-1	1	0	1
Total spending on information services in developed countries increases	1	-1	1	0	1
Money spent by organizations on ICT services increases	0	0	1	0	1
Technological					
There is seamless interoperability between customer access channels.	1	0	1	0	1
Modular and standardized product architecture becomes the dominant solution in ICT	1	-1	1	0	1
Seamless interoperability (roaming) of services between platforms, layers and networks	1	0	1	0	1
It becomes necessary to make technology platforms IP-based	0	-1	1	0	1
Mobile terminals become access agnostic	1	-1	1	0	1
Wimax lowers entry barriers for 3/4-play operators	1	0	1	0	1
Total	**12**	**-15**	**9**	**2**	**19**

Connecting the future drivers into the scenario process could provide a more profound, but complex method to identify and monitor risks in industries. The scenario building exercise requires more in-depth exploration of the tensions and dissonance between different driving forces in the industry. The illustrated future variable listing can be used as background work in a more profound scenario building exercise. In addition to the recognised advantages of scenario work, the presented combination of future variables with the 5-force framework can result in holistic risk analysis of competitive positioning and contribute for the unfolding of more robust and risk-aware strategies.

As can be seen in Figure 3.3, which is a graphical presentation of the example dynamic 5-force analysis presented in Table 3.2, there are several indications of risks that affect the strategic position of the case industry group in a value network. In general, the buyer power and substitute threat are likely to intensify, both arising mainly from the emergence of new service providers in the

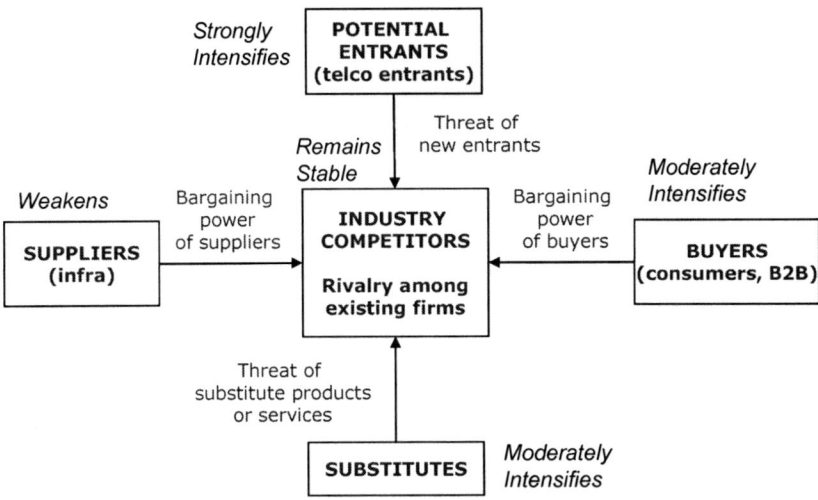

Fig. 3.3 Impact of selected driving factors on 5 competitive forces (adapted from Porter, 1980)

telecommunication industry. In the case example, the most important threat is associated with entry of new actors into the industry. For example, new cheaper IP-based technologies and possibilities to operate as a virtual service provider lower the entry barriers. This is also partially likely to cause better negotiation power with suppliers. In the case example, the competition between the existing industry incumbent firms does not change according to the selected list of future variables.

The changes in the telecommunication service business model illustrate the need to connect industry change variables with risk analysis of the competitive positioning of the company. The purpose of the analysis is to expose whether larger threats or opportunities are about to arise from the power relationships between the industry value network actors.

3.4 Risk Profiles of Interconnected Actors

To continue from the analysis above, we present an approach for analyzing the risk profiles of interconnected actors. It is important to understand the risk profiles of the different actors in the industry. Furthermore, the interconnections between the actors in the value network may result in complex interdependencies impeding risk identification. The primary aim of actor-specific risk analysis is to increase the understanding of the uncertainties of different players in order to explore the future development paths of the industry. Risk and uncertainty analysis may provide an important means for collecting and analyzing relevant data for the scenario development. However, the primary challenge of risk analysis is to address and combine the risks of different levels/units in a value network. This implies the need to consider risks at the organizational, dyadic and value network level.

The general industry analysis framework can be used as the information source when formulating the example risk profiles of industry actors. The risk profiles of selected players in the industry provide a dynamic view on the risks in the industry in general. The analysis can be done by collecting the information regarding the uncertainties of different players, and analyzing the causes and effects of the identified risks. The specific implications of risk factors and their intensity for the different players have been left outside the scope of analysis. The analysis provided here is intended to present examples of generic risk profiles in the industry by using a network infrastructure provider and a service operator as illustrators. The information for the risk analysis has been collected principally from annual reports of companies in these industries.

Tables 3.3 and 3.4 present examples of the risk tables associated with industry and market conditions from the perspective of two actors, a network infrastructure provider and a service operator. These two actor groups have traditionally been highly interconnected in the ICT sector. The primary focus is to explore the risks and uncertainties that arise mainly from the general industrial development. The analysis covers also the potential causes and effects of risk factors. Table 3.3 illustrates the selected risk factors from the network infrastructure provider's perspective and Table 3.4 from the service operator's perspective. Finally, an influence map of the industry development will be outlined as an example of the complex connections of the cause-and-effect attributes of risk factors. The objective is to illustrate the dynamic and systemic interdependencies of the industry development.

The economic and regulatory changes in different countries affect the investment willingness in the industry and cause a substantial threat to the infrastructure providers. The network infrastructure providers' interest in the industry and the end demand growth are principally related to the operators' continuation of making investments in new technologies and services. However, for example the consolidation among network operators may provide a substantial threat to this development. One bottleneck in growth development could also be the availability of lower-cost handsets in lower-income markets. One associated threat in the industry is the intense competition, which is caused by alternative technologies and telecommunication platforms.

From the network service providers' perspective, the regulatory environment causes a substantial threat to the business. In general, competition is increasing in the communication service business and customers may choose among various service providers, which increases the price erosion of services. One essential factor driving the change in competition is the availability of close to free phoning and data services by VOIP (Voice over IP) operators. Consequently, there is reduced interest in the investments to new infrastructure and services. Another risk factor arises from the Mobile Virtual Network Operators' (MVNOs) side. These communication service providers do not own the infrastructure and are therefore able to work with much lower investment risk than traditional operators.

Table 3.3 Selected risks from the network infrastructure provider's perspective

Risk factor	Potential causes	Potential effects
Economic and regulatory changes in various countries	- Economic and political instability - Nationalization of private assets - Price or exchange controls	- Lack of investments - Operating results
Continued growth of mobile communications	- Usage of voice & data - Operators' ability to introduce services - Affordable tariffs by operators	- Decline in markets - Business and operational results
Changes in the regulatory environment for telecommunication Systems and services	- Tariff regulation of pricing services - License fees - Health and safety - Privacy regulations	- Timing and costs of new network construction expansion - Commercial success of these networks
Consolidation among network operators	- International and inter-country consolidation - Competitive pressure	- Less network equipment and associated services required - Delayed network investments - Bargaining position and profit margins
Consolidation among equipment and service suppliers	- Stronger competitors that are better able to compete as end-to-end suppliers - Competitors who are more specialized - Competitors with greater resources	- Material adverse effects on business, operating results, and financial condition
Highly competitive industry	- Price pressure - Rapid technological change - Intense competition from existing companies and new entrants	- Price erosion - Profit margin - Revenue - Operating result
Credit and other risks relating to customers' businesses and operations	- Finance to customers - Customer financing is a competitive factor in obtaining business	- Credit losses - Material adverse effect on business, results for operations or financial condition.
Liability claims	- Potential health risks associated with electromagnetic fields - New scientific findings of adverse health effects of mobile communication	- Adverse effect on business

When considering the example risk profiles separately, it is uncomplicated to identify that they do have several connections with each other. Therefore, when combining the causes and effects of several risk factors, it is possible to determine the connections among several actors in the value network. By this way it is possible to illustrate the larger and more complex risk structures like the map in Fig. 3.4. The example influence map illustrates the cause and effect relationships between elements in an industry system. An arrow represents a positive relationship implying that element 'x' increases the element 'y', and vice versa. Similarly, a negative relationship is illustrated by an arrow with a minus symbol indicating that when element 'x' increases the element 'y' decreases, and vice versa.

Table 3.4 Selected risks from the service operator's perspective

Risk factor	Potential causes	Potential effects
Changes in regulatory environment	- Agencies which regulate and supervise the allocation of frequency spectrum and which monitor and enforcement of regulation	- Operations in geographic areas - Pricing of services - Use of mobile phones
Increased competition	- Customers choose other service providers - Shift from customer acquisition to customer retention - Price erosion	- Reduced market share - Reduced revenue
Delays in the development of handsets, network compatibility, and components	- Number of vendors - Lack of common standards and specifications - Lack of technological interoperability	- Delayed launch of 3 G services - Additional capital expenditures - Profitability
Investments in networks, licenses, and new technology may not be realized	- Acquisition of 3G licenses - Need for new functionality and services - Delayed roll out of 3G services - Delayed building of networks - Additional investment costs	- Demand for services not what expected - Costs of investments do not pay back
Declining revenue/customer	- Introduction of non-voice services - Delayed availability of services and handsets - Higher than anticipated prices of handsets	- Reduced revenue - Price war
Health risks in technology	- Electromagnetic signals emitted by mobile telephone handsets and basses stations	- Ability to retain and attract customers - Adverse effect on results of operations

The map in Fig. 3.4 shows an example of the dynamics and elements connected to the development of new services in the industry. Two basic cycles, either virtuous or vicious can be found in the map. One is related to the cycle which connects the availability of mobile handsets (9) and need to develop new services (1) to the development of new services (2), implying that handsets are capable of performing these services. The other one is related to the investment risk of the operator (4), addressing the lack of financial risk-taking capability of operators (15). This will reduce the operators' willingness to develop and integrate new services (13) and consequently decrease the development of new services (2).

As the example above demonstrates, risk analysis in value networks is a complex process. The primary purpose of risk identification is to provide a complementary view of the key uncertainties in the industry. That way it is possible to address the motives and scope of actions of different actors. It is also essential to analyze the possible strategies for managing the identified risks and uncertainties at the network level. Here, a certain actor may be able to manage a risk which is beyond the control scope of other actors in the value network.

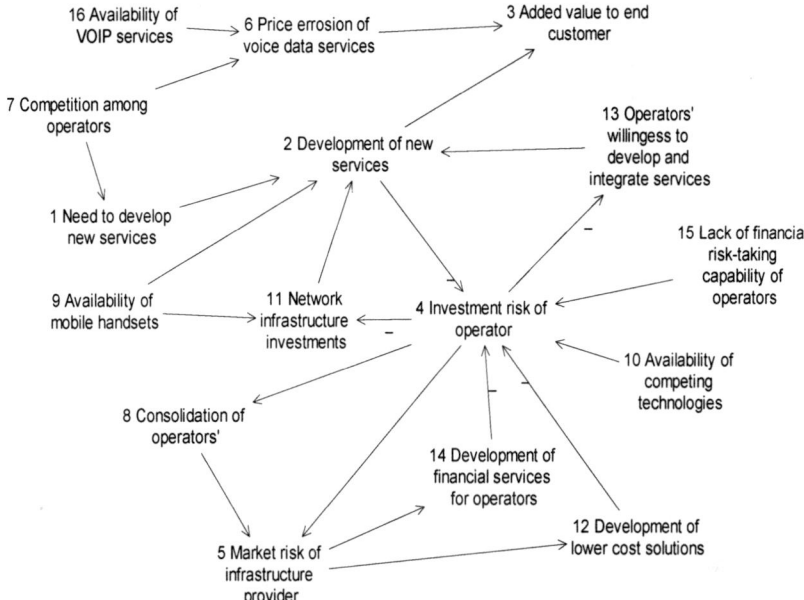

Fig. 3.4 Dynamics in the industry development

The presented risk analysis illustrates the connections associated with the risk profiles of different actors in the value network. The example risk factors in Tables 3.3 and 3.4 are related to rather generic industrial trends and driving forces. An interesting addition would include the analysis of more specific connections with different actors and their possibilities to control the risks and industry change.

3.5 Discussion

An important risk and performance indicator of the value network as a whole is the capability of the end customer and value creating actors to benefit from value creation and its delivery. Thus, value is based on a constellation of relationships formulating value to all the actors in a value creating system. When considering the "hub" or keystone position in the network, the actor has an important role in maintaining the health of the whole business ecosystem. This implies that the risk management capability of keystone actors should be directed towards the orchestration of the whole network, filtering the changes in business environment and customer needs to other actors, as well as providing new value creating opportunities for connected niche actors. In general, the question of risk management is much more about what trends and uncertainties may change in the value creation network, and how interdependent the actors are from the point of view of risks in the network.

As the various frameworks and methods for risk analysis in value networks require in-depth analytical consideration, one important management challenge is to develop a process for monitoring the weak and strong signals to anticipate changes in the industry structure or customer needs. These signals should be further connected to the competitive positioning of the firm in the industry value network.

The actors' risk profiles may provide a comprehensive view of the future industry development. They can also be used as a means for tracing the strategy attributes and motives of different players in the industry. These risk profiles should be connected with each other, since the risks in the value network are connected to each other. This way it is also possible to find ways to manage risk in network relationships collaboratively. Systematic and analytical processes in risk management are needed in collaborative business networks.

References

Agrell PJ, Lindroth R and Norrman A (2004) Risk, information and incentives in telecom supply chain. International Journal of Production Economics 90: 1–16.
Allee V (2000) Reconfiguring the value network. The Journal of Business Strategy 21: 36–39.
Cohen WM and Levinthal DA (1994) Fortune favours the prepared firm. Management Science 40: 227–251.
Contractor FJ and Lorange P (2002) The growth of alliances in the knowledge-based economy. International Business Review 11: 485–502.
Draulans J, deMan AP and Volberda HW (2003) Building alliance capability: management techniques for superior alliance performance. Long Range Planning 36: 151–166.
Fine CH (1996) Industry clock speed and competency chain design: an introductory essay. Proceedings of Service Operations Management Conference, Dartmouth College, Hanover, New Hampshire, June 24–25, 1996.
Gilad B (2004) Early warning: using competitive intelligence to anticipate market shifts, control risk, and create powerful strategies. Amacom, New York.
Hallikas J, Karvonen I, Pulkkinen U, Virolainen V-M and Tuominen M (2004) Risk management processes in supplier networks. International Journal of Production Economics 90: 4758.
Harland C, Brencheley H and Walker H (2003) Risk in supply network. Journal of Purchasing and Supply Management 9: 51–62.
Iansiti M and Levien R (2004) Strategy as ecology. Harvard Business Review, March: 68–78.
McGahan A (2004) How industries change. Harvard Business Review, October: 87–94.
Normann R and Ramirez R (1993) From value chain to value constellation: designing interactive strategy. Harvard Business Review, July–August: 65–77.
Parolini C (1999) The value net: a tool for competitive strategy. Wiley, Chichester.
Peppard J and Rylander A (2006) From value chain to value network: insight for mobile operators. European Journal of Management, 24: 128–141.
Porter ME (1985) The competitive advantage: creating and sustaining superior performance. Free Press, New York.
van der Heijden K (1996) Scenarios: the art of the strategic conversation. Wiley, Chichester.
Varis J (2004) Partner selection in knowledge intensive firms. Doctoral Dissertation, Acta Universitatis Lappeenrantaensis 199. Lappeenranta University of Technology.

Chapter 4: Predicting and Managing Supply Chain Risks

Samir Dani

Wolfson School of Mechanical and Manufacturing, Loughborough University, Loughborough, Leicestershire, LE113TU, UK

4.1 Introduction

Since the start of the new century the world at large has experienced escalating uncertainty as a result of climate changes, epidemics, terrorist threats and an increasing amount of economic upheaval. These uncertainties create risks for the proper functioning of supply chains. This chapter provides an insight into developing a proactive approach to predict risks and manage uncertainties that may potentially disrupt the supply chain.

The aim of the chapter is to present a holistic perspective regarding supply chain risk management and incorporate a methodology to manage supply chain risks proactively. When discussing supply chain risk issues with industry personnel it was noticed that post 9/11, the issue of supply disruption had gained importance within the industry. But the focus on managing these disruptions and sources of these disruptions has been primarily reactive. Supply chain personnel in some instances have remarked that they have in the past researched and presented to their top management proactive risk management solutions which had been subsequently rejected and no investment provided. There is now, however, an increasing interest regarding proactive tools and hence this chapter seeks to present a framework for implementing proactive risk management. The chapter also suggests some tools which may prove useful in predicting supply chain risks.

The chapter begins by defining and discussing the concepts of Uncertainty and Risk as suggested in the literature. A discussion regarding supply chain risks and supply chain risk management leads on to the issues of proactive risk management. The chapter then suggests the use of predictive methods for data gathering and analysis to aid the proactive management of supply chain risks. Such an approach requires that a thorough understanding of the external and internal sources of supply chain risks is gained and that systems are put in place in order to counter these risks. A methodology encapsulating the predictive and proactive approaches is presented, which will also prove useful for initiating further research and applications regarding proactive supply chain risk management.

4.2 Uncertainty and Risk

Does Uncertainty generate risk or does Risk beget Uncertainty? This is a tantalising conundrum. Irrespective of the solution to such a conundrum, either scenario is not one that any business would find appealing, although most inevitably encounter in practice. In his influential work "Risk, Uncertainty, and Profit", Frank Knight (1965) established the distinction between risk and uncertainty. According to Knight a phenomenon which is un-measurable is "Uncertainty" whereas one that is measurable is "Risk".

Risk is defined as uncertainty based on a well grounded (quantitative) probability. Formally,

> Risk = (the probability that some event will occur) × (the consequences if it does occur)

Genuine uncertainty, on the other hand, cannot be assigned such a (well grounded) probability. Furthermore, genuine uncertainty can often not be reduced significantly by attempting to gain more information about the phenomena in question and their causes (Lövkvist-Andersen et al. 2004).

Deloach (2000) has defined business risk as the level of exposure to uncertainties that the enterprise must understand and effectively manage as it executes its strategies to achieve its business objectives and create value. According to the Royal Society (1992, p. 4) "risk is the chance, in quantitative terms, of a defined hazard occurring". Norrman and Jansson (2004) have expressed risk as, Risk = Probability (of the event) × Business Impact (severity). They mention that while risks can be calculated, uncertainties are genuinely unknown.

Holton (2004) has attempted to define risk from an operational point of view. He suggests that risk has two components: exposure and uncertainty. Holton (2004) also suggests that when operationally defining exposure, uncertainty and risk it is only a personal perception that we have of the situation. Chiles and McMackin (1996) observe that a manager's perspectives of risk are associated with the notion of economic loss. Spekman and Davis (2004) suggest that, generally due to the downside effect of the outcome, "risk" tends to have negative connotations. Hence, what exactly is risk, and, is it possible to quantify the risks that can only be perceived? One of the ways to define some aspects of perceived risks may be the development of appropriate risk metrics. The next section will explore the more specific aspects of supply chain risks.

4.3 Risks in the Supply Chain

In today's business world a supply chain may be stretched out across the globe in order to provide the customer with the product at the lowest cost and highest quality. The supply chains are thus exposed to a whole new set of factors, which can create chaos and disruption. Local political turmoil, the ever increasing complexity and uncertainty of weather conditions, terrorism, counterfeiting, and a plethora of other such issues create external risks in the supply chain. But this does not mean that the supply chain is devoid of any risks internally. Supplier issues, strikes, quality problems, and logistics issues are more internal operational risks, which need a different level of mitigation. Zsidisin (2003) suggested that risk in a supply chain context can be defined as the potential occurrence of an incidence associated with inbound supply in which the result is the inability of the purchasing organization to meet customer demand.

Christopher and Peck (2003), taking inspiration from Mason-Jones and Towill (1998), have categorised supply chain risk into five categories:

1. Internal to the firm: **Process, Control**
2. External to the firm but Internal to the Supply network: **Demand, Supply**
3. External to the network: **Environmental**

Peck (2005, 2006) suggests that the sources and drivers of supply chain risk operate at several different levels. These are intricately linked as elements of a system, and are described within four discrete levels of analysis:

1. Level 1 – value stream/product or process.
2. Level 2 – assets and infrastructure dependencies.
3. Level 3 – organisations and inter-organisational networks.
4. Level 4 – the environment.

Each level reflects quite different perspectives but together these levels cover elements of a supply chain and the environment within which they are embedded (Peck 2005). This has also been suggested by Faisal et al. (2006) that risk sources are the environmental, organizational or supply chain related variables that cannot be predicted with certainty and that affect the supply chain-outcome variables.

Spekman and Davis (2004) suggested dimensions for understanding supply chain risks incorporating:

1. Physical movement of goods
2. Flow of information
3. Flow of money
4. Security of the firm's internal information systems
5. Relationship between supply chain partners
6. Corporate social responsibility and the effect on a firm's reputation.

These dimensions were also resonated by Cavinato (2004) when identifying risks and uncertainties in supply chains, focussing on five sub-chains/networks for every supply chain:

1. Physical network
2. Financial network
3. Informational network
4. Relational network
5. Innovational network

In LaFonde (2007 www.manufacturing.net) one of the respondents has mentioned that *"It really is almost impossible to predict when most emergencies will happen... Many companies think, 'It can't happen here' or 'We would never have that problem in our plant,' but then when something does occur, they are caught off-guard and not prepared"*. The concept of "resilience" is related to risk and vulnerability in a perspective that not all "risks" (hazards or threats) can be avoided, controlled, or eliminated. Instead, resilience focuses on the ability of the system to return to its original or desired state after being disturbed, e.g., its ability to absorb or mitigate the impact of the disturbance (Peck 2006).

4.4 Supply Chain Risk Management

Efficient risk management can provide value to various stakeholders of a firm. The compliance with appropriate procedures and corporate governance policies can help to reduce or avoid crisis situations. Risk management entails identifying operational risks and developing mitigation procedures for maintaining operational performance. Along with considering supply chain risk management from an operational viewpoint, it is also beneficial to consider supply chain risk management from a strategic management perspective. Developing the appropriate corporate governance policies to tackle issues of sustainability and ethical sourcing leads to a better corporate reputation and also helps in risk management. Rice and Caniato (2003) report that many firms have developed various risk assessment programmes that are intended to:

1. Identify different types of risks;
2. Estimate the likelihood of each type of major disruption occurring;
3. Assess potential loss due to a major disruption; and
4. Identify strategies to reduce risk.

In considering the risks primarily in the supply chain, Rice and Caniato (2003) and Zsidisin et al. (2000, 2004) suggested that a supply chain risk assessment programme motivates a firm to develop contingency plans, which thus can also be used to meet certain legal requirements such as the Sarbanes-Oxley Act of 2002 and KonTraG.

Research in this area has primarily focussed on the supplier side. Spekman and Davis (2004) have suggested that interdependency carries risk in the supply chain, but these can be managed. Zsidisin et al. (2000) and Zsidisin (2003) present suggestions for minimising risk:

1. Carrying buffer stock and improving inventory management;
2. Using alternative sources of supply;
3. Use of contracts to manage price fluctuations; and
4. Quality initiatives.

These suggestions reinforce research conducted by Smeltzer and Siferd (1998) who concluded that risks associated with poor selection of suppliers can be reduced by developing quality certification programs and auditing the suppliers to assure that they meet the required standards. Lee and Whang (2003) developed a model to show how firms can reduce inventory due to less inspection time. Another aspect of the research conducted around minimising supplier related risk is concerned with the number of suppliers. Both Sheffi (2001) and Kleindorfer and Saad (2005) suggested the use of multiple suppliers as a way to reduce certain supply chain risks.

Since the beginning of the current century, companies are increasingly recognising the importance of risk assessment programs and are using different methods, ranging from formal quantitative models to informal qualitative plans, to assess supply chain risks. Some of the enablers for better supply chain risk management include Lean, Six Sigma and Agile philosophies (Christopher and Rutherford 2004; Chapell and Peck 2005); Event Management software (Malykhina 2005); and Radio Frequency Identification (RFID) (Niemeyer et al. 2003). These provide better visibility, velocity and more effective process control (Christopher and Lee 2001).

According to Norrman and Jansson (2004), the stages of the risk management process can vary from risk identification/analysis (or estimation) via risk assessment (or evaluation) to different ways of risk management. Juttner et al. (2003) suggest that supply chain risk management is the process of identifying and managing risks in the supply chain through a co-ordinated approach amongst supply chain members in order to achieve the supply chain objectives. Researchers have considered supply chain risk management from various perspectives (Gaudenzi and Borghesi 2006): financial and corporate governance perspective (Meulbroek 2002), perspective of business continuity and crisis management (Adams et al. 2002), the ability to react quickly to ensure continuity (van Hoek 2003; Rowbottom 2004), reputation management perspective (O'Rourke 2004), perspective oriented towards the goal of reliability (Moore 2002), and the achievement of the best trade-off between quality controls (through inspections) and process self-control (Svensson 2002), often utilising the Six Sigma approach and tools (Eckes 2001).

Some other approaches to supply chain risk management involve managing risks affecting: specific supply chain levels (Cavinato 2004), systems inside and outside the chain, such as the Information system (Finch 2004), specific projects (Halman and Keizer 1994) with an aim to identify and manage risks that threaten the project's success (Ramgopal 2003) and causes of project failure (Spekman and Davis 2004).

4.5 Proactive Supply Chain Risk Management

Supply chain risk management strategies can be described as being reactive or proactive. Being reactive is a default position when a risk materialises. This is in effect necessary when a supply chain operates without worrying about risks on a day to day basis but reacts to mitigate when the difficulty or disruption strikes. This impacts the supply chain members until the situation is resolved, which needs to be done quickly as a delay can cause serious damage even to a large corporation as per the Philips/Ericsson case (Sheffi 2005). To overcome the need to react after the occurrence of an event, a proactive strategy has been proposed by researchers Norrman and Jansson (2004).

In a proactive strategy, potential risks are identified at the supply chain design stage, their probability and impact are assessed and they are ranked in terms of importance. The focus of this exercise is to target the identified risks in order to avoid them. This may not be possible in all cases and hence there is a need to develop and implement contingency plans to minimise the impact if and when the risk occurs.

This process sounds the most logical thing to do for supply chain managers, but it needs resources upfront in terms of investment and people. Hence, if a risk never materializes, it becomes very difficult to justify the time spent on risk assessments, contingency plans, and risk management (Zsidisin et al. 2000). This also leads to evaluating the total cost of an undesirable event occurring against the benefits realized from having strategies in place that significantly reduce the chance and/or effects of detrimental events with supply. Also, it is not always possible to obtain good estimates of the probability of the occurrence of any particular disruption and accurate measurement of the potential impact of each disaster.

Although the process of proactively managing the risks looks to be fairly familiar to most of the risk management/mitigation strategies, it is not explicitly cited in the supply chain risk management literature. Preston and Smith (2002) have developed a proactive risk management process for controlling the uncertainty in product development. This process uses the following variables in a process map for identifying a proactive risk management strategy:

1. The probability of risk occurring
2. Risk event drivers
3. Risk events
4. The probability of the impact
5. Impact drivers

Norrman and Jansson (2004) describe how Ericsson, the company affected by a fire at a sub-supplier has implemented a new organisation and new processes and tools for SCRM by developing a proactive risk management strategy. These incorporate the inclusion of supply chain management and sourcing functions under the corporate risk management function, development of a risk management tool "Ericsson risk management evaluation tool (ERMET)" and involving its supply network in the risk management process.

One of the most important enablers for a proactive risk management strategy is the presence of a culture and attitude that provides resources and motivates employees to develop risk contingency plans. Smeltzer and Siferd (1998) suggested that the formation of a risk management culture to encompass proactive risk management is extremely necessary as:

1. Employees may feel that no steps are taken towards managing risks identified and reported, but more is done after the impact,
2. Employees feel that top management would look upon them negatively if they are involved in proactively identifying risks,
3. Depicting risks to shareholders may have an adverse effect on the value of the firm.

4.6 Predicting Supply Chain Risks

One of the main requirements for an effective proactive risk management process is to obtain good estimates of the probability of the occurrence of any particular disruption and accurately measure the potential impact. Good estimates of probabilities are obtained by risk prediction techniques. Predicting the occurrence of the risk will vary according to the uncertainty and complexity. It may be possible to predict an occurrence based on historical data but perhaps not possible to predict a one-off environmental event. In supply chains, risk prediction and identification is being helped by early warning systems or satellite tracking systems, and smart containers, for example. Once a risk or a potential risk has been identified it is necessary to do a cost-benefit analysis to ascertain the effect of the risk. This will lead to the selection of the appropriate proactive methods for mitigating the risk.

4.6.1 Tools for Risk Prediction

There are various tools for risk prediction, ranging from complex mathematical models to a less complicated Event Tree Analysis. The following section provides a description of two types of risk prediction tools: "Data Mining" and "Failure Mode Effect Analysis (FMEA)"

4.6.1.1 Data Mining

Risk in general is a term attributable to future loss, and risk management is attributed to the process and resources utilised in order to control the loss (Haimowitz and Key 2002). Hence, the more relevant issue is to ascertain whether the future predictable losses are controllable. In order to predict the losses and reduce the uncertainty it is necessary to be able to look into the future based upon the past capabilities of the system. Data mining is a process that has the ability to use pertinent data to uncover sources of risk exposure that may otherwise remain obscure or unnoticed before prior to the risk being realised. Haimowitz and Key (2002) suggest that a proactive data miner will use their understanding of risk for enterprise advantage through competitive gain or innovation (product, process or service).

There are two dimensions of risk: *frequency* is the rate at which undesirable events exhibit themselves, while *severity* is the magnitude of the loss, once exhibited. According to Haimowitz and Key (2002), data mining generally applies to risk problems in the high frequency and low severity scenario. Data mining can help in making more severe risks less frequent and more frequent risks less severe. The risk or severity is lessened by identifying controllable drivers of severity or risk frequency.

The role of data mining is to analyse historical data, to improve prediction capability. Some of the common analytic approaches used by data miners are:

1. Estimation of the parameters of past performance: Means, Standard deviations, Correlations, and Associations for hypothesis testing
2. Classification: Segmentation, or Clustering of data units to facilitate the modelling process
3. Construction of a functional relationship: or model between responses and explanatory variables.

While the strategic goal of the data mining and modelling exercise is to predict the key phenomenon, the operational goal is to gather and understand the relevant data with the aim of discovering patterns to provide business intelligence. Berry and Linoff (1997) have suggested some specific tools used by data miners for analysis:

(a) Estimation: Tools useful for exploratory data analysis. These tools will not lead to patterns but are more useful in analysing the data to identify the most relevant sets of data to concentrate further analysis. These include the use of statistical tools, Pareto analysis, and graphical analysis.
(b) Clustering/segmentation: This approach is used to logically group observations on the basis of similarity in their characteristics, reducing the level of heterogeneity in the data. These are a precursor to the modelling phase, such as K-means and Distance matrices (Euclidean/non-Euclidean).
(c) Classification/discrimination: The process of assigning observations to a predetermined number of classes. This is performed by dividing the dataset into mutually exclusive groups such that the members of each group are as "close" as possible to one another, and different groups are as "far" as possible from one another. The distance between the groups is measured with respect to specific variable(s) required for prediction such as chi-squared automatic induction, classification and regression trees (Breiman et al. 1993), regression analysis, discriminant analysis, and rule induction.

(d) Prediction: Formal mathematical models are built for the purpose of predicting the occurrence of the phenomenon. Techniques used for this purpose are: Linear/nonlinear regression, Classification and Regression trees (Breiman et al. 1993), Multiple adapted regression splines, artificial neural nets, genetic algorithms, time – series regression models, and stochastic models.

Data mining activities are generally divided into two main types: Predictive data mining and Descriptive data mining. Prediction involves using attributes of the data to predict unknown future values of the dependent variables. Operational data in its raw form is of limited business value when it is mainly used for reporting what has happened. However, if this data is analysed and modelled using Predictive data mining tools it can transform the data into actionable decisions. Predictive data mining is a powerful tool for recognizing patterns and proactively predicting what will happen. Descriptive data mining, however, focuses on trying to obtain insight into the data by finding patterns before trying to predict. Both methods use some of the following core data mining tasks: Classification, Regression, Clustering, Summarization, Dependency, Modelling, Link analysis and Sequence analysis. To conduct a data mining process effectively it is important to ascertain:

1. The fit between the data mining technique and the task, and
2. Conditions under which the identified relationships are valid

4.6.1.2 Failure Mode Effect Analysis (FMEA)

A Failure Mode Effect Analysis can be described as a systematic group of activities intended to

1. Recognise and evaluate the potential failures of a product or process and the effects of that failure.
2. Identify actions, which could eliminate or reduce the chance of the potential failure occurring.
3. Document the entire process.

The fundamentals of an FMEA process are

1. Define scope, functional requirements, design parameters and process steps.
2. Identify potential failure modes: Failure modes indicate the loss of at least one functional requirement. It is the manner in which a failure occurs. This step in the process takes into account a foresight view (based on past experience and any new information) of what could cause a failure to the system or process.
3. Potential failure effect: This step investigates the effect the failure will have on other entities or processes.
4. Severity: "How bad" or "serious" the effect of the failure mode is. Usually severity is rated on a discrete scale from 1 (no effect) to 10 (hazardous effect). Severity ratings of 9 and 10 indicate a potential effect of high

importance and this could typically be a safety or government regulation issue. Critical effects need deeper study for all causes to the lowest level, using a method of Fault Tree Analysis.
5. Potential causes: These are the causes of the failure. In this step, all causes that can be attributed to the failure occurring are investigated.
6. Occurrence: This is the likelihood of the event happening (i.e. failure in the system) on the basis that "the cause occurs". FMEA assumes that if the cause occurs, failure will occur too. The probability of occurrence is ranked from 1 to 10, where 1 signifies a remote probability of occurrence and 10 a very high probability of occurrence.
7. Current controls: The objective of the controls is to identify and detect the deficiencies and vulnerabilities as early as possible. This step looks at the current processes in place to mitigate the failures (if already known).
8. Detection: A subjective rating is assessed corresponding to the likelihood that the detection method will detect the first-level failure of a potential failure mode. This is ranked from 1 to 10, where 1 signifies that it is unlikely to detect and 10 signifies a very high detection potential.
9. Risk Priority Number (RPN): These are used to prioritise the potential failures and are calculated as "Severity × Occurrence × Detection ranking".
10. Actions Recommended: The team should then select and manage subsequent actions needed to locate and control the situation.

4.7 The Predictive – Proactive Methodology

Figure 4.1, depicts the "predictive-proactive" methodology for managing supply chain risks. The methodology is represented by a process map associated with undertaking risk management and, specifically, supply chain risk management. The methodology assumes that for proactively managing risks, it is important to have sufficient information regarding the impending situation to aid the decision process on developing a mitigation plan. The predictive mode is hence reliant on the ability of the organisation to provide sufficient and appropriate information.

Predictive Mode: In the predictive mode the focus is on acquiring data and analysing it to discover meaningful patterns which will aid in identifying risks. The data gathering phase will engage with both the reactive and proactive approaches, as in accessing legacy data for a particular situation and projecting probable future scenarios. Various tools can be used in this phase. Data Mining as described earlier will be used to study legacy data for generating patterns of behaviour causing risk whereas FMEA will be used to project the probable situations for risk. These two tools together will provide sufficient data for the next phase. This is a very important phase as the analysis conducted in this phase will provide the necessary information for risk identification. The predictive mode enables the risk management agencies to form proactive solutions based on quantitatively analysed data.

Proactive Mode: Once the data is analysed in the predictive mode, this data is fed into the proactive phase of the risk management process to enable the identification and quantification (effect) of the risk scenarios. As seen in Fig. 4.1, "strategic objectives" or objectives for the supply chain are at the centre of the process and form the most important part of the complete risk management process. This depicts that the strategic objectives will control the risk management process and will also influence the solutions that will be developed for mitigating the risk. Referring to Fig. 4.1, the event or the performance of the mitigating solution has an impact on the strategic objective for further scenarios of risk management and will also influence the information that the risk management team has regarding sources of risk. This process is an iterative process as the cycle will repeat by studying new issues and risks identified after the analysis of the event. The new data will then be analysed using tools identified in the predictive mode.

In a proactive mode, there are two options for working with a supply chain:

(i) Designing a new supply chain,
(ii) Updating an existing supply chain

When designing a new supply chain or updating an existing one, the proactive mode will benefit in managing risks associated with the new situation. The predictive mode will provide sufficient analysis for identifying risks and quantifying their probability of occurrence and the impact. Also, information regarding the external and internal sources of risk in the form of security issues related to supply

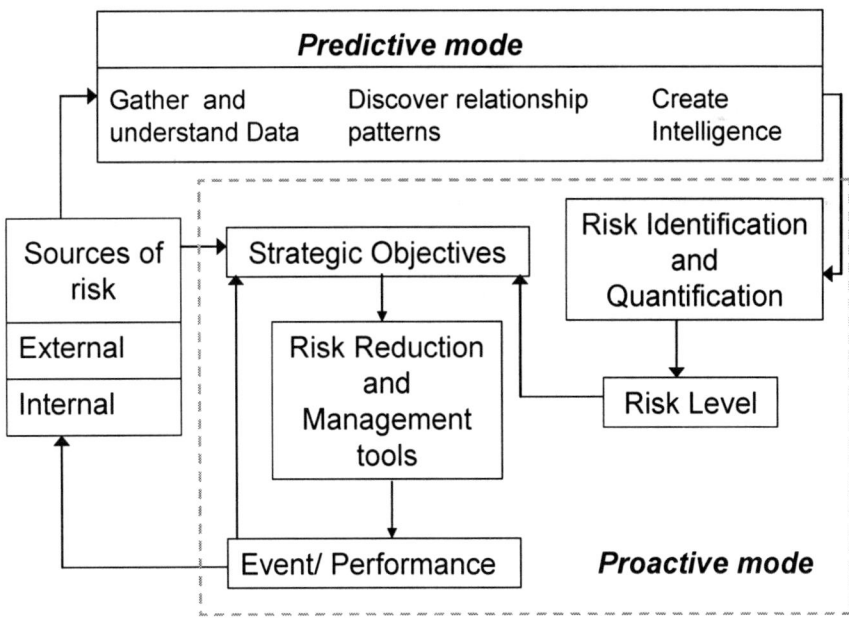

Fig. 4.1 The "predictive-proactive" supply chain risk management methodology © Samir Dani, 2007

chain members, risk profiles for countries in which the chain operates, and risk profiles for individual members in the supply chain will be considered. These sets of information will then be used to determine the strategic objectives for the chain and risk reduction and mitigation plans will be put together to meet these objectives. This, however, does not limit the existence of uncertainty and there will be cases in which the mitigation plan will not be as efficient as required, or that a completely different scenario has materialised. Hence, it is necessary to update the strategic objectives and the data set of risk sources in line with the unexpected event or performance of the mitigation/risk reduction plan.

4.8 Conclusion

This chapter has provided a discussion related to proactive management of supply chain risk. It has been suggested that to be proactive in risk management it is necessary to have an efficient predictive process for data gathering and analysis. The proposed risk management methodology has strategic objectives as a focal point in the process. This will hopefully make it easier for top management to follow the methodology and be more in control of the process.

The methodology as shown in Fig. 4.1 may look simplistic, but it does provide a process map for risk management. The focus of the methodology is to structure the two processes of data analysis and decision making. The stress on including the strategic objectives reflects the importance of providing benchmarks for the firm to base their future risk mitigation plans and to measure the performance of their plan against the possible event. Data Mining as explained previously, can be used as a predictive analysis tool for risks with high frequency and low severity (Haimotwitz and Key 2002), supplemented by other tools (e.g., FMEA) which would be utilised to get an insight into future scenarios of risk having a low frequency of occurrence.

The chapter has sought to generate interest in developing and using proactive strategies for managing supply chain risks. The methodology will be validated in industry in the future. In its current form the methodology can be used as a guiding tool for companies to implement a proactive risk management strategy.

References

Adams, T. J., Austin, S. P., Soprano, R S. and Stiene, L. M. (2002). "Assessing the transition to production risk", *Program Manager*, September, pp. 10–21.
Berry, M. J. A. and Linoff, G. (1997) *Data Mining Techniques for Marketing, Sales and Customer Support*, Wiley, New York.
Breiman, L., Friedman, J. H., Olshen, R. A. and C. J. Stone (1984) *Classification and regression trees*, Wadsworth Inc., Monterey, CA, USA.
Cavinato, J.L. (2004) "Supply chain logistics risks", *International Journal of Physical Distribution & Logistics Management*, Vol. 34, No. 5, pp. 383–388.

Chapell, A. and Peck, H. (2005) The application of a six sigma methodology to military supply chain processes, in *Operations and Global Competitiveness: Proceedings of the Euroma Conference*, pp. 809–818.

Chiles, T. H. and McMackin, J. F. (1996) "Integrating variable risk preferences, trust, and transaction cost economics", *Academy of Management Review*, Vol. 21, No. 1, pp. 73–99.

Christopher, M. and Peck, H. (2003) "Building the Resilient Chain", *The International Journal of Logistics Management*, Vol. 15, No. 2.

Christopher, M. and Rutherford, C. (2004) *Creating Supply Chain Resilience Through Agile Six Sigma.* Critical Eye, Jun–Aug, pp. 24–28.

Christopher, M. and Lee, H. L. (2001) "Supply chain confidence: the key to effective supply chains through improved visibility and reliability", *Global Trade Management*, November, pp. 1–10.

Deloach, J. W. (2000) *Enterprise-Wide Risk Management: Strategies for Linking Risk and Opportunity*, Financial Times Prentice Hall, 284pp.

Eckes, G. (2001) *Six Sigma Revolution*, Wiley, New York, pp. 29–41.

Faisal, M. N., Banwet, D. K. and Shankar R. (2006) "Supply chain risk mitigation: modelling the enablers", *Business Process Management Journal*, Vol. 12, No. 4, pp. 535–552.

Finch, P. (2004) "Supply chain risk management", *Supply Chain Management: An International Journal*, Vol. 9 No. 2, pp. 183–196.

Gaudenzi, B. and Borghesi, A. (2006) "Managing risks in the supply chain using the AHP method", *The International Journal of Logistics Management*, Vol. 17, No. 1, pp. 114–136.

Haimowitz, I. J. and Keyes, T. K. (2002) "Risk Analysis", *Handbook of Data Mining and Knowledge Discovery*, Klösgen, W. and Żytkow, J.M. (eds.), Oxford University Press.

Halman, J. M. and Keizer, J. A. (1994) "Diagnosing risks in product-innovation project", *International Journal of Project Management*, Vol. 12, No. 2, pp. 75–81.

Holton, G. A. (2004) "Defining risk", *Financial Analysts Journal*, Vol. 60, No. 6, CFA Institute.

Juttner, U., Peck, H. and Christopher, M. (2003) "Supply chain risk management: outlining an agenda for future research", *International Journal of Logistics: Research and Applications*, Vol. 6, No. 4, pp. 199–213.

Kleindorfer, P. and Saad, G. (2005) "Managing disruption risks in supply chains", *Production and Operations Management*, Vol. 14, pp. 53–68.

Knight, F. (1965) *Risk, Uncertainty and Profit*. Harper & Row, New York (first published 1921).

LaFonde, A. (2007) "Coping with Disaster,"*Manufacturing.net*, Advantage Business Media, January 30, viewed 14 September 2007, http://www.manufacturing.net/article.aspx?id=138176&terms=coping%20with%20disaster

Lee, H. and Whang, S. (2003) "Higher supply chain security with lower cost: lessons from total quality management", Working paper, Graduate School of Business, Stanford University.

Lövkvist-Andersen, A., Olsson, R., Ritchey, T. and Maria Stenström, M. (2004) Developing a Generic Design Basis (GDB) Model for Extraordinary Societal Events using Computer-Aided Morphological Analysis, presented in SRA (Society for Risk Analysis) Conference in Paris 15–17 November, available from http://www.swemorph.com

Malykhina, E. (2005) "The real time imperative", *Information Week*, 3 January 2005, pp. 1020, 43.

Mason-Jones, R. and Towill, D. R. (1998) "Shrinking the supply chain uncertainty circle", *Control*, pp. 17–22.

Meulbroek, L. (2002), "The promise and challenge of integrated risk management", *Risk Management and Insurance Review*, Vol. 5, No. 1.

Moore, K. G. (2002) "Six sigma: driving supply at Ford", *Supply Chain Management Review*, July/August, pp. 38–43.

Niemeyer, A., Pak, M. and Ramaswamy, S. (2003) Smart tags for your supply chain. McKinsey Q., Vol. 4. Available online at: http://premium.mckinseyquarterly.com

Norrman, A. and Jansson, U. (2004) "Ericsson's proactive supply chain risk management approach after a serious sub-supplier accident", *International Journal of Physical Distribution & Logistics Management*, Vol. 34, No. 5, pp. 434–456.

O'Rourke, M. (2004) "Protecting your reputation", *Risk Management*, April.

Peck, H. (2006) "Reconciling supply chain vulnerability, risk and supply chain management", *International Journal of Logistics: Research and Applications*, Vol. 9, No. 2, pp. 127–142.

Peck, H. (2005) "Drivers of supply chain vulnerability: an integrated framework", *International Journal of Physical Distribution and Logistics Management*, Vol. 35, pp. 210–232.

Ramgopal, M. (2003) "Project uncertainty management", *Cost Engineering*, Vol. 45, No. 12, pp. 21–24.

Rice, J. and Caniato, F. (2003) "Building a secure and resilient supply network". *Supply Chain Management review*, Vol. 7, pp. 22–30.

Rowbottom, U. (2004) "Managing risk in global supply chains", *Supply Chain Practice*, Vol. 6, No. 2, pp. 16–23.

Royal Society (1992) Risk Analysis, Perception and Management, 1992 (Royal Society: London).

Sheffi, Y. (2001) "Supply chain management under the threat of international terrorism" *International Journal of Logistics Management*, Vol. 12, No. 2, pp. 1–11.

Sheffi, Y. (2005) *The Resilient Enterprise: Overcoming Vulnerability for Competitive Advantage*, The MIT Press, 2007

Smeltzer, L. R. and Siferd, S.P. (1998) "Proactive supply management: the management of risk", *International Journal of Purchasing and Materials Management*, Vol. 34, No. 1, pp. 38–45.

Smith, P. G., Merritt G. M. (2002) *Proactive Risk Management*, Productivity Press, New York, USA

Spekman, R. E. and Davis, E. W. (2004) "Risky business: expanding the discussion on risk and the extended enterprise", *International Journal of Physical Distribution & Logistics Management*, Vol. 34, No. 5, pp. 414–433.

Svensson, G. (2002) "A conceptual framework of vulnerability in firms' inbound and outbound logistics flows", *International Journal of Physical Distribution & Logistics Management*, Vol. 32, pp. 110–134.

van Hoek, R. (2003) "Are you ready? Risk readiness tactics for the supply chain", *Logistics Research Network*, Institute of Logistics and Transport, London.

Zsidisin, G. (2003) "Managerial perceptions of risk", *Journal of Supply Chain Management*, Vol. 39, pp. 14–25.

Zsidisin, G. A., Ellram, L. M., Carter, J. R. and Cavinato, J. L. (2004) "An Analysis of Supply Risk Assessment Techniques", *International Journal of Physical Distribution and Logistics Management*, Vol. 34, No. 5, pp. 397–413.

Zsidisin, G. A., Panelli, A. and Upton, R. (2000) "Purchasing organization involvement in risk assessments, contingency plans, and risk management: an exploratory study", *Supply Chain Management*, Vol. 5, No. 4, p. 187.

Chapter 5: Assessing Risks in Projects and Processes

Barbara Gaudenzi

Faculty of Business Economics, University of Verona, Italy

5.1 Introduction

The environment of today's organizations is characterized by continuous changes at an ever-increasing speed. Competition rules change all the time and organizations need to be more flexible as they rise to the challenge of becoming "better, faster and cheaper". This increasingly more dynamic environment is characterized by a huge number of risks, which exist both inside the company, throughout its processes and projects, and at the network level. For this reason, managers should focus their attention on the achievement of the specific objectives of each process and project, depending on the priority that the top management assigns to all of them. However, a best practice gap still exists for analysing risks in projects and processes utilising a systemic perspective.

Each year, natural and manmade disasters cause disruptions amounting to $40 million per day (Nelson et al. 1998). At the same time, based on a survey carried out in 2005 involving a sample of 950 European enterprises with a turnover in the range of 30–300 million Euros, 50% of these enterprises said that they cannot manage the "most significant" risks, where "significant risks" means in particular the inability to offer to the market competitive solutions in terms of value, cost and development time (Marsh and McLennan Companies 2005). Indeed, these risks are related to the ability to manage in an effective and efficient way the relationship with suppliers and clients, the logistic chain and the entire development process of new products. To reach effectiveness in the management of processes and projects, executives should first answer the following questions: How do we identify management priorities? How do we measure performance and the risks that might negatively affect the successful achievement of objectives?

According to the definition given by organization theory, companies are open systems that relate to other external entities – such as suppliers, third parties, customers – and that internally include other sub-systems (Gregory and Rawling 2003; Otto 2003). In a dynamic and competitive environment, these sub-systems can be identified in terms of key processes and key projects. For this reason, it is necessary to deal first with the management of processes and projects in order to react quickly to a competitive environment.

Regarding the processes, these should be considered not only within the company boundaries, but also in relation to other actors involved, as in the supply chain (Bowersox et al. 2002). The term supply chain may be defined as "the network of organizations that are linked through upstream and downstream linkages, in the different processes and activities that produce value in the form of products and services in the hands of the ultimate customer" (Christopher 1998). Apart from the goals of each single organization, all entities should share common supply chain objectives regarding final customers and users. The differing perspectives of the various actors involved should be carefully considered by the managers to prevent different companies or decision-makers from evaluating and assessing risks in different ways, since this might negatively affect the achievement of overall objectives.

It is also worth noting that, more often than previously, organizations carry out developmental work through projects, which have the characteristics of a dynamic system. Projects can be constantly subject to change in terms of scheduling and methods for execution (like a "temporal organization"), have specific sets of goals and are generally task-oriented. This means that project leaders encourage team working involving a diversity of organisational functions, each having different perspectives, in order to solve complex problems. Moreover, a major characteristic of projects is a strong orientation towards personal relationships which occurs at two levels:

- Inside the team: top-down relationships (between top management and project leader; between project leader and the team); horizontal relationships inside the tasks, between all participants.
- Outside the team: with other parallel project teams; with other entities outside the company, like the parties of a larger supply chain.

The characteristics of processes and projects are therefore almost alike: both involve working with multiple relationships, are goal-driven and run in parallel. The definition given by the ISO 10006 Standard, suggests that projects are unique processes. According to this definition, a project consists of a set of coordinated and interrelated activities, which may be part of a larger project structure and is undertaken to achieve a specific objective. As some authors remark (Lundin et al. 2001), projects are devised and scheduled to pursue specific performances, especially regarding the achievement of primary objectives.

A major difference between processes and projects is that processes are potentially continuous and need to reach the objective of being lean and agile, where agility involves the ability to react quickly to the market. On the other hand, projects are start and finish dated by nature. Moreover, projects define objectives and outcomes in a progression, throughout their life cycle, while processes – and their internal and sequential activities – are oriented towards a set of objectives.

A network structure composed of processes and projects may have – like a supply chain – a horizontal and vertical dimension. External partners may join the effort at each stage of these processes and projects in order to contribute to the achievement of the various objectives (Gaudenzi and Borghesi 2006).

For these reasons, it is critical to introduce and develop a project management culture within supply chain management theory as a whole, in order to assess risks in projects involved in long supply chain processes (Asbjørnslett 2003). For instance, Selex Sistemi Integrati is a member of the Finmeccanica Group, a major Italian industrial group globally operating in the aerospace, defence and security sector. Selex is world leader in the provision of Integrated Defence, Air Traffic and Paramilitary Mission Critical Systems. This company works on 400 projects at any one time, each with specific goals, 70% of which are complex and long-term, with over 3,000 dedicated personnel mainly employed in the design and development of high technology. The company can boast enormous capabilities in Systems Integration, Simulation, Engineering, Software Design and Production, combined with comprehensive and advanced customer support solutions along the product's life-cycle. To achieve this goal, Selex closely co-operates with suppliers and external providers, who are treated as real partners. Project management of the company is aligned on a daily basis with the supply chain perspective in order to effectively manage projects and the relationships with suppliers, service providers and customers (Gaudenzi and Gentile 2006). As in the case of Selex, projects are frequently strongly oriented towards the development of complex products. For this reason, project management in the majority of companies is strongly correlated with Product Development Management (PDM), one of the major supply chain processes. When carrying out a project, the challenge lies in the fact that the same supply chain orientation typical of the PDM process should be applied. Companies producing components for the automotive sectors are a good example of coordinating the focus on project management issues and the integration of product development issues within the supply chain. In these cases, upstream and downstream coordination and relationships with other entities – such as suppliers and customers – is particularly critical for the success and value of projects, according to network theory in PDM (Bonaccorsi and Lipparini 1994).

The focus on PDM philosophy and strong relationships with suppliers stresses the importance of considering the supply chain management in terms of its constituent processes and projects. Figure 5.1 provides a schematic representation of both processes and projects in relation to supply chains and the product life cycle. This demonstrates the parallel nature of these two sets of activities and their ultimate fusion in achieving the organisation's goals. The focus of attention in the present chapter is risk assessment and management.

The purpose of this chapter is to explain and evaluate the new challenges faced by supply chain management in adopting such a process- and project-based orientation within the supply chain. The intended outcome is to offer some suggestions on how to assess and manage supply chain risks that otherwise might jeopardize the achievement of supply chain goals whilst, also ensuring profitable supply chain management. First, the chapter will describe the drivers for success in projects and processes. These drivers are: the focus on the objectives, the implementation of change management, the definition of the responsibilities and finally the measurement of performance and risk. These drivers might represent the starting point for the definition of the risk assessment method. It means that the assessment of risks in processes and projects might comprise in its steps these drivers for success. Finally, the chapter will describe the steps comprising the assessment of risks in processes and projects.

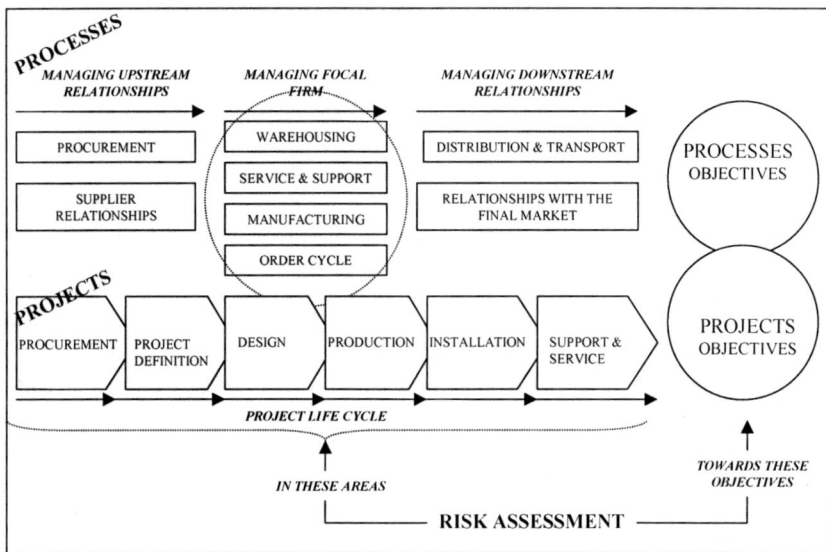

Fig. 5.1 Processes and projects

5.2 Managing Uncertainty and Risks in Projects and Processes

To learn how to cope with the evolution of organizations and how to assess risks, it is necessary to describe some characteristics that apply to both projects and processes. The four drivers for success in projects and processes have been identified. These "win drivers" are:

1. Objectives of projects and processes
2. Change management within processes and projects
3. Roles and responsibilities to achieve success
4. Performance and risk measurement

A brief description of each driver is given below with reference to project and process theory. The analysis of each driver ends with a proposition, and all four propositions will be the starting point for the subsequent section focusing on risk assessment.

5.2.1 Project and Process Objectives

The project objectives should be defined as early as possible during the project life cycle. Objectives are a description of the outcomes (results) expected from the project which in most cases can be easily defined at the start of the work, although defining objectives also means periodically reviewing the project's overall objectives and the objectives of each project phase.

Commitment towards and communication with customers are critical factors in defining a shared set of objectives. Quite often customers are not familiar with the technical implications of the project and their impact on costs, delivery time and other outcomes. This often causes stressful negotiations during the execution of projects. Improving customers' awareness helps in scheduling a good plan and in defining clear goals. On the other hand, organizations should be aware of the technology-related issues and cash flow dynamics involved in the project, in order to achieve customer satisfaction in an efficient and effective way. Goals can change in nature during the project life time, in terms of both characteristics of the outcome and deadlines assigned to each phase. For instance, the so-called open projects typically pursue different goals during the project life cycle (Obeng 1996).

As for processes, according to process management theory and in particular, to some extent, the supply chain management doctrine, supply chain strategy focuses on achieving objectives that should be shared by the entire organization and by the partners in the chain (Christopher et al. 2003). For this reason, supply chain management might be considered as a goal-driven process and, more specifically, a demand-driven process. Debenham (2001) suggested the subdivision of processes into two categories, e.g., activity-based and goal-driven processes. Since, as stated above, supply chain processes and projects are demand-driven, these may be considered as having fairly similar characteristics.

Supply chain objectives can be defined in different ways – in terms of customer service, time compression, and cost reductions – but in each process it is vitally important to prioritize these goals in order to identify the most critical ones.

> Proposition 1: In project as well as process management, multiple goals can exist at the same time. Most importantly, it is necessary to attribute a level of priority to each of these goals in order to define management priorities. Prioritizing goals is of critical importance to the effective analysis of risks.

5.2.2 Project and Process Management

Project and process management is characterised by unremitting change. This is due to the fact that many projects are designed with a set of changing and moving targets, because customers may change the specifications or new technologies may be introduced. Moreover, projects are characterized by a need for flexibility, particularly during project execution (Lereim 2002). This means that project leaders should be ready to change or redefine step by step selected portions or solutions of the project. These problems can also be found in processes, such as for instance when customers require new solutions or when process reengineering is required due to shorter delivery times. In this perspective, managing change implies the risk of losing sight of the goals. For example, continuously redefining technical solutions might cause delays in the scheduled delivery time or additional costs, or both. One should ask: what are customers really interested in? In some cases, failure to comply with a customer's extra request might be more advantageous than causing a delay or additional costs, which customers might later complain about. In this sense, the ability to prioritize objectives helps in the decision-making process.

> Proposition 2: Rapid change and a high correlation between the phases of projects and processes may determine new risks. For example, the risk associated with losing sight of key goals.

5.2.3 Stakeholder Interests

Each stakeholder and supply chain member has specific interests in the project and process (e.g., clients, users and key internal managers) and represent the most critical actors for the success of the project and process.

Project managers (Flannes and Levin 2001) have to achieve project goals within schedule, budget and resources, managing all the risks involved in the achievement of these objectives. Supply chain managers share a common scenario. Customers are interested in getting the best product in terms of quality and performance, in the shortest time and at the lowest price possible. Supply chain managers on the other hand typically focus on the profitability of their activities as their primary goal. When suppliers have to guarantee high flexibility and time compression to accommodate the dynamic needs of their customers, the price for the service rapidly increases.

> Proposition 3: Even though each actor has specific goals, the entire organization should share common supply chain objectives with regard to end customers and users. Failure to do so may result in ineffective decision-making, inappropriate strategies, inadequate measurement of performance and enhanced risk exposure.

5.2.4 Performance Measurement

The famous saying, "We can't manage what we can't measure", has a resonance in many organizations. Failure may originate from a lack of or ineffective performance measurement within processes and projects. Omli et al. (2003 p 163) contend that "Effective performance measurement systems are critical to ensuring project success. Project performance has to be measured systematically and thoroughly, not on an *ad hoc* basis." Uncertainty is also inherent and endemic in project management.

Several factors describe uncertainty contributing to the risk component in projects. Projects simultaneously focus on customers, technology and other actors, resulting in multiple views or perspectives in each single project. Projects often run in parallel resulting in further diversity of perspectives. Since each project plays a different role in the overall business portfolio there is a need to prioritize the most critical items in terms of time, budget and outcome across the entire portfolio. At the same time, the vast amount of information concerning the projects within the portfolio needs to be synchronized, seeking to implement effective methods and models to manage this complexity. This is particularly apparent in a multi-actors dimension, such as in supply chains. Effective upstream and downstream integration and cooperation with suppliers and other actors depends on the ability to manage information exchange. Functions and processes inside the organization should be synchronized with the functions and processes of other partners. When this integration fails, then achievement of critical project goals, such as milestones, on-time delivery and cost control, is at risk.

In this sense, it is fair to say that uncertainty not only affects projects at start-up, when predictions are typically hard to make, but also during other phases, when changes are constantly required, when it is necessary to consider different alternatives and when massive cooperation and exchange of information occurs with other actors. A high level of uncertainty generates risks at all project stages, in terms of both upside risks (opportunities) and downside risks.

In processes, risks can be revealed in many different ways (e.g., business disruptions; deviations from stated service or production standards; and obstacles to the achievement of the goals of the organization and the supply chain as a whole).

> Proposition 4: Organizations that view risk management as an activity that overlaps with all processes and projects may be able to mitigate the negative effect of risks on objectives and boost success and profit.

These four propositions are summarised in Fig. 5.2. They might influence the definition of the appropriate risk assessment method. Particularly, the propositions 1 and 2 might be the starting point for the definition and prioritization of the primary objectives, according to the perceptions and evaluations of managers. The propositions 3 and 4 might be the starting point for the selection of the "areas" where risks should be measured.

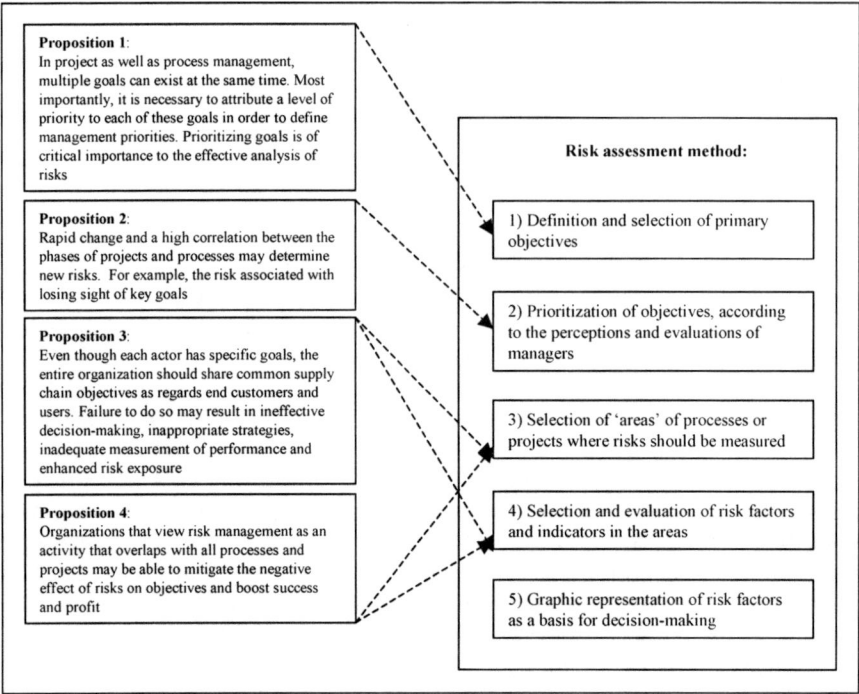

Fig. 5.2 Four propositions and the risk assessment method

5.3 Risk Assessment

The need to manage risks in processes and projects is inherent in most organizations and supply chains. A key characteristic of risks in processes and projects is that selected choices for planning and managing may intrinsically be more risky than others. Assessing risks helps understand how to shift from a risky plan to a less risky plan (Chapman and Ward 1997).

When looking at a supply chain, each process and project inside it seems to be demand-driven. This means that all supply chain members should share a focus on end customers in order to achieve the best customer service, and consequently customer loyalty and profitability (see Proposition no. 4). For this reason, supply chain risk assessment should be linked to specific objectives of the supply chain, such as service quality, timeliness, flexibility and efficiency (see Proposition no. 1). Risks can be considered as a threat or obstacle to achieving the supply chain goal. Risk evaluators should prioritize objectives, assessing the impact of potentially negative events and cause-effect relationships along the supply chain.

Returning to the primary purpose of this chapter to give some suggestions on how to assess risks, the focus is on the identification of major risks along the supply chain processes and throughout the lifetime of specific projects. Risk

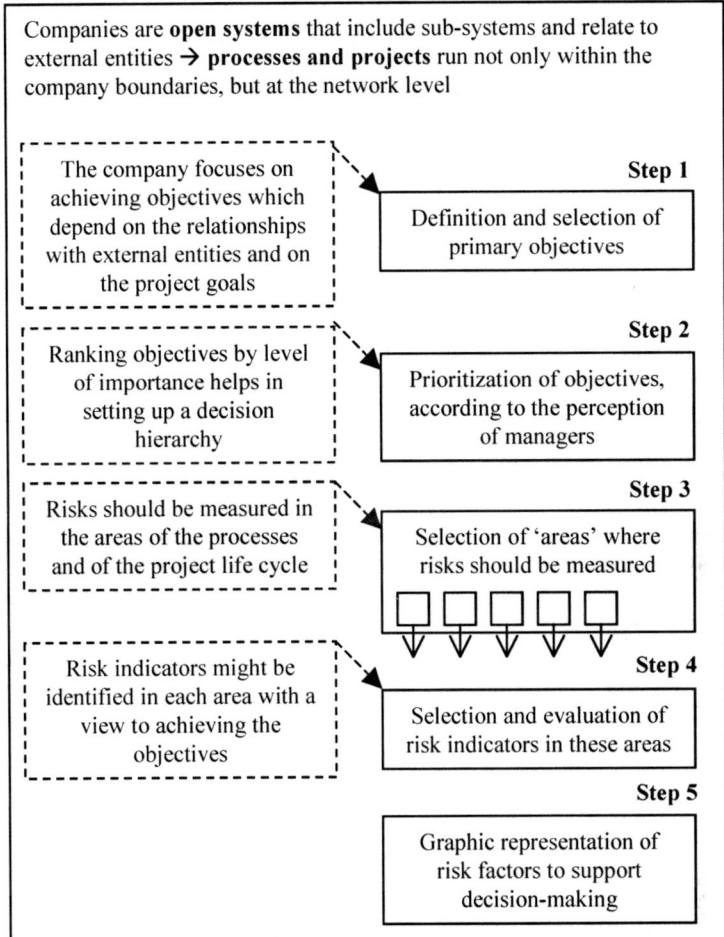

Fig. 5.3 Assessment of Risks in Projects and Processes

factors can always be considered in terms of what drives them, where they lie and what they are associated with.

A risk assessment approach itself may be derived from the philosophy of the performance measurement system employed:

1. The risk assessment method employed should be linked to the achievement of the specific objectives associated with processes and projects
2. The risk factors in the supply chains exist both inside the company and at the external network level. Due to strong correlations between processes inside and outside a focal-company, measurements should be coordinated.

Assessment of risks in processes and projects might comprise the following steps (Fig. 5.3):

1. Definition and selection of the primary objectives
2. Prioritization of objectives, according to the perceptions and evaluations of managers
3. Selection of "areas" of processes or projects where risks should be measured
4. Selection and evaluation of risk indicators in these areas
5. Graphic representation of risk indicators as a basis for decision-making.

5.3.1 Step 1

Step 1 is the definition of primary objectives in projects and processes. Goals depend on the specific upstream and downstream relationships in the supply chain and are a result of the organization's strategic plans, the supply chain scenario and the nature of the relationships with the customers. Some examples of primary objectives are reduced lead-times, improvement of product availability on distribution shelves, timeliness or achievement of high standards in service performance. All of the above objectives often might clash with cost reduction, so the question is: who sets the priorities? It is obvious that decision-makers influence the evaluation with their perceptions. Especially in the case of projects, timeliness might be considered as the ability to reach on time the milestones agreed with the customers. Project-oriented organizations are also strongly oriented towards change, project work and harmonization of different tasks. For this reason, this type of culture has been defined as a "guided missile", where achieving the final goal is crucial and all the resources work toward that end (Trompenaars and Hampden-Turner 1998).

5.3.2 Step 2

Step 2 consists of prioritizing objectives. Ranking objectives by level of importance helps in assessing risk levels and the management's priorities. How do we accomplish that? Each manager in the risk management team should identify the risk factors and problems that might affect his/her specific job objectives. Managers might express different perspectives in this evaluation, depending on their job focus. Decision-making techniques, such as the Analytic Hierarchy Process (Saaty 1994), help in taking into account subjective evaluations and allow decision-makers to participate in the definition of "criticalities" in achieving objectives. These points of view should be used as "drivers" to compare objectives and define a sort of "management's priority". This scale should be used during risk evaluation, in order to define risk levels and the management's priorities. An example relates to a risk assessment analysis carried out for an Italian company in 2006. The goals were prioritized in two totally different ways by the logistics manager and by the marketing manager. The logistics manager believed that keeping stock in the warehouse caused extra-costs and for this reason he preferred a demand-driven production, avoiding high levels of stock. The risk of delay in replenishing stocks would not be viewed by an impartial observer as a "major risk". However, the marketing manager was in total disagreement with the logistics manager believing that service should be the number-one priority. The logic is that higher stock helps reduce delays, making it more likely that stock will be available when demanded and hence ensure higher levels of customer satisfaction. The most appropriate approach probably lies somewhere between the two positions. However, this empirical example illustrates that such diversity of perspectives, objectives and priorities are very common and are not necessarily easily solvable. Such situations may engender higher risk exposure as opposed to resolving risk.

All measures should be consistent. For this reason, all evaluations should be examined by the leader of the risk management team (who might also be the project manager) in order to reach a consistent evaluation (Fig. 5.4).

Step 1: Definition and selection of primary objectives in projects and processes

EXAMPLES OF OBJECTIVES IN PROCESSES:
- Reduced lead-times
- Flexibility and responsiveness
- High standards in service performance

EXAMPLES OF OBJECTIVES IN PROJECTS:
- On-time achievement of milestones
- Cost reduction
- Robustness

Step 2: Prioritization of objectives, according to the perception of managers

→ Inside each project and process team all decision-makers define the 'criticalities' in achieving the objectives
→ The team leader guarantees consistent evaluation of the objectives. He helps the selection and evaluation of the risk indicators (**step 3**)

Fig. 5.4 Step 1 and 2: the definition of objectives and priorities

5.3.3 Step 3

Step 3 is the selection of the "areas" in processes or projects where risks should be measured.

In processes, the supply chain may be broken down into areas such as, procurement, warehousing, order cycle, manufacturing, transport and distribution.

The areas involve all the flows and processes within the chain, both inside and outside the focal company. Risk indicators can be identified in each area with a view to achieving the particular process objective.

The areas involved in the life cycle of projects, also identified within the supply chain (Mentzer 2001) include procurement, project definition, design, production, installation and service. These process and project "areas" may change depending on the specific characteristics of the organizations.

5.3.4 Step 4

Step 4 is the selection and evaluation of risk factors and indicators in the "areas". Risk assessment should focus on identifying threats to the success of processes and projects and factors that may prevent achievement of objectives.

As Einstein stated, "Not everything that counts can be counted, and not everything that can be counted counts." Therefore, managers should ideally be able to identify all these threats. The goal is to identify risks inside defined process areas, with a view to achieving specific objectives. Looking at projects, the issue is no different: risks should be identified inside the project areas with regard to the goals. Moreover, there is a need to recognize and manage similar risk profiles in different projects and to compare different projects in terms of risk, cost and performance.

As stated above, we can borrow some performance measurement principles to define the risk indicators panel. It means that only measurable indicators might be selected. Evaluators should add to the panel indicators that are relevant (providing appropriate information) and objective (not based on opinions) in order to guarantee consistent evaluations within the team and across different functions.

The involvement of managers is always essential in selecting the relevant risk indicators and in evaluating their impact and cause-effect relationships. Since they have a deep knowledge of the processes they manage, only managers can really help in creating an in-depth comprehension of critical issues and interdependencies for a sound risk assessment. An example of risk indicators is given in the next section.

5.3.5 Step 5

In step 5, risk factors should be assessed in terms of their impact ("high", "medium", or "low") on the achievement of objectives. In this phase, it is helpful to create a matrix that takes into account risk factors in the supply chain areas related to the achievement of objectives. An example of this matrix is shown in Fig. 5.5.

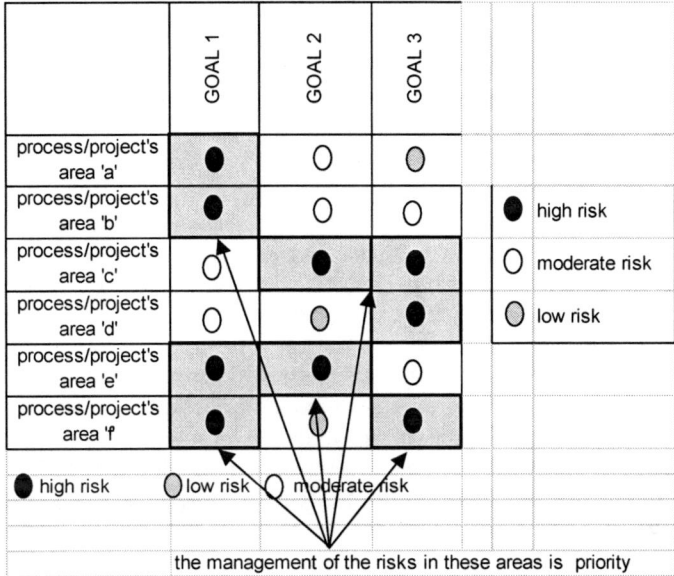

Fig. 5.5 The impact of risk factors on the achievement of objectives

5.4 Some Risk Factors in Processes and Projects

Predicting adverse events that might hinder the efficient and effective management of processes and projects is useful to prevent the occurrence of such events and to implement risk mitigation actions for improved management. A description of some risk factors involved in processes and projects which have the capacity to jeopardize the successful achievement of the overall objectives is provided below. This should be seen as only illustrative and serving as a starting point, since the bulk of the work remains to be carried out by managers during assessment of the risks.

There are numerous adverse events in the "warehousing" and "manufacturing" areas that might cause an extension to the processing time and therefore compromise the efficient and effective management of the overall process. These are, for instance, unexpected down-time, use of damaged or defective materials or semi-finished products along handling and production lines, ineffective management of material handling or business interruptions due to environmental, physical or human factors.

Production planning and management, as well as management of relationships with suppliers, is closely linked to the expected demand, to the point that demands that are higher or lower than expected may cause delay in processing orders, inability to serve some customers or, in case of lower demands, may generate excess finished products remaining unsold and in stock.

In the "order cycle" area, errors during order processing may cause delays in meeting the demand, which in turn negatively affects the downstream service level.

It is fair to conclude therefore that all areas of the logistic process are connected with one another in the achievement of common goals. The occurrence of errors, adverse events or interruptions at selected stages may have a rebound effect and thus seriously compromise the overall performance.

By the same token, when dealing with project management, specific highly critical issues can be encountered in the area of project definition. The need to reduce project execution time, to comply with the request of customers not accepting a standard schedule, may cause the phases of design, production and installation to disalign. Moreover, specific requests from customers may cause additional costs, as in the case of requests for specific product customization or for a higher number of check tests during the work in progress of the project. The additional costs incurred for product customization or to carry out additional tests affect overall project execution time. In some instances, it is necessary to engage external suppliers for the provision of selected components or services in a short time, thus increasing overall costs and running the risk of not being able to synchronize the various project phases.

A high level of correlation between the phases of a project or process increases the probability of the adverse events occurring upstream of the project/process also having a downstream effect, thus compromising the achievement of overall objectives. For this reason, it is of utmost importance to carefully monitor any potential adverse event that might generate such downstream effects and therefore consider its potential ability to negatively affect the entire subsequent flow of activities. The design and production area may in turn generate risks related to delays or design errors, as well as risks connected with failure to coordinate with external partners, when companies opt to outsource certain activities connected with the development of specific components.

Additional risks are related to products' maturity or obsolescence. In fact, if on the one hand a mature product has generated a level of experience, which in part reduces risks correlated with its development, on the other hand in some instances managing the technological refreshment of a product or process may be particularly critical. In fact, extra production costs may ensue, the level of performance of the product may drop or continuity of delivery service to customers may be jeopardized.

5.5 Conclusions

In a dynamic and competitive environment, companies are related to other entities – such as suppliers, third parties, customers – which should share common objectives. Furthermore, companies include sub-systems – such as processes and projects – which are constantly subject to change and have a specific set of objectives.

Projects and processes involve working with multiple relationships, inside and outside each team. At the same time, processes are potentially continuous and projects are start and finish dated by nature. All the activities need coordination in order to achieve the primary objectives. In this context, risks take different forms, such as disruptions, reductions in service or performance levels and delays. It is critical that each organization implements an integrated assessment of risks threatening the performance of the entire network and projects and processes.

Considering the strategic priorities and the nature of the demand, organizations need to define and prioritize objectives to guarantee the ability to achieve them, as required by the stakeholders.

Many decision-makers are involved in all projects and processes. Decision-makers should answer the following questions: what are the priorities of our customers and how can we serve our customers best, avoiding reductions in service levels? Decision-makers should work together and specialize each in their specific area of expertise, although they also need to evaluate in a consistent way the threats to achievement of the best performance. Top managers should coordinate the risk assessment activities inside the teams and prevent different decision-makers from evaluating risks in different ways, since this might negatively affect the achievement of the overall objectives.

For these reasons, risk management can offer an important contribution to effective project management.

References

Asbjørnslett BE (2003) Project Supply Chain Management: From Agile to Lean. Oslo, Norway. (Proceedings of the Conference Project Management: Dreams, Nightmares and Realities.)
Bonaccorsi A, Lipparini A (1994) Strategic Partnership in New Product Development: An Italian Case Study. Journal of Product Innovation Management, 11: 134–145.
Bowersox DJ, Closs DJ, Cooper MB (2002) Supply Chain Logistics Management. McGraw Hill, New York.
Chapman C, Ward S (1997) Project Risk Management. John Wiley & Sons, Chichester.
Christopher M (1998) Logistics and Supply Chain Management: Strategies for Reducing Cost and Improving Services. Financial Time Prentice Hall, New York.
Christopher M, Juettner U, Peck H (2003) Supply Chain Risk Management: Outlining an Agenda for Future Research. International Journal of Logistics: Research & Applications, 6: 197–210.
Debenham J (2001) Agent-Based Process Management. Melbourne, Australia. (Proceedings of the 11th Annual Symposium INCOSE.)
Flannes SW, Levin G (2001) People Skills for Project Managers. Management Concepts. The United States of America.
Gaudenzi B, Borghesi A (2006) Managing Risks in the Supply Chain Using the AHP Method. International Journal of Logistics Management, 17: 114–136.
Gaudenzi B, Gentile I (2006) L'analisi dei Rischi di Progetto Nella Supply Chain: 3D Long Range Early Warning Radar. Roma, Italy (Proceedings of the Project Risk & Opportunity Management forum).

Gregory IC, Rawling, SB (2003) Profit From Time: Speed Up Business Improvement by Implementing Time Compression. New York, Palgrave.

Lereim J (2002) Leadership Challenges in Projects with Moving Targets (Proceedings IRNOP 5 International Research Network of Organising Projects, Renesse, Zeeland, The Netherlands, 28–31 May 2002).

Lundin RA, Söderholm A, Wilson T (2001) On the Conceptualization of Time in Projects. Uppsala, Sweden. (Proceedings of the 16th Nordic Conference).

Marsh and McLennan Companies (2005) MMC Reports 2005 Results. www.mmc.com.

Mentzer JT (2001). Supply Chain Management. Sage, Thousand Oaks, CA.

Nelson D, Mayo R, Moody P (1998) Powered by Honda: Developing Excellence in the Global Enterprise. Wiley, New York.

Obeng E (1996) All is Change, Financial Times. Prentice Hall, New York.

Omli LE, Svendsen EK, Karlsen JT (2003) Aiming for Higher Value Creation of Projects – How to Secure Return on. Oslo, Norway. (Proceedings of the Conference Project Management: Dreams, Nightmares and Realities.)

Otto A (2003) Supply Chain Event Management: Three Perspectives. International Journal of Logistics Management, 14: 1–15.

Saaty T (1994) How to Make a Decision: The Analytic Hierarchy Process. Interfaces, 24: 19–43.

Chapter 6: Risk Management System – A Conceptual Model

Arben Mullai

Lund University, Lund Institute of Technology, Department of Industrial Management and Logistics, Division of Engineering Logistics, Lund, Sweden

6.1 Introduction

This chapter deals with the topic of risk management concerning the transportation of dangerous goods. The evidence presented derives from a European project – the DaGoB (Safe and Reliable Transport Chains of Dangerous Goods in the Baltic Sea Region) project,[1] as well as the author's own research in the field. One of the main objectives of the DaGoB project is to enhance and transfer the knowledge in the field at local, national, regional and international levels.

The literature study shows that risk management is an evolving discipline of science. For a long time, risk assessment and management have been everyday human activities. However, in recent years, this simple perception of risk has changed considerably. Risk management has become a very important topic and a field of study in its own right. In many countries, it has become an increasingly important component of industrial and national decision-making processes concerning many issues including those related to human safety and health, environmental quality, property protection and security (IMO 1997, 2004, 2006; EC 1997, 2006). Contemporary risk management is a cross-disciplinary process that takes a holistic approach and employs a wide range of specific methods, techniques and tools. The process relies on the knowledge of many disciplines of science. However, despite the progress being made, the literature shows that there are still misconceptions, misuse and ambiguities in the field. In addition, as a result of accidents, there is growing public concern about the lack of safety and the

---

[1] The DaGoB project is partly financed by the European Union (European Regional Development Fund) within the BSR INTERREG III B Neighborhood Programme. The project involved numbers of partners from the countries of the BSR, such as Finland, Germany, Sweden and the Baltic States. For more information see: www.dagob.info

consequent pollution caused by the transport of dangerous cargoes. In recent years, in particular after the "9/11" events, security concerning the chemical supply chain has become an important issue for many organisations, industries, governmental authorities and the general public. Such concerns stem mainly from the high and increasing volume of dangerous goods being transported, the potential for deliberate acts, the severe consequences of accidents, and the general belief that risks should be better managed. Therefore, there is a need to further enhance understanding of the field of risk management.

Given the importance, the relevance and the demonstrable need, also reflected in the objectives of the DaGoB project, the chapter seeks to provide a unified understanding of the risk management field in the context of the transportation of dangerous goods. The extensive literature review supplemented by personal research experience (Mullai and Paulsson 2002; Mullai 2004, 2006, 2007) provided the underpinning for the description of the central concepts. The content of this chapter is aimed primarily at risk analysts, risk managers and other members of the scientific communities and practitioners who are interested in risk-related issues, methodologies, research and management practices.

The constituent elements of the risk management system are defined and described in the context of the dangerous goods or the chemical supply chain, focusing in particular on the maritime transport of packaged dangerous goods (PDG). Given the representativeness of the chemical supply chain, the content of this chapter is relevant to many other supply chains or systems. Many industries, sectors or business activities are related to the chemical supply chain. Many risk-related terms, definitions, concepts, methodologies and practices have originally been developed by or on behalf of the actors in the chemical supply chain, including the oil and gas (inland and offshore) industries, the chemical production industry, chemical storage and transportation, the nuclear power production industry and many other related industries and relevant organisations. Some of the world's best risk assessment and management practices, frameworks and techniques may be found in the chemical industry and related organisations. Such practices are adapted and implemented in other industries, sectors or businesses across many countries around the world.

The chapter begins with definitions of the central concepts, namely risk analysis, assessment and management. Then, a unified concept of the risk management system is provided. The main phases, stages and steps presented in the model are explored in some detail.

6.2 Variations in Terms and Definitions

The field of risk management is faced with difficulties in defining and agreeing terms. Risks are dealt with differently across countries, industries and sectors (DCDEP 2000). Terms, definitions and interpretations are as varied as the number of sources providing them (ACS 1998; DNV 1996; EC 1997, 1999; OECD 2001). There are no agreed unified definitions of risk analysis, assessment and management.

There are often misconceptions. Despite their meanings, different terms, for example "risk analysis" and "risk assessment", are often used interchangeably. Further, a single term may be used in different ways, convey different meanings or be applied differently in various contexts. Thus, although the term "analysis" may be narrower than the term "management", the Society for Risk Analysis (SRA 2004), which consists of members from different organisations and countries, has chosen to broadly define the term "risk analysis" as the process that includes risk assessment, risk characterisation, risk communication, risk management, and policy making. The EC Health and Consumer Protection Directorate (EC 2000) defines the term "risk analysis" as the encompassing term used to describe three major sub-fields of the discipline, namely risk assessment, risk management and risk communication. Further, the Comprehensive Risk Analysis and Management Network (CRN 2004), which is a Swiss-Swedish workshop network initiative for international cooperation among governments, academics and industries and sectors, employs a similar definition of "risk analysis" as those stated above.

Variations in risk-related terminology, definitions, concepts, methodologies and practices are attributed to a wide range of factors, including (a) different perceptions, attitudes and values regarding risks in different socio–economic-political contexts; and (b) different needs and specifications of diverse industrial sectors and risks specifications in various countries and regions. Each country has its own priorities, local communities and central authorities with their own interests, and different kinds of legislation (DCDEP 2000). The roots of such variations in the field also stem from the diversity in language, interpretation and the national socio-cultural environmental contexts.

6.3 A Unified Concept of the Risk Management System

A variety of views are adopted in the field of risk management. Although some sources may view the term "risk management system" narrowly, others may treat it as a broad concept, in particular the field of human safety and health, environmental and property protection. Other similar terminologies in use include: "Safety Management System" (SMS), "Integrated Safety Management System", "Risk-Based Decision Making" (USCG 2001), "Risk Policy-Making System", "Social Governance of Risks",[2] "Integrated Socio-Economic Risk Management" (OECD 2000), "Risk Management" (IEC 1995) "Sound Risk Management", "Total Risk Management System" and "Safety, Health and Environmental Management System". The following section seeks to provide a unified understanding of the central concepts related to the risk management system.

The *risk management system* is the overall integrated process consisting of two essential interrelated and overlapping, but conceptually distinct components – *risk assessment* and *risk management*. In recent years, risk communication has become

---

[2] The term is defined by TRUSTNET (2002), which is a pluralistic and interdisciplinary European network involved in the field of risk governance.

an important integrated component of the risk management system. Risk assessment, which is identical to safety assessment, is an element of the system that consists of *risk analysis* and *risk evaluation* (RSSG 1992; IEC 1995). In many cases, the terms "risk analysis" and "risk assessment" are used interchangeably. *Risk analysis* is a scientific process in which, by applying a wide range of methods, techniques and tools, risks are identified, estimated and presented in qualitative and/or quantitative terms (DNV 1995). *Risk evaluation* is the process of comparing estimated risks with established risk evaluation criteria in order to determine the level or significance of risks and appropriate risk management strategies, providing recommendations for the decision-makers (EC 1999). Although risk assessment provides basic inputs for assisting decision makers, it may not necessarily provide answers for many questions, for example questions concerning the level of risks and cost benefit trade-offs in risk control. A wide range of factors influence the issues mentioned. Dealing with these issues also requires consideration of factors other than technical and scientific ones.

6.4 Main Elements of the Risk Management System

The following section describes the main elements of the risk management system. It is based on the study of some of the world's best risk management practices, frameworks and techniques in shipping and other industries, sectors and activities. They include the works of:

1. Institutions or organisations, such as the OECD (2000, 2001, 2002), the USCG (U.S. Coast Guard) (2001), the USEPA (U.S. Environmental Protection Agency) (2000), the UK HSE (Health and Safety Executive) (1991, 1992, 1995, 1999, 2001, 2002), the IMO (International Maritime Organisation) (1997, 2002, 2004, 2006), the EC (European Commission) (1996, 1997, 1999), the German Lloyd's and DNV (Det Norske Veritas) (1995, 1996), the ISO (International Organisation for Standardization) (1999) and IEC (International Electro-technical Commission) (1995), and other organisations (ACS 1998; CCPS 1992; DETRA 1999; RSSG 1992);
2. Researchers in the field (e.g., Erkut 1996; Frewer 2004; Nicolet and Gheorghe 1996; Saccomanno and Cassidy 1993; Weigkricht and Fedra 1993).

Figure 6.1 presents a conceptual model of the risk management system. The risk management system is a stepwise process consisting of two interrelated but distinct phases: *risk assessment* (analysis and evaluation) and *risk management*. Each phase consists of a number of stages, steps and sub-steps that, in principal, are sequential. In many situations, however, this may not necessarily be so. Initiation of the process can be triggered by a combination of different factors at any given time, including the seriousness of events, threats, issues and concerns faced, the availability of resources and data, and the improvements or developments of more advanced analytical methods and tools. The process can start at any point and involve any individual element of the system. The literature shows that each component of the system can be considered a specific domain of science in its own right.

The model (see Fig. 6.1) represents a dynamic model. The overall risk management process has a hierarchical structure that consists of different levels, in which the highest levels are further broken down into stages, steps and sub-steps. The processes are interactive, responding to change, re-evaluation and refinement. Although shown in a sequential and seamless order, e.g., risk analysis, risk evaluation and risk management, some stages and steps can be carried out and accomplished simultaneously. Skipping processes and returning back to the earlier processes are also possible. This is due to a variety of factors, including the availability and accessibility of additional and/or new risk-related data and information, the breadth and depth of the analysis, the results of the study, re-evaluation and redefinition, and decision-makers' needs. In many situations, it may be considered unnecessary to go through all the phases, stages and steps. The process can be suspended at any given phase, stage or time. For example, the risk analysis process can be suspended from further detailed analysis if risks are found to be at a low or negligible level and further study may be deemed unnecessary and cost inefficient.

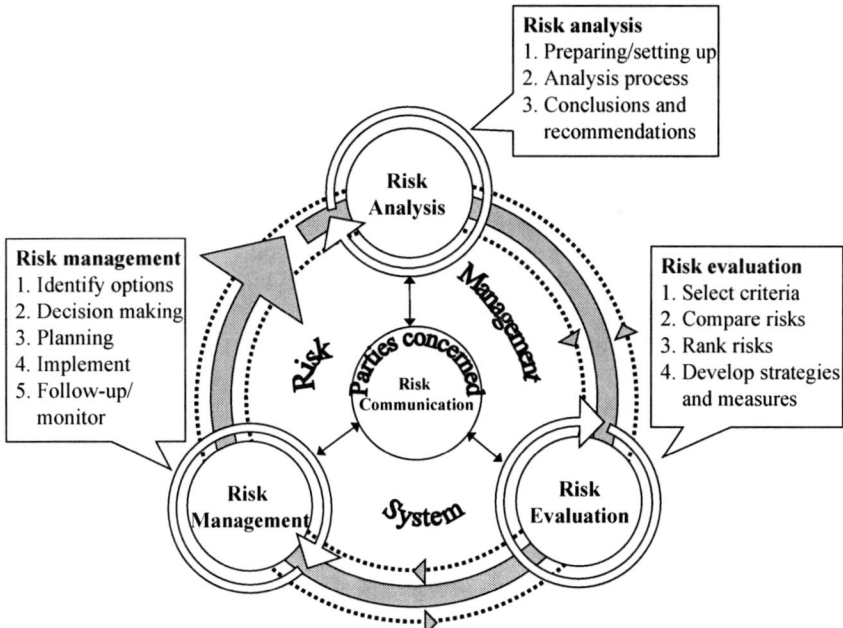

Fig. 6.1 Main elements of the Risk Management System (RMS)

The following sections describe in some detail the key elements of the Risk Management System. Each of the three phases is discussed in turn, starting with the Risk Analysis phase. Within each phase the key stages and steps are explained and illustrated with examples relating to the PDG context.

6.5 Risk Analysis

Risk analysis is the process in which risks are examined at various degrees of detail to determine the extent of the risks, how risk elements are related to each other, and which ones are the most important to deal with. This may not necessarily involve any consideration of the significance of the estimated risks (DNV 1996). The main stages of the risk analysis are (1) *preparations for analysis*, (2) *risk analysis process* and (3) *conclusions and recommendations*. These main stages consist of a number of steps and sub-steps or tasks, which are identified and further developed for ready application in the risk analysis of the maritime transport of PDG based on combination of the literature study (DETRA 1999; DNV 1995; HSE 1991, 1999, 2001; IEC 1995; IMO 1997, 2004, 2006; ISO 1999; OECD 2000, 2001; USCG 2001; Weigkricht and Fedra 1993) and the author's research experiences (Mullai and Paulsson 2002; Mullai 2004, 2006, 2007).

In order to inform about the risks, three interrelated fundamental questions must be answered: "*What has gone and can go wrong?*" "*What are the consequences?*" and "*How likely is that to happen?*" – known as "the triplet definition" of risks (Kaplan and Garrick 1981). The concept of the triplet definition has become widely applicable as an element of standardisation (ACS 1998; IEC 1995). The process that facilitates the answers to these three fundamental questions constitutes the core of the risk analysis. The entire process builds on these "simple" questions, which often require considerable effort, time and resources to provide answers. Depending on a number of factors, including the requirements of the decision makers and the data and resources available, the answers to the questions could be given in a qualitative or quantitative form, or a combination of both. These questions lead to other important questions that, in turn, require additional answers, and subsequently additional effort, time and resources.

6.5.1 Preparing/Setting Up

Prior to answering the triplet questions, a number of important activities are carried out in preparation for the risk analysis, including the following key steps:

1. *Background*: Establish the particular context base on which risks associated with the maritime transport of dangerous goods will be analysed and evaluated and which decisions will be taken. Without the context, without knowing how dangerous goods risks can be compared to other risks, it is hard to put these risks into perspective.

2. *Perform a preliminary or screening risk analysis*: If necessary, perform a preliminary risk analysis in terms of types of marine accidents, ships, dangerous goods, vessel traffic, activities and geographical locations.
3. *Determine who should conduct the risk analysis*: Set up a team of risk analysts, whose members are familiar with the maritime transport system of dangerous goods, risks and risk analysis methods and techniques, including other knowledgeable persons with a variety of relevant expertise in the field.
4. *Identify interested parties*: Identify parties that are concerned with the risk issues and affected by decisions, such as decision or policy makers, ship-owners, cargo interests, employees and many other parties interested in the maritime transport of PDG.
5. *Identify risk generating activities*: Identify risk generating activities, such as packing, handling, stowage, loading and unloading, and the transport of dangerous goods.
6. *Identify and formulate problems*: Some generic issues include human safety and health, marine environment pollution, property damage, security and economic aspects.
7. *Set the objective(s) for risk analysis*: A principle objective of every risk study is to enhance the understanding of risks involved and to provide decision makers with relevant information and tools and recommendations for improving risk management in the maritime transport of dangerous goods.
8. *Define boundaries of the study*: Define the system or physical and analytical boundaries of the study.
9. *Select appropriate methods and techniques*: There is a wide range of methods and techniques for the collection and analysis of risk-related data to choose from. Based on the amount, type and quality of data, the time and resources available, and the legal requirements, if any, select the most appropriate methods and techniques for collection and analysis of relevant risk-related data and information.
10. *Collect relevant risk-related data and information*: Identify the data sources and collect relevant risk-related data and information sets.

6.5.2 Analysis Process

Risk analysis varies from simple to very complex and detailed. A preliminary analysis may be conducted prior to a detailed risk analysis. The stage of risk analysis includes the following key steps:

1. *System definition*
    (a) Define and describe the system whose risks are to be analysed and managed, including the means and objects of transport, dangerous goods traffic, dangerous goods-related activities, and the transport infrastructure.

(b) Review and evaluate the current state-of-the-art regulatory system governing the maritime transport system of dangerous goods. The system is highly regulated.

2. *Hazards identification*
   (a) Define top events, including the wide rage of breaches and failures of packages.
   (b) Explore transport/distribution hazards, including their cause and contributing factors and sequences of events that have or can lead to loss of containment and/or involvement of dangerous goods.

3. *Exposure and consequences analysis*
   (a) Dangerous goods and their hazards: Explore the list/inventory of dangerous goods and their hazards that have or are likely to cause consequences to the risks receptors.
   (b) Dangerous goods release-dispersion-concentration: Explore sequences of events following the release, dispersion, concentration and/or involvement of dangerous goods that can lead to consequences for the risk receptors.
   (c) Modes of contact – the routes of exposure: Explore the ways and routes through which dangerous substances and/or their hazards come into contact and affect the risk receptors.
   (d) Dose-effect assessment: Explore and assess dose-effect relationships.
   (e) Risk receptors exposure: Explore categories of risk receptors exposed to dangerous goods hazards.
   (f) Consequences analysis: Explore the nature of the actual consequences to the risk receptors due to dangerous goods hazards.

4. *Likelihood estimation – quantification*
   (a) Quantify top events, transport/distribution hazards and their causes and contributing factors.
   (b) Exposure estimation: Estimate the size/extent of risk receptors exposed to dangerous goods hazards along with the magnitude, duration, and the spatial extent of exposure.
   (c) Consequence estimation: Estimate the magnitude of the actual consequences to the risk receptors due to dangerous goods hazards, including influencing factors and conditions. Explore the relationships between consequences and other system and risk elements.
   (d) Explore the relationships among the system and risk elements.

5. *Risk estimation and presentation*
   Risks are estimated by combining: a) the likelihood and the severity of consequences; or b) the consequences measured relative to (or averaged over) the population exposed to dangerous goods hazards, which are estimated per one year. Risks are presented in various formats that depend on how the risks are estimated and evaluated, and reflecting the decision makers' and legal requirements, if any. Some alternative risk presentation formats are: risk index (e.g., in the form of a single number

index 1/100,000), tabular format (e.g., tables with bands of fatalities or injures 1-10, 11-100 and 101-1,000), graphs or diagrams (e.g., F N curves, bars, lines and pie diagrams), and maps (e.g., risk contour plot).

6. *Sensitivity analysis*

The Management Index (e.g., Risk Index × Sensitivity Index) provides further ranking for those risks with equal Risk Indexes. Given its scope, sensitivity analysis may not necessarily constitute an integrated step of the risk analysis process.

6.5.3 Conclusions and Recommendations

Synthesize the main results of risk analysis, including key information concerning the main system and risk elements, such as top events, transport hazards and their causes and contributing factors, the inventory and hazardous properties of dangerous goods, the likelihood, the categories and magnitude of exposures and consequences due to dangerous goods hazards, relationships among elements, and the estimated risks. One important objective of every risk study is to develop *a list of recommendations*. In this context, it is necessary to suggest risk management strategies and measures for improving human safety and health and protection of the environment and property in the maritime transport of dangerous goods. A wide range of factors that shape the needs for future research, including system dynamics and interrelations, new and more advanced frameworks, techniques, tools or models, as well as more data and resources may become available. Therefore, it is important to *suggest relevant areas and questions* for future research.

A detailed description of the risk analysis framework is provided in Mullai (2004). The framework is validated based on the large amounts of diverse datasets, including incident data (~600,000 incident cases covering the period 1990–2004) collected from the U.S. and the world's largest hazmat incident databases. The study (see Mullai 2007) may be one of the largest of its kind, and some of the results may not be found elsewhere.

6.6 Risk Evaluation

Risk evaluation consists of the following key stages:

6.6.1 Select Criteria

In many countries and industries, there is a wide range of established qualitative and quantitative risk criteria or standards for the evaluation of dangerous goods risks, including human safety and health, environmental and property risks. For example, risk criteria are developed in the chemical supply chain, including oil and gas inland and offshore industries, hazardous facilities (e.g., chemical plants and storages), nuclear power plants, transport (air, road, rail and sea) of dangerous goods and in general, and health and safety sectors (HSE 1991; IMO 2004, 2006; OECD 2002). Risk criteria represent views, usually of the regulators, of how much risks are acceptable (HSE 1995). In many countries, however, risk criteria reflect, to a large extent, the broad acceptance of society (HSE 1992), because, by law, a wide range of interests have a say in shaping criteria. It is an important task to identify and select relevant specific risk criteria for specific estimated risks in a specific country or industry. Responsible authorities or organisations, for example the Austrian Commission for Tunnel Safety (Knoflacher and Pfaffenbichler 2004), the UK HSE (2001), the IMO (IMO 2004; Spouge 1997), and the USEPA (2000), have established certain sets of principles serving as guides for designing risk criteria in a specific country or industry. These principles cover a wide range of issues concerning risks and risk management, such as the basis for establishing threshold values of tolerable/intolerable risks, the frequency and magnitude of consequences of undesirable events, concentration of risks on particular individuals, locations and territories, the balance between risks/costs and benefits and many more. In many countries and industries, risk criteria may be nonexistent.

Selection of risk criteria may also depend on the results of the risks analysis and how risks are estimated. In cases involving aggregated risks, which combine two or more individual risks, the right risk criteria for evaluation of these types of risks are selected. Not all risk criteria available may be suitable for the evaluation of the aggregated risks.

6.6.2 Compare Risks

In order to determine the significance or the level of estimated risks, at this stage estimated risks are compared against the selected risk evaluation criteria. Risk evaluation may involve different parties concerned with dangerous goods risks, including decision makers at senior levels. Risk evaluation also takes into account a wide range of additional factors and procedures other than scientific and technical ones. The theory on risk perception maintains that the concept of risk is strongly shaped by human minds and cultures (HSE 2001). Numerous studies (e.g., IMO 2004; Johnsson 2004; Vrijling et al. 2004) have shown that risk perception, evaluation, management and attitudes towards risks are affected by a wide range of interrelated factors, including: types of risks involved, benefits of risk sources, the type and the sensitivity of risk receptors and systems and activities exposed, the ability of people to control risks, familiarity with risks, the equality of exposure to risk sources, concentration of risks, the type and the severity of consequences, and scientific uncertainty about the consequences. For example, large-scale disasters weigh more seriously in the public's mind than small-scale individual events. Society generally has a strong aversion to multiple casualty accidents (IMO 2004). In the Netherlands, attempts have been made to express risk aversion mathematically in the form of a risk aversion index and to integrate this into the overall risk evaluation (Vrijling et al. 2004).

6.6.3 Rank Risks

In cases involving various types of risks, the results of risk evaluation may show that risks have various degrees of significance. An important task in quantitative risk analysis is to relate risks to various system elements and risk receptors. In order to prioritize risk management strategies and measures and, subsequently, resources and efforts for managing risks, risks are ranked and prioritized according to their significances as well as sensitivity. The management index, which could be obtained from the sensitivity analysis, can be used for ranking risks with equal indices.

6.6.4 Develop Strategies and Measures

At this stage, based on the results of risk analysis and evaluation, it is possible to develop and present a detailed list of risk management strategies and measures to deal with the present level of risks. There is a large array of approaches and means for dealing with the risks. Although the choices may be endless, there are generally a few principal management strategies, namely *avoidance/elimination, reduction, transfer and acceptance* (USCG 2001; Knight 1999).

Table 6.1 Taxonomy of risk management strategies and measures

		Risk management strategies	Categories of measures	
			Regulatory Command/ control	Non-Regulatory -Voluntary
A	Avoid	- Eliminate		
R	Reduce	- Reduce the frequency of causes (prevention) - Eliminate some causes - Reduce the frequency of consequences - Reduce or mitigate consequences (mitigation)	- Technological - Operational - Managerial - Training/education - Knowledge/ information - Methodological - Financial - Legal - Others	
T	Transfer	- Transfer by contract - Transfer by insurance - Physical transfer - Risk sharing		
A	Accept	- Retain		

The term "risk management measure" can be used as the most generic term representing the wide range of methods, techniques, approaches, or tools, which, in contrast to risk management strategies, are employed for managing risks at a more operational or tactical level. Table 6.1 presents a taxonomy of risk management strategies and measures. A single measure can be enacted to affect one or several risk or system elements. On the other hand, multiple measures can be designed to affect a single element. Often, there is no single solution to guarantee a high degree of efficiency and effectiveness in risk management. As one single measure may not be sufficient, several measures are often combined to achieve risk management strategies. For example, mandatory technological and procedural measures can be combined with financial measures, such as levies or subsidies.

Certain risk criteria contain principal risk management strategies for various risk levels. In order to identify the most effective strategies and measures and to prioritise these, it may be desirable to formulate the set of strategies and measures then submit these again for further scrutiny, including detailed risk analysis and cost-benefit analysis.

6.7 Risk Management

Risk management attempts to provide answers to the questions on how best to deal with risks, such as (USCG 2001): What can be done? What options are available and what are their associated tradeoffs? What are the effects of current decisions on future options? This process, which is distinct from risk assessment, involves the key stages and steps presented below (USCG 2001; Weigkricht and Fedra 1993). Although a large part of this process concerns the decisions of policy makers, risk assessors provide useful information and practical propositions for dealing with risks in a most effective and efficient manner. The key activities of risk management include:

6.7.1 Identify Options

1. *Identify key interests*: Identify and solicit involvement from key interests who will be involved in the decision-making and affected by actions resulting from it.
2. *Risk management strategies*: Identify and determine which risks are important to deal with and what key strategic decisions must be made to avoid/eliminate, reduce, transfer or retain risks.
3. *Risk management measures - options generation*: Identify choices available to the decision makers and factors that will influence the decisions and risk factors, as decisions are rarely based on one single factor alone.
4. *Select methods and tools*: Select the appropriate methods and tools for the analysis of alternative options. Some relevant cost-benefit analysis methods include Cost-Benefit Analysis (CBA), Cost-Effectiveness Analysis (CEA), Input-Output Models (I-O), General Equilibrium Models (GE), and Multi-Criteria Analysis (MCA).

5. *Option analysis and evaluation*: In the light of the results of risk assessment and other relevant evaluation, conduct specific analyses including cost-benefit analysis and comparison and weighing of available options. Almost every implementation of risk management strategies and measures, in particular large and sensitive decisions, involve costs. It may be nearly always possible to take measures that would reduce risks further, but the costs may outweigh the expected benefits. In many cases, a balance between the benefits and costs is needed. In economists' terminology, this means that risks should be reduced until the "marginal cost equals the marginal benefit." Estimation and evaluation of costs and benefits require a common unit of measurement. The monetary value is suggested as the common unit of measurement. For more information about cost-benefit analysis (see Mullai 2006).
6. *Option selection*: Select and recommend appropriate alternative approaches for implementation of risk management strategies and measures.
7. *Residual risks and recommendations*: Identify residual risks and provide recommendations for managing them.

6.7.2 Decision-Making

The decision-making process is a central element of the risk management system. It is a discipline in its own right. This stage concerns decisions on implementation of selected risk management strategies and measures. In consultation with all parties concerned, weighed alternatives are selected and decisions are made for their implementation. The decision may involve implementation of measures to reduce or eliminate unacceptable risks. When appropriate, risks are eliminated, reduced or transferred in the most cost effective manner. When they are justified, risks are retained or accepted.

In many industries or businesses, because of the wide range of complex factors and conditions, decision-makers at all levels are faced with difficult decisions. The process involves not only consideration of technical factors, but also political, social, economic, and many other factors. Further, the process is complicated by the variety and complexity of choices and the environment in which they are made, multiple and often conflicting objectives, different perspectives on risks, the uncertainty and the sensitivity of decisions. It is, therefore, important to provide decision makers with valid, reliable and sufficient information to ensure that they have taken decisions to their best knowledge.

6.7.3 Planning

Preparation and communication of action plans to deal with risks include:

1. Documentation of strategies, actions, goals, and schedule dates;
2. Emergency response and contingency planning;

3. Chemical supply chain including transport and transport-related activities planning;
4. Providing supporting information needed to implement risk management strategies and measures.

6.7.4 Implement

The implementation of risk management strategies and measures include:

1. Implementation of risk management strategies and measures for different risk and system elements;
2. Emergency response procedures and means;
3. Education and training of all persons involved;
4. Supervision, inspection and monitoring to verify compliance with regulations;
5. Measures to compel compliance;
6. Safety management audit.

6.7.5 Follow-Up and Monitoring

Follow-up and monitor the effectiveness of planned actions and the continuous update of all assessments as they change due to the implementation of strategies and measures and changes in the transport system and surrounding environment with the passage of time. Due to the wide range of outcomes, risk management strategies and measures are often difficult to compare and evaluate. The best decisions are those that yield the greatest expected values. The USCG (2001), for example, has designed three principal criteria for the evaluation of risk management strategies and measures, such as: a) *efficacy*, which is the degree to which the risks will either be eliminated or minimized by the proposed actions; b) *feasibility*, which is the acceptability of implementing the proposed preventative action; and c) *efficiency*, which is the cost-effectiveness of the proposed actions in terms of potential dollars lost if no action is taken versus the cost of the actions.

6.8 Risk Communication

Risk communication has become an important integrated element of risk management. Risk-related information generated at each phase, stage or step should be communicated continuously and effectively to all parties concerned.

The literature shows that risk communication can be considered a specific domain of science in its own right. Risk communication and its role in attitudes towards risks, risk assessment and risk management have been explored in several studies (Bender et al. 1997; Bickerstaff and Walker 1999; Frewer 2004; HSE 2001; Leiss 2004; OECD 2002; Reid 1999). For example, minor risks can sometimes produce massive reactions, while major risks often may be ignored

(HSE 2001). This is partly attributed to risk communication approaches. The public responses to risks can be amplified or reduced depending on how risk communication interacts with psychological, social, cultural, and institutional processes (HSE 2001). These and other issues have been the subject of risk communication studies.

Risk communication is an interactive process involving the exchange of information and opinions among risk assessors, managers, decision makers and other parties, including individuals, groups and institutions concerned. The interface among parties is a critical element for ensuring that risk assessment results are used to support the decision-making processes at all levels. Risk communication covers a wide range of activities directed at increasing the knowledge about risk issues and participation in decision making and management. The process includes discussions about the nature and level of risks, risk management strategies and measures. In this process, people express their concerns, opinions and reactions to legal and institutional bodies responsible for risk management. The public prefers clear information regarding risks and associated uncertainties, and the nature and extent of disagreements among different experts in the field (Frewer 2004).

Effective risk communication is an important responsibility of many industries and governments (Leiss 2004). The fundamental requirements of good risk communication practices include undertaking "science translation", addressing uncertainties, and dealing with science and policy interfaces (Leiss 2004). The OECD has also been working to identify practical ways to make risk communication an integral and effective part of the risk management process. The OECD Guidance Document on Risk Communication for Chemical Risk Management, Sect. 2 (pp. 19–26), contains guidelines on a risk communication programme concerning (OECD 2002): (1) designing the strategy for a risk communication programme; (2) designing an effective risk communication message; (3) rules addressing specific risk issues and (4) communication in crisis situations.

In the transport of dangerous goods, risk communication encompasses a wide range of activities, such as dissemination of risk-related issues, data and information, research results, sharing of best practices and experiences in risk methodologies and management, holding public hearings on risk and risk management issues, providing information and warnings about dangerous goods hazards, and developing publicly accessible dangerous goods risk-related databases.

Public information concerning dangerous goods risks has become a norm in many countries and industries. Risk assessment processes and outcomes are required to be opened to greater participation and scrutiny by all parties concerned. This, in turn, has required the need to help the public understand information and to help decision makers understand the public's risk perceptions and responses. Perceptions and responses are complex, multidimensional and diverse, as "the public" consists of many publics with diverse values and interests. Understanding public concerns must be the basis for an effective risk management strategy (Frewer 2004).

6.9 Re-Assessment – A Continuous Cyclic Process

The literature shows that risk-related studies are often carried out on an ad-hoc basis. But, the system and risks associated with it require continuous attention and re-assessment of new situations. Although presented at the "end" of the cycle, the re-assessment or re-analysis can take place at any given phase or stage and at any given moment. A proactive management process is to be viewed as a continuous and cyclical process, because of the wide range of interrelated influential factors, including (a) system dynamics and constant changes; (b) more and better risk-related data and information, and risk assessment and management methods, techniques or tools become available and accessible; and (c) increasing concerns and demands for more frequent and thorough risk studies.

The aforementioned factors are also valid for the maritime transport system of dangerous goods and risks associated with it. The system elements are very dynamic and constantly changing. These include changes in the regulatory system, ships, dangerous goods, dangerous goods traffic, packaging systems, and dangerous goods-related activities. In addition, with the implementation of risk management strategies and measures, one or several system elements may change. Re-assessment of risks on a regular basis is especially important as it keeps decision makers continuously updated about changes in the system and risks, and provides feedback on the effectiveness and efficiency of risk management strategies and measures.

6.10 Summary

Despite the progress being made, there are still variations, misuses and misconceptions in the field of risk management, which are attributed to many different factors, including differences in perceptions, needs, specifications, and even differences in languages. Based on some of the world's best risk management methodologies and practices, attempts have been made to provide a unified understanding of the field. In this Chapter, a conceptual model of the risk management system is presented. Each element of the system is explored in some detail. The risk-related concepts, methodologies and practices employed in the chemical supply chain could be adapted with some adjustments for application in other supply chains. In summary, the latter can learn from the best practices in the chemical supply chain.

References

ACS (American Chemical Society) (1998) Understanding Risk Analysis, Guide for Health, Safety and Environmental Policy Making.
Bender M.J., Swanson S., Robinson R. (1997) On the Role of Fuzzy Decision Support for Risk Communication among Stakeholders. IEEE Proceeding, pp. 317–322.

Bickerstaff K. and Walker G. (1999) Clearing the Smog? Public Response to Air Quality Information. Magazine of Local Environment, Oct. issue, 1999.

CCPS (Center for Chemical Process Safety) (1992) Guidelines for Chemical Process Quantitative Risk Analysis. American Institute of Chemical Engineers, New York.

CRN (Comprehensive Risk Analysis and Management Network) (2004) http://www.isn.ethz.ch/crn/, 2004.

DETRA (Department of the Environment, Transport and the Regions, UK) (1999) Identification of Marine Environmental High Risk Areas (MEHRA's) in the UK. Report prepared by Safetec for the Department of the Environment, Transport and the Regions.

DCDEP (Directorate for Civil Defence and Emergency Planning, Norway) (2000) Risk Assessment in Europe. A summary report from the EU Workshop on Risk Assessment, Part 2, Oslo 25–26 November 1999, ISBN: 82-993462-8-2.

DNV (Det Norske Veritas) (1995) Feasibility Study for Safety Assessment of RoRo Passenger Vessels. Report prepared by the Det Norske Veritas Limited, Technical Consultancy, London, UK.

DNV (Det Norske Veritas) (1996) Safety Assessment of Passenger/RoRo Vessels. Summary report for the North West European Project on Safety of Passenger/Ro-ro Vessels, Det Norske Veritas Ltd., Technical Consultancy, London.

Erkut EV (1996) A Framework for Hazardous Materials Transport Risk Assessment Insurance. Elsevier, Journal of Mathematics and Economics, Vol. 18, No. 2, p. 135.

EC (European Commission) (1996) Recommendations on the development and implementation of Environmental Agreements. COM (96) 561 Final and Official Journal (OJ) No L333/59, 21.12.1996.

EC (European Commission) (1997) Working Paper on Risk Management. Directorate General III of the European Commission, Directive 76/769/EEC.

EC (European Commission) (1999) The Concerted Action on Formal Safety and Environmental Assessment of Ship Operations. Report by Germanischer Lloyd and Det Norske Veritas, project funded by the European Commission under the Transport RTD Programme of the 4th Framework Programme, 1999.

EC (European Commission) (2000) Report on the Harmonisation of Risk Assessment Procedures, Part 2, European Commission Health and Consumer Protection Directorate, October 2000.

EC (European Commission) (2006) The New EU Regulatory Framework (REACH) for Chemicals, European Commission, http://ec.europa.eu/enterprise 2006.

Frewer L. (2004) The Public and Effective Risk Communication. Elsevier Ireland Ltd., Toxicology Letters, Vol. 149, pp. 391–397.

HSE (Health and Safety Commission/ Executive, UK) (1991) Major Hazardous Aspects of the Transport of Dangerous Substances. Report of Advisory Committee on Dangerous Substances, Health and Safety Executive, UK, London, HMSO.

HSE (Health and Safety Executive, UK) (1992) The Tolerability of Risk from Nuclear Power Station. Report of Health and Safety Executive, United Kingdom, HMSO, 1992.

HSE (Health and Safety Executive, UK) (1995) Generic Terms and Concepts in the Assessment and Regulation of Industrial Risks. Discussion Document DDE2, Health and Safety Executive, UK, HSE Books.

HSE (Health and Safety Executive, UK) (1999) Reducing Risks, Protecting People. Discussion Document, Health and Safety Executive, UK.

HSE (Heath and Safety Executive, UK) (2001) Reducing Risks: HSE's Decision-Making Process Protecting People, Report of Heath and Safety Executive, UK, HMSO.

HSE (Health and Safety Executive, UK) (2002) Marine Risk Assessment: Offshore Technology. Report 2001/063, prepared by Det Norske Veritas for the Health and Safety Executive, UK, ISBN 0 7176 2231 2.

IEC (International Electro-technical Commission) (1995) Dependability Management – Risk Analysis of Technological Systems (International Standard IEC 300-3-9), IEC, Geneva.

IMO (International Maritime Organisation) (1997) Interim Guidelines for the Application of Formal Safety Assessment (FSA) to the IMO Rule-Making Process. Marine Safety Committee MCS/Circ.829, International Maritime Organization, London.

IMO (International Maritime Organisation) (2002) Guidelines for Formal Safety (FSA) for use in the IMO Rule-Making Process, MSC/Circ. 1023 and MEPC/Circ. 392.

IMO (International Maritime Organisation) (2004) Formal Safety Assessment: Risk Evaluation. Report submitted by the International Association of Classification Societies (IACS), Maritime Safety Committee, 78th session, Agenda item 19, MSC 78/19/.

IMO (International Maritime Organisation) (2006) Goal-Based New Ship Construction Standards, Safety Level Approach – Safety Level Criteria. Report submitted by Japan, Maritime Safety Committee, 81st session, Agenda item 6, MSC 81/6/10, 21 March 2006.

ISO (International Organisation for Standardization) (1999) Petroleum and Natural Gas Industries – Offshore Production Installations – Guidelines on Tools and Techniques for the Identification and Assessment of Hazardous Events. Draft International Standard ISO 17776.

Johnsson B.B. (2004) Varying Risk Comparison Elements: Effects of Public Reactions. Journal of Risk Analysis, Vol. 24, No. 1.

Kaplan S. and Garrick B.J. (1981) On the Quantitative Definitions of Risks. Journal of Risk Analysis, Vol. 1, No. 1, pp. 11–27.

Knight W.K. (1999) An Introduction to the Australian and New Zealand Risk Management Standards: AS/NZL: 4360: 1999. Report to ISO Working Group.

Knoflacher H. and Pfaffenbichler P.C. (2004) A Comparative Risk Analysis for Selected Austrian Tunnels. Paper on 2nd International Conference "Tunnel Safety and Ventilation" 2004, Graz, Institute for Transport Planning and Traffic Engineering, Vienna University of Technology.

Leiss W. (2004) Effective Risk Communication Practice. Elsevier Ireland Ltd., Toxicology Letters, Vol. 149, pp. 399–404.

Mullai A. and Paulsson U. (2002) Oil Spills in Öresund – Hazardous Events, Causes and Claims. The report on the SUNDRISK Project conducted within the Lund University Centre for Risk Analysis and Management (LUCRAM), Department of Industrial Management and Engineering Logistics, Institute of Technology, Lund University, Lund, Sweden.

Mullai A. (2004), A Risk Analysis Framework for Maritime Transport of Packaged Dangerous Goods (PDG). In: Brindley, C (2004) Supply Chain Risk. Ashgate Publishing Company, UK, Chapter 9, pp. 130–159.

Mullai A. (2006) Risk Management System – Risk Assessment Frameworks and Techniques. Safe and Reliable Transport Chains of Dangerous Goods in the Baltic Sea Region (DaGoB) Project Publication Series 5:2006, Turku School of Economics, Logistics, Turku, Finland.

Mullai A. (2007) A Risk Analysis Framework for Maritime Transport of Packaged Dangerous Goods – A Validating Demonstration, Volume II. Doctorial Thesis, Department of Industrial Management and Logistics, Engineering Logistics, Lund Institute of Technology, Lund University, Sweden.

Nicolet-Monnier M. and Gheorghe A.V. (1996) Quantitative Risk Assessment of Hazardous Materials Transport System. Kluwer Academic Publishers.

OECD (Organisation for Economic Co-operation and Development) (2000) Environmental Health and Safety Publications Series on Risk Management: Framework For Integrating

Social-Economic Analysis in Chemical Risk Management Decision Making Nr. 13 ENV/JM/MONO (2000); Environment Directorate OECD Paris 2000.

OECD (Organisation for Economic Co-operation and Development) (2001) The Chemical Accident Risk Assessment Thesaurus (CARAT) http://www1.oecd.org, 2001.

OECD (Organisation for Economic Co-operation and Development) (2002) Guidance Document on Risk Communication for Chemicals Risk Management, Environment Directorate Joint Meeting of the Chemicals Committee and the Working Party on Chemicals, Pesticides and Biotechnology, ENV/JM/MONO (2002)18, 25-Jul-2002.

Reid S.G. (1999) Perception and Communication of Risk, and the Importance of Dependability. Elsevier Science Ltd., Structural Safety, Vol. 21, pp. 373–384.

RSSG (Royal Society Study Group, UK) (1992) Risk: Analysis, Perception and Management. Report of the Royal Society Study Group, UK, London 1992, Based on British Standards 4778 1991.

SRA (Society for Risk Analysis) (2004) SRA Represents Members, Individuals and Organisations, from Different Countries http://www.sra.org/ 2004.

Saccomanno F. and Cassidy K. (1993) Transportation of Dangerous Goods: Assessing the Risks. Proceedings of the First International Consensus Conference on the Risks of Transport of Dangerous Goods, 1992 Toronto, Institute for Risk Research University of Waterloo, Ontario Canada.

Spouge J. (1997) Risk Criteria for Use in Ship Safety Assessment. Report of the DNV Technica UK, the Institute of Marine Engineers, UK.

USCG (U.S. Coast Guard) (2001) Risk-Based Decision Making Guidelines, 2nd Edition, http://www.uscg, 2001.

USEPA (U.S. Environment Protection Agency) (2000) Water Quality Standards Handbook. EPA 823–B94–005a, Section 304(a) of the Clean Water Act (CWA).

Vrijling J.H.K., van Gelder P.H., Goossens L.H.J., Voortman H.G., Pandey M.D. (2004) A Framework for Risk Criteria for Critical Infrastructures: Fundamentals and Case Studies in the Netherlands. Journal of Risk Research, Vol. 7, No. 6, pp. 569–579.

Weigkricht E. and Fedra K. (1993) Computer Support for Risk Assessment of Dangerous Goods Transportation. Institute for Risk Research.

Chapter 7: Using Simulation to Investigate Supply Chain Disruptions

Steven A. Melnyk*, Alexander Rodrigues, and Gary L. Ragatz

*Corresponding author: The Department of Supply Chain Management, The Eli Broad Graduate School of Management, Michigan State University, East Lansing, MI 48824–1122

7.1 Introduction

Managers and researchers are coming to realize that Supply Chain Disruptions (SCDs) constitute a real and significant threat – a threat that has to be better understood. However, the challenge facing many researchers is that of developing an understanding of these disruptions, what causes them, what factors moderate or influence the disruptions, and of identifying, and comparing alternative strategies and policies for dealing with such disruptions.

Past research into this topic has drawn extensively on either anecdotal or case based research (Chopra and Sodhi 2004; Zsidisin et al. 2005). Prior research has been helpful in developing a better understanding of the need for improved management of supply chain disruptions and in developing initial frameworks and sets of "effective practices" when dealing with such disruptions. However, such research is limited because the researchers are constrained to the experiences of the respondents. Furthermore, when dealing with empirical data, it is difficult to evaluate how an event taking place in a second tier supplier affects the performance of the firm since we have to identify and account for the impact of any policies being used and actions taken by the first tier supplier. If we are to develop a better understanding of supply chain disruptions, how to describe them, what factors influence them (and in what manner), and what policies/strategies can be used to deal with them, then an alternative approach is needed.

This chapter introduces such an approach – computer-based discrete event simulation. Simulation has long been used in Operations Management, Logistics, and Supply Management to study problems such as scheduling (job sequencing, production scheduling, order release, delivery reliability), capacity planning, process design-service, cellular manufacturing, and resource allocation (Shafer and Smunt 2004). It is now being used as a vehicle for studying supply chain related problems (Bowersox and Closs 1989; Levy 1995; Parlar 1997; Ridall et al. 2000; van der Vorst et al. 2000; Holweg and Bicheno 2002; Shafer and Smunt 2004; Terzi and Cavalieri 2004; Venkateswaran and Son 2004; Allwood and Lee 2005).

However, using simulation to study problems involving supply chain disruptions brings its own set of unique problems and challenges. These challenges and problems are most evident in four areas: (1) describing and modeling the events triggering the supply chain disruption (e.g., how to describe the SCD and its associated critical traits, and the location of the SCD); (2) the building of the simulation model itself; (3) identifying and setting appropriate policies and parameters (e.g., determining what is to be treated as a given and what are the management policies that can be used to affect either the occurrence of the event triggering the disruption or the impact of the disruption on the affected organization); and, (4) analyzing the resulting data generated from the simulation runs. This chapter explores these four challenges. Moreover, it strives to provide researchers with frameworks and guidelines that can be used when studying SCD using discrete event simulation.

As will be shown, the study of SCDs presents the researcher with an attractive trade-off. On one hand, the simulation-based study of SCDs is inherently more complex than the simulation-based study of a job shop (as an example). In return, the researcher is rewarded with "richness" in terms of the resulting problem settings and the insights gained. Ultimately, it is our hope that this paper encourages researchers to explore the topic of SCDs and draw on the tools and capabilities offered by discrete event simulation.

7.2 Understanding Supply Chain Disruptions: Background

Supply Chain Disruptions, or situations where there is a physical problem with product deliveries, is one of the major categories of supply risk. Consequently, to understand a SCD, we must first understand the concepts related to supply risk.

The study of supply risk has a long history, initially starting with the study of inventory models (Lee et al. 1997; Sodhi 2005) and multiple sourcing policies (Anupindi and Akella 1993; Berger et al. 2004) to buffer organizations from the effects of supply chain disruptions. However, there has been a recent surge in research investigating organizational behaviors and responses to various facets of supply chain risk. These studies include the implementation and use of supply risk assessment tools (Hallikas et al. 2002; Zsidisin et al. 2004; Kleindorfer and Saad 2005), supply risk perceptions (Mitchell 1995; Zsidisin 2003), supply chain security (Prokop 2004; Lee and Whang 2005), supply risk management (Zsidisin and Ellram 2003; Juttner 2005; Kleindorfer and Saad 2005), and the financial effects of supply chain risk (Hendricks and Singhal 2003, 2005).

Supply risk involves the probability that an incident occurs with supply that has a detrimental financial effect on the firm. These incidents fall into one of four categories: (1) significant price increases; (2) adverse impacts on firm reputation, (3) loss of intellectual property; and, (4) supply chain disruptions. The first category of supply risk includes uncontrollable price increases due to commodity price volatility (Seifert et al. 2004; Zsidisin 2005) and currency rate fluctuations from pursuing global sourcing strategies (Carter and Vickery 1988; Kazantzis and Tessaromatis 2001). With this form, the firm is faced by uncertainty in terms of the price to be paid for incoming supplies or the price (and profit) obtained from sales to customers in the downstream segment of the supply chain. The second category of supply risk involves its detrimental effects to a firm's reputation from its supply chain activities, such as environmental performance, labor practices at its supplier organizations, and overall ethical practices and philosophies associated with its supply chains (Carter and Jennings 2004; Magnan and Fawcett 2006). The third is concerned with the loss of intellectual property developed within the firm as the firm outsources production to the upstream supply chain.

The fourth category of supply risk and the focus of this study, supply chain disruptions, can occur from problems associated with suppliers being unable to provide products or services. Examples of sources of these disruptions include (but are not limited to) the following:

- *Natural*: Disruptions caused by nature. These include hurricanes, hail, fire, dust storms, lightning, and tornados.
- *Demand Shifts*: Disruptions due to demand shifts occur when the demand generated by customers exceeds the available capacity. Because it takes time to add capacity, demand shifts (due to factors such as market forces or increase in demand due to product standardization) create situations where the existing capacity is insufficient to meet demand.
- *Supplier Problems*: The supplier, for whatever reason, is unable to provide goods and services that satisfy the demand requirements of the firm. This inability can be due to factors such as problems in producing goods satisfying the customer's minimum quality requirements or problems with delivery reliability.
- *Human/Organizational Behavior*: Any disruption that occurs directly as a result of a human or organizational action (either deliberate or accidental). Included are such actions as terrorism, arson, human error, strikes and slowdowns.
- *Information/Technology*: Any disruption due to a breakdown in the information or technology systems. This could come from such factors as a system crash, corrupted data, or a computer virus.
- *Financial*: Those disruptions caused by adverse changes in the financial condition of any party involved in the supply chain (e.g., bankruptcy, or liquidation of a supplier).
- *Legal/Regulatory*: Those disruptions caused by legal/regulatory problems (e.g., health and safety violations, product liability law suits, government mandated shut downs).

It must be recognized that these sources often do not occur independently of each other. For example, consider a situation where we are dealing with a supplier that is on the edge of financial insolvency. That supplier experiences an initial disruption caused by an information problem. That initial disruption triggers another disruption caused by the supplier's subsequent bankruptcy.

7.3 Modeling Supply Chain Disruptions: A Framework

Before any researcher can model a supply chain disruption, they must first understand what it is. Consequently, a good starting point for the development of the SCD framework is an operational definition of a supply chain disruption:

> ... *the outcome of a process whereby one or more events (to be referred to as the "triggering event") taking place at one point in the supply chain adversely affect the performance of one or more components located elsewhere in the supply chain.*

A SCD can be viewed as the output of a chain of events. This chain begins with an event that triggers the start of the disruption. This event is then transmitted through the supply chain from the source to the firm. As it moves through the chain, its impact is shaped and influenced by such factors as the location of the source of the disruption within the supply chain, the inventory, ordering, and buffering policies in use by the various supply chain partners, the amount of visibility/warning regarding the disruption, the availability of alternative sources of supply, and the lead times (production, transportation).

Given the preceding discussion, we can identify four factors that influence the process linking the triggering event to the disruption suffered by the supply chain component: (1) the specific traits associated with the triggering event; (2) the structure of the supply chain (which identifies the nature of the linkages linking the various components of the supply chain); (3) the policies, procedures, and parameters in use by the various components of the supply chain; and, (4) the performance measures used in assessing performance (at either the organizational or supply chain levels). These four factors (triggering event and its trait; supply chain structure; policies, procedures, and metrics; and, performance measures) form the major foundations of the framework proposed in this chapter for modeling and studying supply chain disruptions.

7.3.1 The Triggering Event

When dealing with supply chain disruptions, it is important to first recognize that every supply chain experiences such disruptions. In most cases, these disruptions are minor and represent nothing more than a momentary "hiccup." They can also originate from issues encountered either at the customer side (downstream) or supply side (upstream) of the supply chain. In this study, attention is on disruptions originating in the Supplier Side (e.g., *Supply Side Supply Chain Disruptions* or SS-SCD) as compared to *Customer Side Supply Chain Disruptions* (CS-SCD).

All SCDs, whether they are SS-SCD or CS-SCD, begin with some form of triggering event. This triggering event can be the result of factors such as technological or information breakdowns (e.g., a supplier experiences a breakdown in their information system) or a natural event (our supplier is unable to ship product because of a storm). Without a triggering event, there is no SCD. Yet, a triggering event is more than simply the onset (or lack) of a problem. It consists of a number of attributes that describe the triggering event in detail. Five of these traits are illustrated in Fig. 7 1.

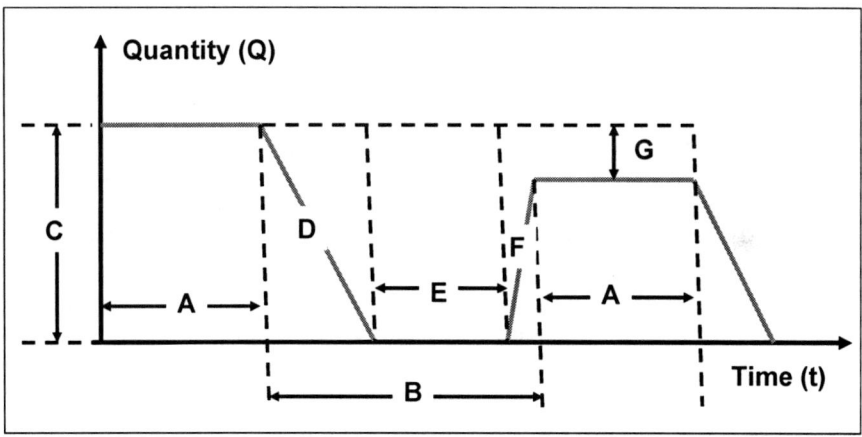

Fig. 7.1 Decomposing a supply chain disruption – the major elements

- *Disruption Periodicity* (DP): Denoted by A in Fig. 7.1, the Disruption Periodicity is the interval between the disruptions. This trait can be viewed as a range running from long to short. With a long DP, there is sufficient time between successive disruptions so that the supply chain can move back to steady state. Alternatively, when the interval is short, the system has not yet achieved steady state before it is subjected to another disruption. With a short DP, the researcher must deal with data that not only reflects the main effects of the disruptions but also the possible interaction between disruptions.
- *Disruption Time Period* (T_d): Denoted by B in Fig. 7.1, the Disruption Time Period is the time period over which the triggering event is present AT THE SOURCE. That is, in the case of a SS-SCD, the disruption time period begins with the onset of the disruption of supply at the supplier. It ends when the supplier has been able to correct the conditions causing the disruption and the problems with supply have been corrected.
- *Disruption Quantity Loss* (Q_L): Represented by C in Fig. 7.1, the Disruption Quantity Loss stands for the number of units that the supplier is no longer able to provide as a result of the disruption. It is important to recognize that not all disruptions result in a total loss of supplier output. It is reasonable to

expect partial loss of production, especially under conditions where the total demands placed on the supplier exceed that supplier's capacity. Under such conditions, the supplier may prefer to provide each customer only a portion of their demands (such as what happens when the supplier places the customer on allocation).

- *Disruption Profile* (DPr): Consisting of segments D, E, F in Fig. 7.1, the Disruption Profile refers to the exact shape of the disruption. From Fig. 7.1, we can see that any supply disruption can consist of up to three stages: (1) the **onset** (segment D), (2) the **nadir** (segment E), and, (3) the **recovery** (segment F). Both the onset and the recovery can assume a spectrum of shapes ranging from "gradual" (which is what is displayed in Fig. 7.1) to sudden. Sudden disruptions occur when the loss of supply goes from pre-disruption quantity level Q to the lower post-disruption quantity in one step. This "step" behavior can occur for a number of reasons, such as the supplier experiencing a catastrophic loss of output because of a man-made event such as terrorism, or because of a natural event such as a plant fire or an earthquake/hurricane. Gradual disruptions, in contrast, are the result of such factors as the ability of the supplier to keep producing at some level, the presence of buffer stocks at the supplier's location, or the ability of the supplier to reallocate production from one form of demand (e.g., cancel production of orders for safety stock or for forecast needs and redirect the production to actual customers). The Disruption Profile can also affect and be affected by the types of risk management policies selected and their relative effectiveness. It can be argued that what works in an environment where the onset and recovery is gradual might not work when faced by a sudden onset and recovery.

- *Post-Recovery Output Level* (Q_{PD}): Denoted by G in Fig. 7.1, the Post-Recovery Output Level refers to the level of supplier output after the onset and recovery from the disruption. It is tempting to assume that the *Post-Recovery Output Level* (Q_{PD}) is identical to the level of supplier output before the disruption. This does not always occur. Under some conditions, the Q_{PD} may be less than the output levels prior to the disruption (the supplier has recovered a significant but not all of the output levels). In still other cases, the Q_{PD} can be higher. This occurs when the supplier, hoping to avoid a similar problem in the future or taking advantage of the opportunities offered by a disruption and replaces older equipment with newer, more efficient machinery.

With these five traits (summarized in Fig. 7.1.), it is now possible to describe and model any specific SCD. Yet, the researcher must also consider the following two factors: Disruption Breadth and Disruption Location.

- *Disruption Breadth* (DB). This refers to whether the triggering event is **single** (there is only triggering event present) or whether the triggering events are **multiple** (there are multiple triggering events). From a simulation perspective, with a single disruption breadth, the triggering event is present

at only one node in the supply chain. With a multiple disruption breadth, triggering events are present at two or more nodes in the supply chain.
- *Disruption Location* (DL). This factor refers to where in the upstream supply chain the disruption event occurs. Disruptions can occur at the first, second, or third tiers of the supply (e.g., DL = 1,2,3 where DL = 1 represents a disruption taking place at the first tier in the upstream supply chain). The farther away that the disruption occurs, the less visible it is to organizations ultimately affected by it (Fine 1998). Further, as you move away from the firm, the ultimate control exercised by the firm on its second, third, or higher tiers can be expected to fall exponentially. Finally, it is important to recognize that the further away from the firm in the supply chain the disruption is located, the more likely that the ultimate impact of the disruption will be influenced by the actions taken at intervening stages.

Table 7.1. Supply chain disruptions – major traits and values

TRAITS	DEFINITION	MAJOR LEVELS
Disruption Periodicity (DP)	The time interval between SCD events.	Long – sufficient time between disruptions so that the system can return to steady-state. Short – system has not returned to steady-stated before experiencing next disruption.
Disruption Time Period (DTP or T_d)	The time interval over which the disruption is present at the source.	
Disruption Quantity Loss (DQL or Q_L)	The amount of production measured in units lost at the outset of the disruption.	
Disruption Profile (DPr)	The shape of the disruption loss from beginning to end.	Onset Nadir Recovery
Post-Recovery Output Level (Q_{PD})	The level of output reached in steady-state after the end of the disruption.	
Disruption Breadth (DB)	The extent to which the observed SCD is a result of a single triggering event or multiple triggering events.	Single Multiple
Disruption Location	At which tier in the upstream supply chain the triggering event is present.	DL = 1,2,3... (The larger the DL value, the further away is the triggering event).

7.3.2 The Simulation Model

One way of studying how these various components come together and interact is to build a computer simulation model of a supply chain and to expose it to a disruption. We can also use this same simulation model to assess how alternative policies and actions can affect this chain of events or, if possible, control/eliminate it (and by doing so prevent the SCD from occurring).

Computer simulation is the process of designing and building a model of a real or representative system and then using this system as an environment for carrying out controlled experiments (Law and Kelton 2000). As noted by several researchers, computer simulation offers the researcher several important advantages (Kelton et al. 2004). It allows the researcher to study events that would be potentially disastrous for most firms and to evaluate and understand processes that would take long periods of time to complete and that would be potentially confounded by external factors such as human intervention. Simulation encourages active and complete experimentation with various possible policies under a variety of different settings. It also enables researchers to develop insight into how the observed outcomes are generated.

One critical decision facing the researcher is selecting the specific type of simulation approach to use (e.g., static vs dynamic; deterministic vs stochastic; continuous vs discrete; systems dynamic vs discrete event vs complex systems). While some researchers have strongly presented the advantages offered by

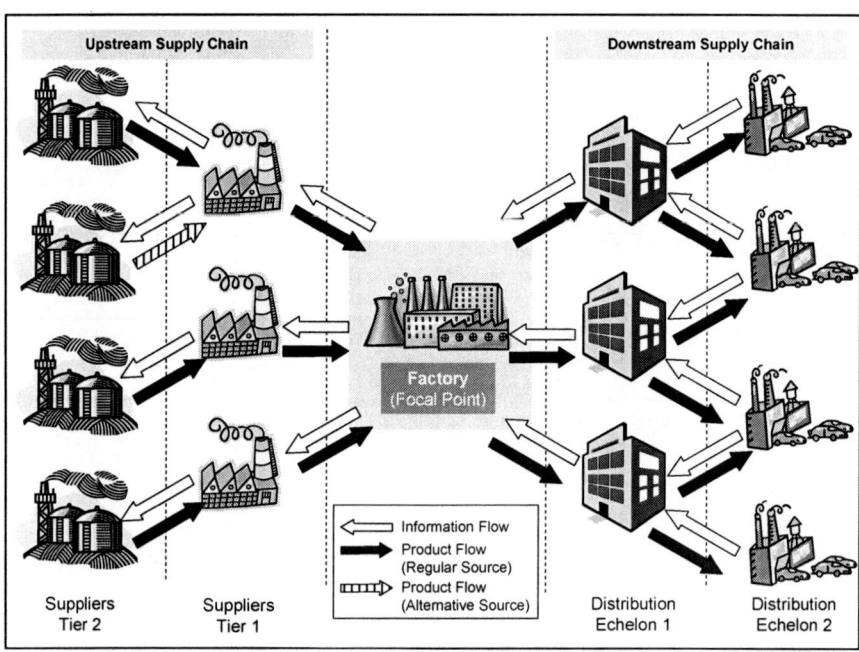

Fig. 7.2 Structure of the "typical" supply chain model – a simulation framework

autonomous agent/complex system and object oriented models (van der Zee and van der Vorst 2005), this study focuses its attention on the dynamic stochastic discrete event simulation model. Such models have been widely used and are supported by simulation programs such as Arena (Kelton et al. 2004).

When building a simulation model of a supply chain, two considerations must be noted. First, the system must be a dynamic, multi-echelon system that brings together multiple tiers of suppliers and multiple tiers of customers (Fig. 7.2). Second, while the overall simulation model deals with the entire supply chain, any analysis of this data will tend to focus on the performance of only one node or entity in the model – the focus of interest (as shown in Fig. 7.2). While simulation does allow us to look at the performance at multiple points, typically, the question that we most often addressed takes the form of "How does a supply chain disruption taking place at some upstream point in the supply chain affect the performance of this firm?"

7.4 Parameters and Experimental Factors

Simulation provides the researcher with a very attractive vehicle for generating data. It is attractive in that it reports all the data that you want with continuous regularity and 100 percent accuracy. Driving this vehicle is the experiment. Within the experiment, the researcher must specify the *parameters,* and the *experimental factors*. Parameters are those elements that describe elements exogenous to the simulation model and outside of the control of the researcher. Parameters represent the "givens" or the system constraints under which the simulation model operates. In contrast, experimental factors are those elements that are under the control of the researcher. Experimental factors incorporate the policies, tools, and procedures that can be used to deal with SCD. In this section, we will focus our attention primarily on the experimental factors.

7.4.1 Experimental Factors

Within the experimental designs that drive simulations, these form the endogenous factors that are modeled. The various experimental factors can be assigned to one of several categories. It should be noted that these categories should be viewed as an initial set; they are not intended to be comprehensive:

- *Information-related Policies*: These are policies and strategies focusing on information flows within the supply chain. Central to these policies are considerations of lead times of information flows, information quality (Chatfield et al. 2004), information sharing, and the degree to which advanced warnings or distress calls are given. For information lead times, our concerns focus on the speed and the variance; for information quality, the concerns shift to that of variability in the quality or the degree to which the information provided is complete. When dealing with the disruption, we have to differentiate between advanced warnings, which are any communications that are given by the partner experiencing the disruption in advance of the disruption and distress calls, which are any communications given after the onset of the disruption.

- *Buffer-related Policies:* These are policies that deal with the use and positioning of the three major types of buffers available to firms (e.g., inventory, lead times, and capacity). It is here that we would naturally expect to see the impact of such policies as Lean Systems (Imai 1986; Ohno 1988; Flynn et al. 1995; Levy 1995; Imai 1997; Levy 1997; Liker 1997; Womack and Jones 2003). By applying lean system approaches, tools, and procedures, we would expect to see a reduction in these buffers since these buffers are slack and slack becomes viewed as "waste" (Demchak 1996). The goal of any lean system is to reduce and/or eliminate all forms of waste.
- *Alternate Sourcing*: This factor consists of actions taken by the firm in developing and implementing alternate sources of supply. These sources of supply can be secured, such as when the processes in alternate supply sources have been evaluated and proven previously, with the result that any inefficiencies incurred in bringing them on line would be relatively low. Alternatively, these sources can be unsecured (e.g., unproven) with a resulting high penalty in terms of efficiency considerations.
- *Component Substitution*: When faced by a disruption in supply, one option available is that of identifying and using as a substitute another item that is in greater availability.

7.4.2 Performance Measures

The final element of the simulation model involves the specification and measurement of performance. As noted by Scheffe (2005), SCDs affect system performance (as measured at either the organizational or supply chain levels) in one of three ways:

- *Financial*: These are effects that primarily influence the cost of doing business and managing the supply chain. Performance is measured typically in terms of costs. Examples of this category include inventory levels, lost sales, revenue/contribution margin by time period, expediting, penalty clauses, and investments in goodwill.
- *Strategic*: In this category, we find effects that influence the firm's ability to achieve its strategic objectives. Strategic impacts are more difficult to evaluate. They deal with the impact of the SCD on the ability of the firm to achieve its strategic objectives, as stated in terms of the four major components of value: lead time, cost, quality, and flexibility. Strategic effects can be viewed as a "lens" that influences the specific types of performance metrics selected for analysis. For example, a firm focused primarily on competing on lead times would be most interested in those

measures that deal directly with lead times: average lead times, lead time variance, and order tardiness/lateness (to name a few). It would also have less interest in any measures dealing with the other dimensions of performance (e.g., quality, cost or flexibility).
- *Operational*: These are the effects on operations that occur as a result of operating problems in the supply chain. Operational measures focus on such measures as lead time/flow time (mean and/or variance), as measured in hours; or fill rates, as measured in percentage of orders placed; or inventory levels, as measured in units. In contrast to strategic measures that are linked to and driven by corporate objectives, operational measures are frequently generic. They are captured to help the researcher and the manager better assess how the system is running and to flag operating problems.

When dealing with performance measurement, the critical issue that must be addressed involves the level of analysis. When studying SCDs, the impact can be evaluated at two different levels: at a local or organizational level or at the supply chain level. In the first case, you select a specific organization to act as the focal point of interest and measure the impact of the SCD in terms of how it affects the operation of the specific organization. Alternatively, you can measure the performance at the overall supply chain level. In this case, the goal is to identify in quantitative terms the impact of the triggering event and the resulting disruptions on the performance of various tiers. If the triggering event takes place at a tier 2 supplier, then the intent of this measurement system is to evaluate the impact of this event on the tier 1 supplier, the organization and its customers.

7.5 Generating and Analyzing Simulation Data

In many cases, our interests in SCD are fairly straight forward. We want to answer certain questions such as:

- Has the disruption significantly affected system performance?
- Has the absence or presence of certain experimental factors affected the impact of the disruption on the supply chain?
- Is there a significant difference in the effectiveness of two or more policies in terms of their impact on the performance of a system experiencing a disruption?
- To what extent are the performance effects observed the result of main effects or interactions (or both)?
- Is the post-disruption steady-state mean significantly different from the pre-disruption steady-state mean?

How we address these and other related questions is strongly dependent on the type of statistical analysis that we use.

When analyzing the data generated by a series of simulation-based experiments, there are at least two possible approaches available: (1) classical statistical analysis; and, (2) time series analysis using outlier detection. Each approach has strengths and weaknesses. Further, each approach provides some but not all of the insight needed to understand how SCD affects system performance and how various policies and system parameters affect not only the response of the supply chain to the disruption but also the impact of the disruption on system performance. Consequently, when using these approaches, methodological triangulation should be promoted. That is, researchers should be encouraged to use multiple approaches since the insight provided by each approach, when combined, provide a more complete picture of what is taking place within the simulated system. However, before the concept of methodological triangulation can be better understood, it is first necessary to review the two potential approaches to the analysis of data.

7.6 Classical Statistical Analysis of Simulation-Based Experimental Data

Most researchers who have worked with discrete event simulation are familiar with classical statistical analysis. By "classical," we mean those tests that deal with assessing differences in means or that perform correlation analysis. Included in these tests are statistic procedures such as t-tests (paired and unpaired), analysis of variance (univariate and multivariate), factor analysis, linear regression (in its various forms: ordinary least squares, LOGIT, PROBIT, and robust regression) and non-parametric tests.

In most of these tests, the researcher creates an experiment, which is defined in terms of factors (main issues of interest) and levels (specific values applied to the factors). The result is an experimental design consisting of cells or specific combinations of factors and levels. The interest of the researcher is in determining if there is a significant difference between cell means and if the factors associated with the cells have had a significant impact on the recorded means.

This approach has several advantages. First, it is well known to most researchers (especially those working with simulation studies). These procedures are extensively discussed by Law and Kelton (2000) in their text book on computer simulation. Second, these procedures allow the researcher to assess not only main effects (are there differences in impact between various independent variables) but also interactions. Finally, these procedures are widely implemented in computer statistical packages such as Minitab©, SPSS©, SYSTAT©, JMP©, and STATA©.

However, classical statistical analysis, when applied to SCD, does suffer from an important limitation directly attributable to the transient response created by the supply chain disruption itself. The disruption introduces a form of data variance. Traditionally, in the case of steady-state simulations, the method proposed for dealing with any transient variance is to delete the data associated with the transient from the dataset. In most cases, this is a reasonable approach since the transient data is typically generated by the simulated system getting to steady-state.

This "ramp-up" data is not representative of the performance of the system once it has reached steady-state.

Within the SCD simulation, deleting transient data associated with the startup is appropriate; deleting transient data associated with the disruption is inappropriate. This latter type of data cannot be deleted because it represents information of greatest interest to the researcher. This is what both the researcher and the manager want to better understand. However, by including this transient in the dataset, the result is an increase in variance that is directly due to the SCD. This rise in variance is likely to create an increased probability of encountering a Type II error – a situation where the alternate hypothesis is rejected when, in fact, the null hypothesis should be rejected. In the end, the amount of variance introduced by the disruption is dependent on two factors: the impact of the disruption and time horizon for the simulation. That is, the longer the time horizon simulated and the smaller the disruption, the less variance introduced by the disruption. Conversely, the shorter the simulated time horizon and the greater the disruption, the greater the variance observed.

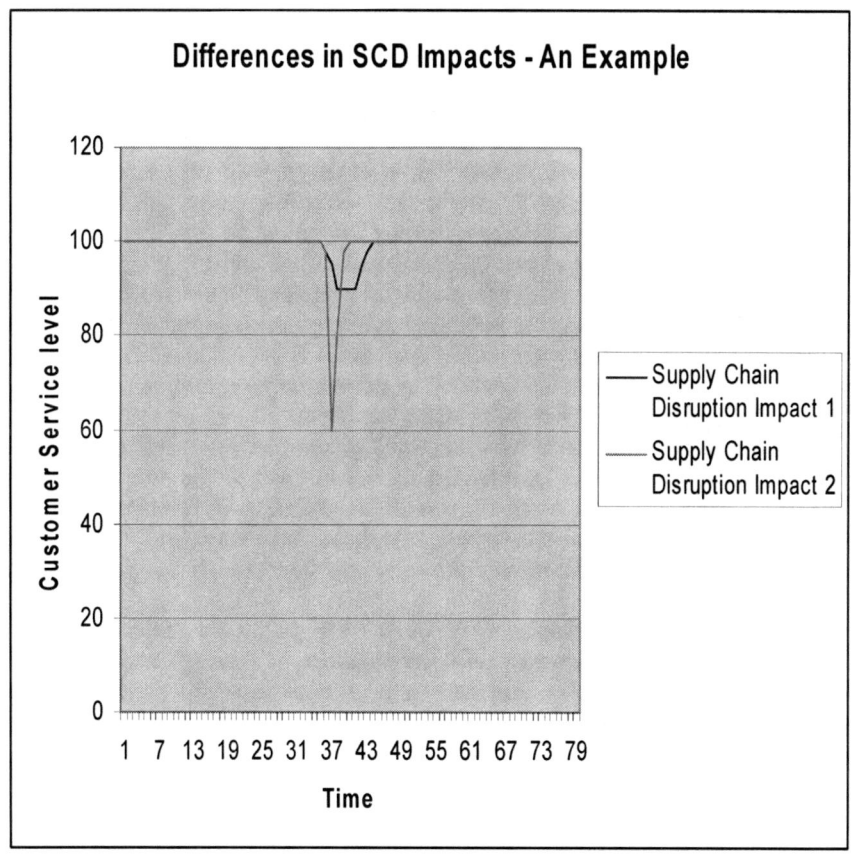

Fig. 7.3 Comparing two supply chain disruptions

Ultimately, classical statistical analysis enables the researcher to deal with questions pertaining to changes in means and factors affecting these changes (e.g., policies, parameters). Yet, the insights gained by this technique are limited since they do not deal with the time dimensions of the disruption. We can potentially be faced by two disruptions that have the same net impact but that exhibit very different time patterns (see Fig. 7.3). Even though the net impact is identical, how these effects manifest themselves may have different levels of attractiveness. How a disruption unfolds over time is often of critical interest to the researcher and the manager. To provide the required insight demands a different approach – an approach that recognizes that the transient is essentially a times series. As such, it is most appropriately studied with statistical techniques appropriate for dealing with time series. In this instance, the proposed approach is that of *Intervention Analysis*, in general, and *Outlier Detection*, specifically.

7.7 Intervention Analysis: Evaluating the Transient Response

What differentiates the SCD simulation from the conventional simulation study is the central role played by the *transient* behavior of the process being simulated. It is this behavior that captures the impact of the triggering event on the performance of either the firm or the supply chain. Yet, in focusing on this element of performance, the SCD simulation study emphasizes something that most conventional simulation studies seek to discard or minimize. As is evident from Fig. 7.1, when studying the impact of a SCD on firm performance, we are essentially assessing the impact of an external event, such as a major supplier shutdown or a disruption to the logistics network, on a time series representing firm performance. Interest in assessing how such external events affect time series is not unique to SCD studies. In time series, such studies fall under the umbrella of *intervention analysis* (or *impact analysis*) (Box and Tiao 1965, 1975; Wei 1990).

Traditionally, if a time series was subjected to a known intervention occurring at a specific time, its effect in changing the mean level of the time series was evaluated using a two-sample t-test (Liu and Hudak 2004). However, the t-test is not appropriate in the case of serially correlated data (Box and Tiao 1965). Further, this test may not be appropriate where the intervention is a pulse lasting some t time periods.

Box and Tiao (1975) subsequently developed a procedure for analysis of a time series in the presence of known external interventions. In their approach, there are two types of interventions – pulses and steps. A pulse is an intervention with a finite duration (typically one time period), while a step involves a permanent change or intervention (e.g., the introduction of a new governmental regulation or the permanent loss of a supplier). Intervention analysis is a statistical procedure that enables the researcher to evaluate the impact on a time series, as represented by an ARIMA (Autoregressive Integrated Moving Average) process.

There are several components in an intervention model: a deterministic component describing the intervention(s), the associated response of the system to the intervention, and a stochastic disturbance term. The overall modeling strategy is to obtain reasonable initial representations for these components and to iterate to a final model based on intermediate estimates, diagnostic checks, and model interpretations (Liu and Hudak 2004).

7.8 Outlier Detection

When the timing of the intervention and its impact on the system are known in advance, intervention analysis is the preferred approach. However, in many cases, especially in a simulation where the triggering event takes place at the second or third tiers, the timing of the impact of the disruption on the firm's performance may be difficult to identify in advance. Under these cases, the researcher is better served by turning to procedures for *Outlier Detection* in the simulated time series. It can be argued that Intervention Analysis is simply a special form of Outlier Detection.

Outlier Detection recognizes four types of outliers (Pankratz 1991; Liu 2005):

- *Additive Outlier* (AO): An event that affects the series for only one time period.
- *Innovational Outlier* (IO): An event whose effect is spread according to the ARIMA model of the process so that the event affects all values observed after its occurrence. The IO often represents the onset of an external cause (Liu 2005) – a situation similar to a disruption being suffered in the upstream supply chain.
- *Level Shift* (LS): This represents an event that affects a series at a given time and whose effect then becomes permanent.
- *Temporary (or Transient) Change* (TC): This is an event that has a brief or transient impact on the system performance. That is, after the initial impact, the effect of a TC decays exponentially according to some dampening factor such as δ (Tsay 1988).

Procedures do exist that allow a researcher to jointly estimate model parameters and outlier effects (Chen and Liu 1993b, 1993a). Currently, there are only a few statistical packages available that offer the capability of analyzing such outliers or of assessing these aspects of the outlier. One such package is SCA Workbench$^{©}$.

7.9 Dealing with Replications: Time Series Considerations

One of the major features of any simulation study is its ability to replicate results easily and quickly. With a sufficient number of replications, we can approximate the underlying system parameters (e.g., means, variances). Replication is a critical element of conventional simulation since it enables the researcher to increase the power of analysis by generating a sufficiently large number of observations per cell. Yet, within the context of a SCD simulation, there is the question of whether replications have a role. The simple answer is yes. However, how they are used is different when compared to conventional simulation studies.

When replicating results for a given combination of parameters and experimental policies, we must realize that we are faced by an M x N matrix (M replications by N time periods). Rather than carrying out ARIMA and Outlier Detection for each replication, an alternative approach is to generate a new 1 x N matrix, where each cell consists of the mean value for the values from the M replications. This "Mean Time Series" (MTS) becomes the focal point for application of time series analysis. Further, with sufficient replications, we can assume that the values contained within this MTS are unbiased estimators of the true system values for that time period.

7.10 Other Applications

With the combined simulation and analysis procedure outlined in this chapter, it is now possible to address the following types of questions:

- Under what conditions do "lean" supply chains become fragile supply chains, when faced by a SCD?
- How do the policies used by intermediate suppliers (suppliers located between the source of the disruption and the firm) affect the traits of the SCD?
- To what extent can early warnings provided by the supplier help us deal with the effects of a SS-SCD?
- Can we develop operational definitions and measures of such SCD-related terms as resilience (Peck 2005)?

Addressing these and other questions will do much to enhance our knowledge of SCDs and how to cope with their effects.

7.11 Methodological Triangulation

These two methodologies can and should be regarded as complementary. By using them together, the researcher (and the manager) can develop a more complete, richer, and more insightful "picture" of what is taking place and how the disruptions can affect performance. Classical experimental statistics can help the researcher determine if the disruption has had an impact and to determine if the experimental factors have had any significant impact on the resulting performance. However, by itself, classical experimental statistics do not give the researcher the "entire" picture of the disruption. What is ignored is the time dimension (as noted in Fig. 7.1). To capture this dimension, we need time series analysis. Combining these two techniques, we have a form of triangulation – methodological triangulation.

With methodological triangulation, the researcher is using a combination of appropriate methodologies in such a way as to overcome the weaknesses of one methodology and the problems inherent in its application. By combining the methodologies, it is possible that two sets of policies can generate the same end results but generate very different time series. One time series could show the disruption having a rapid onset. It is also possible to use time series analysis to address questions that the classical experimental statistics cannot (e.g., has the system returned to the original level of performance?).

7.12 Concluding Comments

Supply chains and the problems and challenges affecting the management of this new operating environment are becoming increasingly more important. One of the major challenges facing today's supply chains is that of dealing with the impact of supply chain disruptions. Increasingly, researchers and managers are now recognizing that they are dealing with supply chains that can be best described as fragile. These are systems that once exposed to a supply chain disruption break down and are limited in their ability to recover. As awareness of the supply chain fragility and supply chain disruptions has developed, there is a need to present researchers and managers with a method of exploring these issues and of identifying and evaluating alternative methods for coping with the effects of supply chain disruptions. This chapter has proposed one such method – that of the dynamic discrete event simulation.

Discrete event simulation is not presented as a substitute to empirical research; rather it is a complement. It enables researchers to investigate alternative approaches and to experiment with environmental conditions and with management policies.

This chapter has presented the use of this approach within the context of various frameworks. These frameworks have been used to structure and organize our understanding of supply chain disruptions and to facilitate our modeling of both the disruptions and the simulation environments in which such disruptions take place. We believe that dynamic discrete event simulation is an appropriate vehicle for studying supply chain disruptions. It has also identified and addressed those simulation and analysis issues that are unique to the study of supply chain disruptions. Finally, this chapter has explored two alternative statistical approaches to the analysis of the data generated by such a simulation – classical statistical analysis and time series analysis using Intervention Analysis/Outlier Detection. Both approaches have their own strengths and weaknesses. Time series analysis is potentially attractive because it explicitly enables the researcher to explore the effects of a supply chain disruption as they manifest themselves over time.

It is hoped that this chapter and the frameworks, guidelines, and procedures contained within it will encourage other researchers to undertake research into this increasingly important topic. Such research offers the potential of helping managers better prevent SCDs, and, if unable to prevent such disruptions, better remediate the effects of such SCDs on performance.

References

Allwood, J.M., and Lee, J.H. (2005). The design of an agent for modeling supply chain network dynamics. *International Journal of Production Research*, 43(22), 4875–4898.
Anonymous (2005). Business: When lightning strikes; business continuity planning. *The Economist*, 377(8450), 83.
Anupindi, R., and Akella, R. (1993). Diversification under supply uncertainty. *Management Science*, 39(8), 944–963.
Berger, P.D., Gerstenfeld, A., and Zeng, A.Z. (2004). How many suppliers are best? A decision analysis approach. *Omega*, 32(1), 9–15.
Bowersox, D.J., and Closs, D.J. (1989). Simulation in logistics: A review of present practice and a look to the future. *Journal of Business Logistics*, 10(1), 133–148.
Box, G.E.P., and Tiao, G.C. (1965). A change in level of a non-stationary time series. *Biometrika*, 52, 181–192.
Box, G.E.P., and Tiao, G.C. (1975). Intervention analysis with applications to economics and environmental problems. *Journal of the American Statistical Association*, 70, 355–365.
Carter, C.R., and Jennings, M.M. (2004). The role of purchasing in corporate social responsibility: A structural equation analysis. *Journal of Business Logistics*, 25(1), 145–186.
Carter, J.R., and Vickery, S.K. (1988). Managing volatile exchange rates in international purchasing. *Journal of Purchasing and Materials Management*, 24(4), 13–20.
Chatfield, D.C., Kim, J.G., Harrison, T.P., and Hayya, J.C. (2004). The bullwhip effect impact of stochastic lead time, information quality, and information sharing: A simulation study. *Production and Operations Management*, 13(4), 340–353.
Chen, C., and Liu, L.-M. (1993a). Joint estimation of model parameters and outlier effects in time series. *Journal of the American Statistical Association*, 88(421), 284–297.
Chen, C., and Liu, L.-M. (1993b). Forecasting time series with outliers. *Journal of Forecasting*, 12(1), 13–35.
Chopra, S., and Sodhi, M.S. (2004). Managing risk to avoid supply chain breakdown. *MIT Sloan Management Review*, 46(1), 53–61.
Crowe, T.J., Fong, P.M., Bauman, T.A., and Zayas-Castro, J.L. (2002). Quantitative risk level estimation of business process reengineering efforts. *Business Process Management Journal*, 8(5), 490–511.
Demchak, C. (1996). Tailored precision armies in fully networked battlespace: High reliability organizational dilemmas in the information age. *Journal of Contingencies and Crisis Management*, 4(2), 93–103.
Fine, C.H. (1998). *Clockspeed: Winning industry control in the age of temporary advantage*. Reading, MA: Perseus Books.
Flynn, B.B., Sakakibara, S., and Schroeder, R.G. (1995). Relationship between jit and tqm: Practices and performance. *Academy of Management Journal*, 38(5), 1325–1360.
Hallikas, J., Virolainen, V.-M., and Tuominen, M. (2002). Risk analysis and assessment in network environments: A dyadic case study. *International Journal of Production Economics*, 78(1), 45–55.
Hendricks, K.B., and Singhal, V.R. (2003). The effect of supply chain glitches on shareholder wealth. *Journal of Operations Management*, 21(5), 501–522.

Hendricks, K.B., and Singhal, V.R. (2005). Association between supply chain glitches and operating performance. *Management Science*, 51(5), 695–711.

Holweg, M., and Bicheno, J. (2002). Supply chain simulation a tool for education, enhancement and endeavour. *International Journal of Production Economics*, 78(2), 163–175.

Imai, M. (1986). *Kaizen: The key to Japanese competitive success.* New York, NY: Random House Business Division.

Imai, M. (1997). *Gemba kaizen: A commonsense low-cost approach to management.* New York, NY: McGraw-Hill.

Juttner, U. (2005). Supply chain risk management: Understanding the business requirements from a practitioner perspective. *International Journal of Logistics Management*, 16(1), 120–141.

Kazantzis, C.I., and Tessaromatis, N.P. (2001). Volatility in currency markets. *Managerial Finance*, 27(6), 1–22.

Kelton, W.D., Sadowski, R.P., and Sturrock, D.T. (2004). *Simulation with arena.* New York, NY: McGraw-Hill.

Kleindorfer, P.R., and Saad, G.H. (2005). Managing disruption risks in supply chains. *Production and Operations Management*, 14(1), 53–68.

Law, A.M., and Kelton, W.D. (2000). *Simulation modeling and analysis.* Boston, MA: McGraw-Hill.

Lee, H.L., and Whang, S. (2005). Higher supply chain security with lower cost: Lessons from total quality management. *International Journal of Production Economics*, 96(3), 289–300.

Lee, H.L., Padmanabhan, V., and Whang, S. (1997). Information distortion in a supply chain: The bullwhip effect. *Management Science*, 43(4), 546–558.

Levy, D.L. (1995). International sourcing and supply chain stability. *Journal of International Business Studies*, 26(2), 343–360.

Levy, D.L. (1997). Lean production in an international supply chain. *Sloan Management Review*, 38(2), 94–102.

Liker, J.K., ed., *Becoming lean: Inside stories of U.S. Manufacturers* (Portland, OR: Productivity Press, 1997).

Liu, L. (2005). *Time series analysis and forecasting.* River Forest, IL: Scientific Computing Associate Corporation.

Liu, L., and Hudak, G.B. (2004). *Forecasting and time series analysis using the sca statistical system.* River Forest, IL: Scientific Computing Associate Corporation.

Magnan, G.M., and Fawcett, S.E. (2006). Supplier codes of conduct: Organizational and implementation challenges. *Proceedings of the 17th Annual North American Research Symposium on Purchasing and Supply Management*, San Diego, CA,

Mitchell, V.-W. (1995). Organizational risk perception and reduction: A literature review. *British Journal of Management*, 6(2), 115–133.

Ohno, T. (1988). *Toyota production system: Beyond large-scale production.* Cambridge, MA: Productivity Press.

Pankratz, A. (1991). *Forecasting with dynamic regression models.* New York, NY: John Wiley and Sons.

Parlar, M. (1997). Continuous review inventory problem with random supply interruptions. *European Journal of Operational Research*, 99(2), 366–385.

Peck, H. (2005). Drivers of supply chain vulnerability: An integrated framework. *International Journal of Physical Distribution and Logistics Management*, 35(3/4), 210–232.

Prokop, D. (2004). Smart and safe borders: The logistics of inbound cargo security. *International Journal of Logistics Management*, 15(2), 65.

Ridall, C.E., Bennet, S., and Tipi, N.S. (2000). Modeling the dynamics of supply chains. *International Journal of Systems Science*, 31(8), 969–976.

Seifert, R.W., Thonemann, U.W., and Hausman, W.H. (2004). Optimal procurement strategies for online spot markets. *European Journal of Operational Research*, 152(3), 781–799.

Shafer, S.M., and Smunt, T.L. (2004). Empirical simulation studies in operations management: Context, trends, and research opportunities. *Journal of Operations Management*, 22(4), 345–354.

Sheffi, Y. (2005). *The resilient enterprise.* Cambridge, MA: The MIT Press.

Sodhi, M.S. (2005). Managing demand risk in tactical supply chain planning for a global consumer electronics company. *Production and Operations Management*, 14(1), 69–79.

Terzi, S., and Cavalieri, S. (2004). Simulation in the supply chain context: A survey. *Computers in Industry*, 53(1), 3–16.

Tsay, R.S. (1988). Outliers, level shifts, and variance changes in time series. *Journal of Forecasting*, 7(1), 1–20.

van der Vorst, J.G.A.J., Beulens, A.J.M., and van Beek, P. (2000). Modeling and simulating multi echelon food systems. *European Journal of Operational Research*, 122(2), 354–366.

van der Zee, D.J., and van der Vorst, J.G.A.J. (2005). A modeling framework for supply chain simulation: Opportunities for improved decision making*. *Decision Sciences*, 36(1), 65–95.

Venkateswaran, J., and Son, Y.J. (2004). Impact of modeling approximations in supply chain analysis an experimental study. *International Journal of Production Research*, 42(15), 2971–2992.

Wei, W.W.S. (1990). *Time series analysis: Univariate and multivariate methods.* Redwood City, CA: Addison-Wesley.

Womack, J.P., and Jones, D.T. (2003). *Lean thinking: Banish waste and create wealth in your corporation.* New York, NY: Free Press.

Zsidisin, G.A. (2003). Managerial perceptions of supply risk. *Journal of Supply Chain Management*, 39(1), 14–26.

Zsidisin, G.A. (2005). "Managing commodity spend in turbulent times," CAPS Center for Strategic Supply Research.

Zsidisin, G.A., and Ellram, L.M. (2003). An agency theory investigation of supply risk management. *Journal of Supply Chain Management*, 39(3), 15–27.

Zsidisin, G.A., Ellram, L.M., Carter, J.R., and Cavinato, J.L. (2004). An analysis of supply risk assessment techniques. *International Journal of Physical Distribution and Logistics Management*, 34(5), 397–413.

Zsidisin, G.A., Melnyk, S.A., and Ragatz, G.L. (2005). An institutional theory perspective of business continuity planning for purchasing and supply management. *International Journal of Production Research*, 43(16), 3401–3415.

SECTION TWO - SUPPLY CHAIN DESIGN AND RISK

Chapter 8: Single Versus Multiple Sourcing: A Supply Risk Management Perspective

Constantin Blome* and Michael Henke

*Corresponding author: Supply Management Institute (SMI), European Business School EBS, Oestrich-Winkel, Germany

8.1 Introduction

The general trend to focus more on core competencies and the ongoing shift of value creation to suppliers has caused the dependency of purchasing companies on their suppliers to grow (Ellram and Birou 1995; Boutellier and Corsten 2000). Together with this development, more and more cooperative relationships with suppliers have evolved, and as a consequence more single sourcing relationships have emerged: e.g., strategic supplier relationships, strategic partnerships, and strategic alliances. This leads to the fact that the purchasing companies are confronted with new risks involving the risk of supply disruption. These risks are often seen as a risk of single sourcing.

Single and multiple sourcing are the more frequently discussed approaches in Supply Risk Management (Khan et al. 2007). However the risk analysis of both these approaches follows stereotypes. Whether single or multiple sourcing reduces the risk, is disputed. Furthermore, one other important fact is often missing: in many cases, there is no opportunity to choose between these two alternatives. The discussion of these two approaches will be developed in the remainder of this chapter, employing a differentiated view of risk to develop the basis for a more contingency-based method for managing the associated risks. It is subsequently argued that it is necessary to integrate this method into the process model of Supply Risk Management to enable a sustainable and company-wide Risk Management approach.

8.2 The Decision of Single and Multiple Sourcing

In addition to other sourcing approaches, (e.g., global and local sourcing or systems, modular and unit sourcing), multiple and single sourcing have been discussed for a long time as measures or strategies in Supply Management. The decision concerning single and multiple sourcing entails the decision about the number of suppliers with whom a purchasing company wants to have supplier relationships for a certain good or service, or possibly for a segment of goods or services. Whilst this present discussion will not debate whether single and multiple sourcing are best considered as a measure or a strategy, it is necessary to define the scope of the two approaches. Multiple sourcing exists when several, distinctly independent sources are available for a good or a service and the buying company purchases these goods or services from more than one supplier. Special types of multiple sourcing such as dual sourcing (purchasing from two suppliers) will not be separately analyzed in this chapter. For single sourcing the decision to opt for one supplier for a service or good – although other adequate suppliers are available and a choice is possible – is crucial. Sole sourcing is also based on one supplier per good or service, but it differs from single sourcing in one important respect: there is only one single supplier available (e.g., due to patents). The possibility to choose is therefore not available, at least not on a short- or mid-term basis (Newman 1989).

The discussion of the number of suppliers and the discussion on supplier relationships are often mixed up. This is especially the case in parts of the English supply management literature, where one can find the terms single sourcing and cooperative supplier relationships synonymously used. This combination is self-evident because single sourcing and cooperative supplier relationships as well as multiple sourcing and transactional supplier relationships are linked in an advantageous manner. But it is also possible that single sourcing is combined with a transactional supplier relationship in a beneficial way (e.g., operational demand with the aim to reduce processes and the use of bundling). Hence, the two terms, number of suppliers and supplier relationships, are not used synonymously in the present discussion. Even though the selection of a type of supplier relationship (transactional vs cooperative) predetermines the decision of single and multiple sourcing (Gadde and Hakansson 1993), the two decisions are separate and different decisions.

Moreover, the assumption that pursuing the approach of single or multiple sourcing leads automatically to the positive outcomes of a transactional or cooperative supplier relationship is a fallacy. The actual advantages of single and multiple sourcing are listed in Fig. 8.1. The disadvantages, although not detailed in the figure, may be taken as the equivalent counterparts. The supply risks of each approach are the only disadvantages that will be discussed in this chapter.

Pros of cooperative supplier relationships	Pros of transactional supplier relationships
• Use of strategic cost-reducing potential • Commitment of suppliers • Use of suppliers' Know-how • Faster development of new products • Improved planning possibilities and information exchange with suppliers • Earlier detection of misleading developments • Higher quality levels • Simpler sourcing processes • Better use of resources • Reduction of stock • etc,	• Lower supplier relationship costs • Lower prices due to higher competition possible • Higher flexibility due to lower switching costs • Lower dependence on single suppliers • No ompending loss of know-how • No decline in supplier motivation due to long-term contracts • etc,
Pros of single sourcing	**Pros of multiple sourcing**
• Cost reduction through bundling • Cost reduction through standardisation • Smaller number os suppliers and interfaces • Lower transaction costs • Easier quality assurance • Higher specialisation • Easier sourcing processes • etc,	• Lower prices due to higher competition • Lower dependence on single suppliers • Lower dependence on single technologies • Flexible change of suppliers • etc,

Fig. 8.1 Advantages of certain types of supplier relationships[1]

When discussing decisions of single and multiple sourcing one fact is often – especially in theory – missing, the decision has always to be made in a particular context and not detached from it. This point of departure of practice from theory – especially given the market conditions – affects the choice of single and multiple sourcing significantly. The assessment of the terms and context of the decision varies completely with the underlying paradigm in purchasing and supply management. While in the traditional paradigm (purchasing as an operational role) purchasers acted on the assumption that there is always an adequate number of capable suppliers available, the modern paradigm (supply management as a strategic role) takes a more realistic perspective: capable suppliers are a scarce resource, although this is a hotly contested view even in boom situations (Dobler and Burt 1984; Gadde and Hakansson 1993).

Although a real choice between single and multiple sourcing is possible for standard products like office equipment, the modern supply management paradigm shows that a real choice between these two options is not always possible. In the case of strategic goods and services, it is possible that due to advantages in know-how,

---

[1] According to Arnolds et al. (2001); Ellram and Birou (1995); Monczka et al. (1998); Baily et al. (1994).

prices, distribution, innovation, services etc. only one single supplier comes into question (e.g., sole sourcing). This may lead you to suspect that sole sourcing situations are more frequent for strategic goods and services than for example the number of patents might suggest. The use of an alternative supplier with a lower level of capability would limit the competitive decision of the buying company. Only on a long-term basis would the development of an adequate second supplier be possible (e.g., supplier development). In many situations the option to develop a second supplier may not be available in practice due to duplication of resources, technology competition, know-how advantages, image etc. Building up a second supplier can also lead to a higher corporate risk for the buying company because of slower innovation processes due to splitting the total demand across more than one supplier.

8.3 Evaluation of the Single and Multiple Sourcing Decision from a Risk Perspective

In the US literature, the following definition of supply risks is established:

"The potential occurrence of an incident or failure to seize opportunities with inbound supply in which its outcomes result in a financial loss for the [purchasing] firm." (Zsidisin 2005, p. 3)

Hitherto, supply risks have not been systematically measured, properly analyzed, and assessed (Zsidisin 2005). However, a systematic and structured picture of supply risks is the basis for deriving risk minimizing measures in general (Wildemann 2006) and especially in the single and multiple sourcing decision. Risks can be expressed in different styles. An all-embracing and selective list of all supply risks does not exist. But exemplarily, Wildemann's allocation (Wildemann 2006) of risks into supplier risks, demand risks and market risks should be mentioned, as well as the following classification of Zsidisin (2005), presented in Table 8.1.

A pragmatic way to distinguish between all supply risks is the following classification[2]:

- capacity risks (e.g., quantity and time risks),
- technology/technical risks (e.g., development risks),
- quality and service risks (e.g., specification risks),
- financial risks (e.g., price, liquidity and currency risks),
- location risks (e.g., off-shoring risks),
- management risks (e.g., embezzlement and fraud risks),
- strategy/market risks (e.g., behaviour of competitors),
- contractual risks (e.g., infringement of intellectual property rights),
- "force majeure"/environmental risks (e.g., war and terrorism).

---

[2] Based on Wildemann (2006); Weerd (2003); Jahns (2003); Atkinson (2003); Gleißner (2005); Basler Ausschuss für Bankenaufsicht (1994).

Table 8.1 Supply risk characteristics

Characteristic	Definition or Description
Cost reduction capabilities	The act of lowering the cost of the same goods or services.
Cycle time	The time between purchase request to a supplier and receipt.
Disasters	Any occurrence that causes great harm or calamity.
Environmental performance	Activities such as selecting materials used, product design processes, and process improvements.
Financial health of suppliers	Profitability trends in cash flow and the existence of financial guarantees.
Inbound transportation	Methods to distribute, handle, and transport inputs.
Information system compatibility and sophistication	Information system capability of suppliers to transfer timely, accurate, and relevant information to buyers.
Inventory management	Supplier ability to manage raw materials, work-in-process, and finished goods and inventories.
Legal liabilities	Legally enforceable restrictions or commitments relating to the use of the material, product, or service.
Management vision	Supplier management attitude and ability to foresee market and industry changes.
Market price increases	Trends, events, or developments that may increase prices.
Number of available suppliers	The existence of monopoly or oligopoly conditions in the supply market.
Process technological changes	The frequency of new ideas and emerging technology.
Product design changes	The unpredictability of changes in product technology.
Quality	The ability of suppliers to conform to specifications.
Shipment quantity inaccuracies	The gap between the actual demand requests and the quantity shipped.
Supply availability	Availability of strategic materials in terms of quality and quantity, and the relative strength of suppliers.
Volume and mix requirements changes	Demand fluctuations in quantity and type for a component or service.

Source: Zsidisin (2003).

A single source situation is itself a capacity risk. This perspective will be extended by contrasting a single sourcing decision with a multiple sourcing decision in terms of measures for Supply Risk Management.

In the practitioner's world, single sourcing is seen as a riskier alternative in comparison to multiple sourcing, although the advantages of single sourcing are indisputable. Continental, the German automotive company, states in its 2006 annual report: "Nevertheless, single sourcing cannot always be avoided" (Anonymous 2006). There are numerous practical examples illustrating the negative effects of single sourcing, often with disastrous consequences, which confirm the stereotype: e.g., Ericsson and Philips in 2000 (a fire at Philips resulted in a stoppage of production at Ericsson with € 400 m damages) (Christopher and Peck 2004); Ford in 1998 (computer problems of a door- and trunk-locking mechanism led to a three-day production downtime with € 100 m damages) (Anonymous 1998); Toyota and Aisin in 1997 (a fire at an Aisin plant, being linked to Toyota via JIT, resulted in $300 m damages) (Nishiguchi and Beaudet 1998). Equally, in many parts of the literature, one can find the stereotype of assuming single sourcing as risky and multiple sourcing as risk-free. In the following, it will be shown that there is no such dichotomy as a general rule.

Single and multiple sourcing as a source of risk but also as a risk reducing measure will be discussed later in the chapter. People have argued for a long time whether single or multiple sourcing leads to a reduction in supply risk. Some highlight that single sourcing can reduce the risk (e.g., Ellram 1991; Zsidisin 2003; Treleven and Schweikhart 1988), and others like Smeltzer and Siferd (1998), Pilling and Zhang (1992) and Lonsdale (1999) suggest that single sourcing can lead to an over-dependence on a supplier. This over-dependence implies the potential outcome that the buying company will be unwillingly exploited by the supplier. The dependence on a supplier is often seen as the main source of supply risk in single sourcing.

Although single sourcing relationships are doubtlessly linked with higher degrees of dependency, a higher dependency does not mean inevitably a higher supply risk. While one of the main assumptions of the traditional purchasing paradigm was that dependencies on suppliers have to be avoided generally, this has slightly changed in the modern paradigm. Today, the main advantages of cooperative – and as a consequence dependent – partnerships are undisputed. Furthermore, the awareness in examining dependencies and risks has grown such that not only the purchasing company's view, but also the supplier's view has to be included.

The decision concerning single or multiple sourcing alternatives has to take into account the situations in which risks come into existence due to dependencies. Dependencies lead to risk especially in such situations where dependency is asymmetrically allocated (one-sided dependency, Fig. 8.2). This means that either the supplier is much more dependent on the buying company, or the converse effect exists. This entails not only the risk of exploiting the dependent side, but also higher risks overall, e.g., in terms of time and quality risks. When the dependent buying company counts on the reliable and accurate delivery from the independent supplier, in buoyant trading conditions, it is possible that the supplier may serve other clients first. This can have significant negative consequences for the buying company.

If supplier and buying company are dependent on each other in a similar way, both parties will attribute the same importance and will arrange their resources accordingly. Two cases have to be distinguished:

1. The goods and services as well as the supplier relationship and the associated supply risk are marginal.
2. The importance of the goods and services as well as the supplier relationship and the associated supply risk are high.

Case 1 is of minor interest, but case 2 shows that both parties need to act in a cooperative way. Both are highly dependent on the other side's performance. To draw detailed conclusions for single and multiple sourcing decisions, a closer examination of the situation is necessary. Therefore, we separate strategic and non-strategic demand.

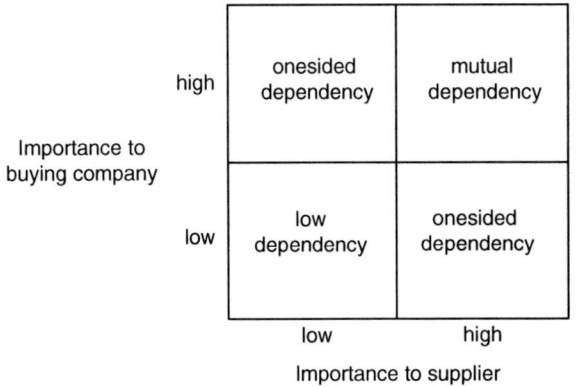

Fig. 8.2 Matrix of dependency

In the case where a company uses a single sourcing approach for a strategic good or service, it would be advantageous for the buying company if the supplier also regards the relationship as important. A cooperative approach would be helpful. From a risk perspective, the mutual dependency can result in higher flexibility of the supplier in fulfilling the buying company's needs. The consequence could be a lower capacity risk (especially time based capacity risk). Also a higher motivation for strategic cost reducing initiatives and a higher quality level can reduce cost and quality risks (quality and service risk as well as management risk). Of course, the high dependency bears the risk of painful consequences in the event of a disruption. That is also the reason why a proactive risk management approach is necessary, especially in single sourcing situations. However, the supply risks of dependency on a single supplier can be overcompensated by the greater opportunities and the lower corporate risks.

Multiple sourcing for strategic demand means high supply risk potential. The consequence of splitting the total demand may reduce the interest for the single supplier to develop new innovations because of a smaller possibility to amortise the expenses (e.g., higher innovation risk). Additionally, the time for developing innovations can increase due to the number and variety of small and independent R&D departments instead of one large one. This can lead to significant competitive disadvantages for the buying company. Equally, the risk of a supply disruption can increase. In the situation of buoyant demand all suppliers will serve their most important customers first, but due to the splitting of demand it can happen that the buying company is not a top client for any of its suppliers. The service and flexibility can be lower in comparison to single sourcing, too. In conclusion, this shows that a lower dependence is not inevitably linked to a lower risk for the buying company.

For non-strategic demand it is of minor importance for the buying company in terms of supply risk whether the supplier sees the relationship as important or unimportant. In both situations, a fast change of suppliers is possible.

An overview, that shows which supply risks are related to single sourcing and which supply risks are related to multiple sourcing, will be omitted for two reasons. Firstly, supply risk is very much based on the specific situations and hence generalisation is difficult. The second reason is that supply risk consists of two main factors: probability of occurrence and outcome of risk occurrence (e.g., financial damage). A third factor – the so-called Mean Time to Repair (MTTR) (Henke et al. 2006) – will for complexity reasons not be analyzed separately. The probability of occurrence and the outcome of risk occurrence act for multiple and single sourcing, very often in a contrary way. The difficult quantification of these factors is only possible on the basis of the particular situation in a substantiated manner. Although an abstract presumption is not possible, a pattern for the alternatives of single and multiple sourcing can be identified as:

1. Single sourcing:
   - *Probability of occurrence:* The number of potential risk sources is smaller for single sourcing. But risk sources can be better recognised and be better proactively managed if applicable. The probability of occurrence can be lowered through a proactive Supply Risk Management approach.
   - *Damage:* The possible damage in the case of single sourcing increases, because alternative sources are not available for strategic demand on a short-term basis. The damage can be even higher for sole sourcing. But single source suppliers can change their set-up in such a way that production capacity is independently positioned, as in a multiple sourcing environment. In any case, most multiple sourcing suppliers are not independent sources because of oligopoly or the dependence on raw materials. In the case that multiple sourcing suppliers are mainly supplied by the same pre-supplier and the production of this pre-supplier ceases, the result is similar to a single sourcing situation.

2. Multiple sourcing:
   - *Probability of occurrence:* The probability that a risk occurs at all, grows with the number of risk sources. The higher complexity of a high number of risk sources makes it more difficult for the supply manager to manage all risks as well as he/she might in a single source environment.

According to this, the probability of occurrence can increase. But a systematic risk management system can reduce this effect again.
- *Damage:* Because of spreading the demand to several suppliers, the damage of a disruption of one source is smaller in comparison to a single source. Furthermore, if different suppliers can help out then the potential damage will decrease.

Although the effects of single and multiple sourcing based on the factors discussed seem to be contrary, this does not mean that a decision for single or multiple sourcing is risk-neutral. The possibilities with which the risk-minimizing effects can be achieved in the two approaches are totally different depending on the point of departure. The possibility to manage risk in general varies a lot with the single or multiple sourcing approach. Additionally, the costs of the risk minimizing activities as well as the positive outcome of the two alternatives differ very much from each other. A multi-level decision for the assessment of single or multiple sourcing as a risk-reducing measure, based on a process model is necessary.

The different measures for managing risk have to be evaluated with regard to the risk strategy, the results of the risk assessment and the economy (Diederichs et al. 2004). Subsequently, adequate measures can be chosen having considered the advantages and opportunities of both the single and multiple sourcing approaches.

In particular, the aim is to reduce total cost in regard to risk or chance. A measure to manage risk makes economic sense only when its positive effects (in terms of risks: e.g., avoided costs) are bigger than its negative effects (e.g., incurred costs) (Rogler 2001). This suggests that it is also necessary to evaluate supply risks not only from a supply management viewpoint, but from a corporate viewpoint as well.

The general conclusions of this and the foregoing discussion for multiple and single sourcing decisions are the following:

- On a short- and mid-term basis there is often no real choice between single and multiple sourcing.
- Multiple sourcing does not result automatically in lower supply risks, especially for strategic demand multiple sourcing can lead to higher supply risks.
- Single sourcing can have risk-reducing effects for strategic demand.

8.4 Integrating Supply Risk Management into an Enterprise Risk Management System

Risk-reducing measures are segregated by Zsidisin and Ellram (2003) into two categories, behaviour-based and buffer-oriented management. While behaviour-based management makes use of quality improvement programs, certification of suppliers, supplier development, etc., buffer-oriented management includes safety stock, multiple sources and building up supplier inventory. The behaviour-based approach tries to aim for the optimum performance of the existing supply chain through managing the risk directly. The buffer-oriented approach seeks to reduce risks through the use of cushions or buffers. The degree of security grows but at the cost of expensive redundancy.

The contrast of cooperative partnerships involving behaviour-based management compared with transactional relationships and buffer-oriented management are obvious. As a consequence, buffer-orientation is good for multiple sourcing in combination with transactional supplier relationships, and behaviour-orientation is beneficial for single sourcing situations in combination with cooperative supplier relationships. For a transactional and single sourcing relationship only some measures of the buffer-orientation approach can be applied.

The analysis of the methods of Supply Risk Management based on earlier, primarily English and US research indicates the existence of some shortcomings when viewed in terms of single and multiple sourcing. The development of a process-oriented and integrated approach of Supply Risk Management and the integration of that approach into corporate risk management can achieve a significant contribution in closing this research gap. In the future, it will be very important not to analyse supply risks and their management solely from the perspective of supply management, but more importantly from a more strategic and corporate viewpoint: It is necessary to develop integrated systems of Supply Risk Management (Jahns 2004), that are closely linked with the so-called "Enterprise Risk Management (ERM)" (The Committee of Sponsoring Organizations of the Treadway Commission 2004). Supplier relationships and supply risks can have a significant strategic impact on the company's success. Therefore, single and multiple sourcing decisions should always take into account the company wide consequences of such decisions.

References

Anonymous 1998 Supply failure won't change Ford policy. Professional Engineering June 1998.
Anonymous 2006 Contintental Jahresbericht. See http://report.conti-online.com/de/02_unternehmensprofil/kapitel_2_11_de.html.
Arnolds H, Heege F and Tussing W 2001 Materialwirtschaft und Einkauf. Wiesbaden.
Atkinson W 2003 Supply chain risk management. Riding out global challenges. Purchasing 132(14):43–47.
Baily P 1994 Purchasing Principles and Management. London.
Basler Ausschuss für Bankenaufsicht 1994 Richtlinien für das Risikomanagement im Derivativgeschäft. Basel.
Boutellier R and Corsten D 2000 Basiswissen Beschaffung. Munich and Vienna.
Büschemann KH 2005 Eine Dieselpumpe bringt Autohersteller in Verlegenheit. Süddeutsche Zeitung 04.02.2005:19.
Christopher M and Peck H 2004 The five principles of supply chain resilience. Logistics Europe 12(1):16–21.

Diederichs M, Form S and Reichmann T 2004 Controlling-Special, Standard zum Risikomanagement. Controlling 16:189–198

Dobler DW and Burt DN 1984 Purchasing and Supply Management, New York.

Ellram LM 1991 Supply chain management: the industrial organization perspective. International Journal of Physical Distribution and Logistics Management 21(1):13–22.

Ellram LM and Birou LM 1995 Purchasing for bottom line impact: improving the organization through strategic procurement. Chicago.

Gadde LE and Hakansson H 1993 Professional Purchasing. London, New York.

Gleißner W 2005 Beschaffung und Einkauf. Das Management von Beschaffungsrisiken – eine Einführung. In: Gleißner W (ed) Risikomanagement im Unternehmen, Praxisratgeber für die Einführung und Umsetzung. Augsburg pp 4–14.

Henke M, Jahns C and Geisler P 2006 Das China-Risiko. In: Online-Portal forum gelb August 2006.

Jahns C 2003 In interview: Vom Einkauf zum Supply Management. Es gibt nichts Gutes, außer man tut es. Beschaffung Aktuell 31(10):30.

Jahns C 2004 Der Paradigmenwechsel vom Einkauf zum Supply Management (Teil 14). Es gibt ihn, es gibt ihn nicht, es gibt ihn, ... Beschaffung Aktuell 32(8):29.

Khan O, Burnes B and Christopher M 2007 Risk and supply chain management – creating a research agenda. Currently in review, see http://www.som.cranfield.ac.uk/som/research/groups/scrf/Articles.asp.

Lonsdale C 1999 Effectively managing vertical supply relationships: a risk management model for outsourcing. Supply Chain Management: An International Journal 4:176–83.

Monczka R, Trent R and Handfield R 1998 Purchasing and Supply Chain Management. Cincinatti

Newman RG 1989 Single sourcing: Short-term savings versus long-term problems. Journal of Purchasing and Materials Management 25(2):20–25.

Nishiguchi T and Beaudet A 1998 Case study: the Toyota Group and the Aisin Fire. Sloan Management Review 40(Fall):49–59.

Pilling BK and Zhang L 1992 Cooperative exchange: rewards and risks. International Journal of Purchasing and Materials Management 28(2):2–9.

Rogler S 2001 Management von Beschaffungs- und Absatzrisiken. In: Götze U et al. (eds) Risikomanagement. Heidelberg pp 211–240.

Smeltzer LR and Siferd SP 1998 Proactive supply management: the management of risk. International Journal of Purchasing and Material Management 34(1):38–45.

The Committee of Sponsoring Organizations of the Treadway Commission, 2004 Enterprise Risk Management – Integrated Framework. Jersey City.

Treleven, M and Schweikhart SB 1988 A risk/benefit analysis of sourcing strategies: single versus multiple sourcing. Journal of Operations Management 7(4):93–114.

Weerd E 2003 Alles unter Kontrolle. Risikomanagement. Initiativbanking, 2003(3):21.

Wildemann H 2006 Risikomanagement und Rating. Munich.

Zsidisin GA 2003 Managerial perceptions of supply risk. The Journal of Supply Chain Management. A Global Review of Purchasing and Supply 39(1):14–25.

Zsidisin GA 2005 Approaches for Managing Price Volatility. In: 5th International Research Seminar on Risk and the Supply Chain. Centre for Logistics and Supply Chain Management, School of Management, Cranfield University 12th September 2005.

Zsidisin GA and Ellram LM 2003 An agency theory investigation of supply risk management. The Journal of Supply Chain Management 39(3):15–27.

Chapter 9: The Role of Product Design in Global Supply Chain Risk Management

Omera Khan*, Martin Christopher, and Bernard Burnes

*Corresponding author: Cranfield School of Management, Cranfield University, Bedford, UK

9.1 Introduction

Recent years have seen a rapid decline in the UK textile and clothing manufacturing industry and those companies that have survived have been forced to outsource or move production offshore to low labour cost economies as demands from retailers for lower prices puts margins under pressure. The way in which these companies manage their design capability in a global supply network becomes critical, because of the increased lead times that offshore manufacturing presents. In a dynamic, fast moving sector, where quick response and agile supply chain management is the recipe for success, design has become an increasingly important capability in the search for competitive advantage. The purpose of this chapter is to address the role of design in global supply chain risk management.

This chapter recognises the strategic role of product design in global supply chain risk management. It provides a framework for design-led risk management and thus presents a case for recognising design as more than a creative function in the supply chain but as a platform to manage risk in supply chains. Whilst there is growing literature in the field of supply chain risk there is less empirical evidence providing practical examples of managing supply chain risk. Design-led risk management offers a novel approach to mitigating supply chain risk. The empirical research reported in this chapter is specific to the clothing manufacturing and fashion retail industry, although the findings support industries and supply chains where product design has an integral role and plays an important part in the competitiveness of the final product. There would be benefit in extending the research into other sectors. The chapter begins by summarising the key trends in the textile and clothing industry and discussing one of the major trends of global sourcing. We then introduce the research methodology, which is based on an in-depth longitudinal case study of a major UK retailer. Data collection tools included observation of supplier meetings/workshops, semi-structured interviews and access to key company documentation and archives. This is followed by a detailed case analysis of the company from which we propose strategies for managing risk in a global market.

9.2 Trends in the Textile and Clothing Industry

Increased competition at the international level and the globalisation of markets has radically changed the competitive context even in a highly traditional sector such as textiles and clothing (Elliott 2005). Strategies based on reducing the time required to respond to the market and faster introduction of new products as well as improvement in the quality of the service offered to the final customer are all fundamental to achieving competitive advantage in this sector.

The UK's textile and clothing industry has always had to respond to change in order to stay competitive, but in the last few decades the industry has had to keep pace with the swift rate of change caused by the trends towards global sourcing, outsourcing and off-shoring (Warburton and Stratton 2002; Howell and Soucy 1991). However many firms in this sector still have traditional configurations, such as the sequential approach to product and process design, functionally based organisational structures, a lack of communication both internally and externally and the tendency to see external relations in a confrontational and opportunistic light rather than from a collaborative perspective (Warburton and Stratton 2002). It can be argued that such characteristics slow down the speed to respond to market requirements and may not offer the best service to the final customer.

The spread of technology, the opening of world trade markets and the industrialisation of developing economies have multiplied the sources of competition to UK manufacturers (Bhagwati 2004). At the same time it is suggested that markets have become unstable and more unpredictable and that this has changed the degree of risk (Christopher et al. 2004). As more companies seek to develop a network of suppliers globally, the impact of risk in the textile and clothing sector has increased for a number of reasons:

1. Product complexity has increased due to the demand for variety and technical performance
2. Consumer tastes fluctuate and dictate trends more frequently which requires shorter lead times and quick response
3. Forecasting design trends in an international supply network is a riskier process when lead times are longer

Off-shoring has led to supply chains becoming more complex (Fitzgerald 2005). The traditional configurations of supply chains have in fact become supply networks which are complex webs of interactions with a focus on speed of response and new ways of adding value. Thus, supply chain management (SCM) or more specifically supply network management (SNM) has become an intrinsic process to ensure a firm's survival (Barry 2004).

A fundamental process in the textile and clothing supply network is the design of products. Traditionally the design of products in this industry has often been the responsibility of external agencies or the garment manufacturers and has often been a lengthy process. Design activities in the past have taken place sequentially and frequently involve many changes and modifications adding significantly to total lead-times. The design process has become even more critical with the trend to global sourcing and in particular offshore manufacturing (Walsh et al. 1991; Warburton and Stratton 2002; Howell and Soucy 1991). This significant transformation which potentially brings considerable benefits to a firm but also possible risks is not always fully recognised and companies have failed to mitigate design associated risks. These risks include increased product lead times and incoherent designs/products. Therefore, off-shoring the design of products may mean that a retailer loses control over a vital process in the supply network (Khan 2005).

9.3 The Trend to Global/Off Shore Sourcing

Whilst there are different definitions of global sourcing, a common feature is that they refer to materials, products or services being sourced from or performed by a provider from a location outside the company's home country (Trent and Monczka 2005; Brown and Wilson 2005; Embleton and Wright 1998). Similarly there are a number of terms used to describe the purchasing process from suppliers outside the buying firm's country (Zeng 2000) such as international sourcing, worldwide sourcing, global sourcing, off shoring and outsourcing and many practitioners and academics use the terms interchangeably. In this chapter we take the view that the term global sourcing encompasses both offshore sourcing and manufacturing and we argue this has significant implications for supply chain risk management.

A review of the literature indicates that there are two conflicting views of global sourcing. Here we examine both the positive and negative impacts of global sourcing. Those that take the positive view of global sourcing refer to it as a source of competitive advantage (Frear et al. 1992; Samli et al. 1998). Fagan (1991) highlights the obvious benefit of low labour cost as well as product availability, quality, technical supremacy and penetration of growth markets (e.g. access to new markets before competitors). Taking a similar view on the positive benefits are Ettlie and Sethuraman (2002) who describe global sourcing as providing access to new technology which is uniquely available in specific regions of the world and offers opportunities for improved product design and quality and increased manufacturing flexibility. In addition Trent and Monczka (2003) suggest that access to process and product technology and the ability to introduce competition to the domestic supply base can have benefits. A survey amongst UK and USA retailers by Lowson (2001) indicates that the two main advantages of global sourcing are considered to be cost and quality. However, interestingly, the survey also revealed that quality, response time, design and shorter lead times are considered as the key advantages for domestic sourcing.

The negative impacts of global sourcing are referred to in the literature as both risks and costs or 'hidden costs' of global sourcing. Many have argued that the globalisation of supply chains and the trend towards outsourcing has increased the exposure to risks (Christopher and Lee 2004; Jüttner et al. 2003). Others mention that risks have increased as a result of global sourcing because of extended material pipelines, longer lead times, relying on new and unfamiliar sources of supply and total costs that may go far over the unit purchase costs, increased rules and regulations, currency fluctuations, customs requirements, language, cultural and time differences. In addition to these are the problems of managing cross functional and cross-locational coordination (Trent and Monczka 2005). James (1990 p 87) states that 'many firms have moved into globalisation without a real appreciation of the environmental risks that they face with dispersed sourcing. The more links in the chain the more room for tension and greater exposure to risks'. James argues that many companies have globalised their supply chains without developing structures and systems designed to reduce their vulnerability to supply chain risks.

Fitzgerald (2005) provides a list of risks related to low-cost country sourcing:

1. Supply disruption due to poor infrastructure, communication etc.
2. Long lead times
3. Poor quality
4. Security issues (political instability, potential terrorist activities)
5. Hidden costs due to changes in tariffs, duties and taxes etc.

The above list is mirrored by Braithwaite (2003) who presents five types of risks related to global sourcing as above, but he includes the point that valuable know how is given away, which could be catastrophic for some high risk industries such as pharmaceuticals. Managing different cultures, currencies, languages and business practices, foreign exchange rates, duty/customs regulations, quality assurance and transportation delays are all cited in the literature as risks associated with global sourcing (Cho and Kang 2001; Howell and Soucy 1991). Another cost related, more strategic risk of global sourcing is discussed by Hendricks and Singhal (2003). Based on investigations of 519 supply chain glitches (e.g. disruptions in the supply chain such as inaccurate forecasts, poor planning, part shortages, quality problems, production problems, equipment breakdowns, capacity shortfalls, and operational constraints) in the companies studied, the authors argue that there is a strong link between the glitches and shareholder value. On average the glitches decreased the shareholder value by 10.82%.

Warburton and Stratton (2002) present a case of a North American apparel manufacturer, in which a dual sourcing model was developed where 20% of the production was kept in the USA and the rest was manufactured offshore. The case study concluded that even if the labour cost was dramatically less expensive offshore, a dual sourcing solution made sense. First, a number of hidden costs were associated with the offshore production, making it less attractive than it seemed at first sight. Second, fluctuations in consumer buyer patterns made the long lead times very costly, both in terms of lost sales and redundant inventory.

Ritter and Sternfels (2004) found in a survey that offshore manufacturing turned out to be less successful than expected, due to the fact that the importance of direct labour costs as a proportion of the overall costs is decreasing. In addition to this, keeping production plants near the end consumer (in their case in the USA), increased responsiveness to changing market conditions. This can be an important competitive weapon. Furthermore, the authors argue that in the high-tech electronics industries the long lead times as a consequence of producing overseas can result in price declines between 2–6% during that lead time.

The increasing trend towards off-shore manufacturing has become evident, as more companies are concentrating on their 'core competence' and outsourcing non-core activities to specialist suppliers. This trend has arisen from the increase in the number of different technologies a company now needs in order to develop innovative products that can compete in a volatile and fast changing market. As a result, companies are increasingly relying on innovation and technologies provided by suppliers. It can be argued that global sourcing and offshore manufacturing appears to bring potential benefits to a firm, specifically cost savings driven by lower labour costs. As a consequence in many cases, companies are exposed to greater risk and uncertainty in their supply chains since they are too focused on costs and cost reduction at the expense of other considerations.

9.4 Research Method

To explore these issues further a major British retailer, Marks and Spencer (M&S), was chosen as a case example. The case study approach was adopted for the empirical investigation of this research as it is useful for exploring areas where theory is still developing. It enables the researcher to gain in-depth understanding of a situation which is difficult to investigate using other techniques such as surveys. It is useful for examining how and why questions so that new insights and knowledge may be gained. In-depth, semi-structured and unstructured interviews were undertaken in this study to gather detailed information. These types of interviews allowed discussion between the interviewer and the interviewee and for ideas and thoughts to flow more freely in a conversation rather than a structured interview that would not encourage an open discussion. Also an in-depth interview allowed the researcher to probe deeper into the interviewee's thoughts and ideas and to collect more detailed data that a structured interview may not allow. Interviews were tape recorded and transcribed to save the data. According to Flynn et al. (1990) transcriptions can be used to improve interviewing techniques and to detect the presence of leading questions on the part of the interviewer. They may also be used in conjunction with content analysis, e.g., the interview is taped and a transcript is prepared. Content analysis then codifies the transcript, noting recurrent usage of a phrase or concept of interest and hypotheses may be developed or tested, based on the content analysis of the transcript (Flynn et al. 1990; p 259).

In addition to conducting interviews, all observations and meetings were documented. The importance of documentation is to 'maximise recall and to facilitate follow up and filling of gaps in the data' (Voss et al. 2002). As well as transcriptions, other forms of documentation include the collection of material collected in the field or through other sources. It also includes documenting ideas and insights that surfaced during or subsequent to the field visit, In addition, submitting draft reports to the respondents to review and verify the information increased the accuracy of the documents (Voss et al. 2002). The following list summarises the research methodology that was undertaken to investigate the aims of the study:

1. Access to Marks and Spencer was gained by contacting a senior manager in the field of supply chain management through emails followed by phone calls.
2. Once an interest in the research project was established, the research synopsis highlighting the aims of the study was sent to the manager along with an interview agenda.
3. Subsequent interviews were arranged with the senior manager at the head quarters of M&S.
4. Whilst shadowing the senior manager for a week, observations at meetings were conducted and further samples/interviewees were established, and consequently contacted in the same way as steps 1–3.
5. A close relationship was developed with key samples/interviewees to develop trust between the researcher and the sample. This was an integral part of the research process. As a lack of trust may have hindered the exclusive access to proprietary information which was both sensitive and confidential. During the early stages of the investigation the researcher recognised that interviewees were reluctant to provide certain information, hence, it was necessary to develop a good relationship and trust with the interviewees to overcome this barrier.
6. Step 5 enabled the researcher to gain access to M&S top ten suppliers, who were also contacted in the same way as steps 1–3.
7. Gaining access to M&S core suppliers was integral as one of the aims of the study was to investigate the correlation of risk and supply networks, and the supply chain was a key focus of the study.
8. All interviews were transcribed and sent back to interviewees for verification. Observation notes, documents and archival records were collected to substantiate the primary information.
9. This research process continued for a period of 2 years due to the longitudinal nature of the case. The samples evolved over the course of the investigation, some samples were interviewed several times because of their relevance to the study, and others were interviewed only once, but nevertheless provided key information and relevant links to other samples.

10 Data collection and data analysis were a parallel process and both evolved over the course of the investigation. This process enabled the researcher to identify new samples and explore emerging themes in more depth during the investigation

The longitudinal single case study proved to be an effective methodological approach to conduct this research and as a consequence a rich insight into the phenomenon of risk in textiles and clothing has been collated. The case was exploratory as there were no previous studies of risk conducted in textiles and clothing and focusing the case on a single organisation meant that more time could be invested in developing the research, which was required because of the sensitivity of the research subject. The researcher embraced a somewhat 'native' approach, blending into the organisation 'as one of them' rather than being seen as an investigator. This was an advantage, as respondents were less reluctant to share information, and as trust between the researcher and respondents developed so did the access to sensitive information increase. The longitudinal aspect of the case was appropriate to a single case, as more time was invested in researching the phenomenon over a longer period. Although the first access to M&S was not until 2001, data was collected prior to this to develop a longitudinal case study. This data collecton covered the period 1998–2004, representing M&S's most critical period.

Using a case study example of M&S, this chapter examines how the company significantly changed its sourcing policies and how this changed the company's risk profile. The case is presented in two parts to show the strategic impact of offshore sourcing and the implications for risk management between the years 1998–2002 (Part One) and 2002–2004 (Part Two). Discussing the results in two distinct parts enables us to understand the transformation and degree of change that took place in the company throughout the investigation, and enables us to identify how supply chain risk management evolved and gained in significance.

9.5 Case Study of Marks and Spencer: Part One

Part One introduces the company and analyses the impact of global sourcing between the years 1998–2002 and discusses some of the key challenges that the company faced in global sourcing decision making. The four key areas that are analysed in this part of the case are: (1) A Shift in Global Sourcing, (2) Direct Sourcing, (3) Central Procurement and (4) The Impact of Change on Supplier Relationships.

9.5.1 Company Introduction

Marks and Spencer is regarded as a UK retailing institution and is one of Europe's leading retailers. M&S built its reputation on a combination of reliability, quality and service and for decades this proved to be a recipe for success. Underpinning this was a close, more or less paternalistic, relationship with its supply base, which until late in the twentieth century was almost exclusively made up of UK companies. However, when the retail arena began to change in the 1990s, M&S found that its commitment to its traditional UK suppliers turned from being an asset to a liability. A major problem the company faced was that their previously loyal customer base began to desert them because they saw their clothing as expensive and unattractive. By the end of the 1990s, sales plummeted and profits dwindled. Nevertheless, since then, M&S has rebuilt itself and has won back many of its lost customers, but this has been a costly and painful exercise. As this case study will show, its renewed success is based on offering clothes that its customers want to buy at prices they are willing to pay. As in the past, underpinning this is its close relationship with its supply base. However, the supply base is now a global one, the relationship is based on partnership and not paternalism, and a key element in its success has been M&S's ability to manage the risk associated with global sourcing.

9.5.2 A Shift to Global Sourcing

As a business M&S felt disadvantaged in having so much manufacturing in the UK as most of their competitors shifted sourcing abroad. The company was more production-led than design-led and had built many close relationships with their suppliers but with increased competition from low-cost suppliers abroad, M&S were disadvantaged compared with their competitors. As a result their supply base strategy changed significantly from having a majority of suppliers in the UK and few in the Far East, to greatly increased sourcing from such countries as Mauritius, Sri Lanka, Cambodia, China, and Bangladesh. Sourcing from different parts of the world enabled M&S to significantly reduce the risk of becoming over-dependent on certain suppliers. Being able to switch production to a different supplier acts as a contingency and mitigates the impact of the risk to the company.

In 2001, M&S expanded its supply chain networks by collaborating with suppliers in new product development. M&S also introduced a new supply chain strategy by implementing direct sourcing. Direct sourcing refers to the practice of taking responsibility for raw material sourcing. This way the raw material suppliers are also responsible for manufacturing, therefore reducing the product development lead time. This has enabled M&S to achieve better margins as buying direct from the source of manufacture has obvious cost benefits compared to buying via a garment supplier who takes its share from the margin. For M&S this meant cutting out a number of tiers of suppliers in the supply network, thus, streamlining the network considerably.

Direct sourcing is intended to bring significant long-term benefits to M&S and mitigate the risk of offshore manufacturing. Buying directly from suppliers has many obvious benefits such as quick response, increased flexibility, shorter lead times but it also enables M&S to develop close personal relationships by dealing directly with the supplier. Direct sourcing allows the procurement process to be quicker and mitigates the risks of miss-communication in the textile supply chain, such as merchandise not arriving on time and suppliers not being briefed properly. By accessing external capabilities and increasing their knowledge and competence of the supply network, M&S reduce the risk of buying direct. M&S are far more aware of the prices of raw materials, the technical qualities of fibres and fabric constructions, finishes and innovations and the relevance of quotas and duties to price and transportation costs, than they were before.

9.5.3 Central Procurement

In response to the risks faced by the company as a result of the poor buying strategies in 1998, one of the critical decisions made to improve the business strategy was to implement a central procurement department. Sourcing and procurement was previously generally confined to UK suppliers with whom a close relationship had been nurtured over several years, so supply-side risk was thought to be low. However, when this strategy fell apart in 1998, M&S recognised that several parts of their business model needed updating. This was mainly a response to the fact that M&S had a large number of suppliers, which could only be managed by reducing the number of suppliers and by strengthening the relationships with the remaining and new suppliers. The central procurement department was created to leverage scale and to gain some consistency in M&S's approach to procurement, regarding introducing a new supplier or eliminating an existing supplier. The main function of the procurement team was to identify new manufacturers who were able to supply M&S in the most cost effective manner and influence and advise the buying teams, which prior to the department being set up were in charge of the buying for their department but lacked consistency of approach with other departments and with the company strategy.

The procurement team identified new potential suppliers and considered the possible risks of procuring from specific suppliers. In terms of design risk, the procurement department recognised the risks of procuring designs from some suppliers that did not sit well with designs from other suppliers. Thus the procurement team tried to manage this risk by using selected suppliers for specific products.

9.5.4 The Impact of Changed Supplier Relationships

The relationship between M&S and its suppliers went through many changes between 1998 and 2004. There were many companies who no longer supplied M&S and there were new suppliers with whom M&S developed a relationship. A significant change in M&S's strategy was the increase in the number of seasons per year and the increased product variety. This made it difficult for the suppliers to procure raw materials in advance due to an extensive number of components involved and high-risk levels associated with holding high levels of inventories. By developing relationships with first and second tier suppliers M&S achieved better collaboration and integration across the supply chain, hence reducing risks in the product development process. Some of the risks identified in the supply chain that could impact on the product development process were a lack of communication between raw material suppliers, garment manufacturers and M&S; lack of resources in the supply base; political instability of the supply base; inefficient transportation and logistics of goods; and a lack of design skill and technology in the supply base.

However these risks were mitigated by fostering close relationships across the supply chain, visiting suppliers regularly to ensure performance and standards were at the highest level. Communication was the key to developing relationships to enable M&S and its suppliers to have a better understanding of each other's requirements. By fostering close partnerships with a smaller number of suppliers and implementing direct sourcing as a means of working closer with their suppliers M&S mitigated and managed risks to its supply chain.

9.6 Case Study: Part Two

Part Two examines the company between the years 2002–2004 and analyses the various strategies that M&S implemented to sustain recovery and the new initiatives M&S planned to undertake to improve efficiency. The three key areas highlighted in this part of the case are, (1) Intelligent risk taking, (2) Consolidation of M&S supply network and (3) Internalising the design process.

9.6.1 Intelligent Risk Taking

M&S have created a more risk aware culture within the organisation through the application of a risk matrix for strategic decision-making. This is a standard 3 by 3 matrix (example shown in Fig. 9.1) which measures the likelihood of risks (horizontally on the matrix) versus the impact of risk (vertically on the matrix). Risks which are assessed through the matrix range from product proposition e.g., the failure to meet customer needs, through to space planning, failure to drive profit and quality, failure to meet standards, inappropriate choice of suppliers and loss of market leadership.

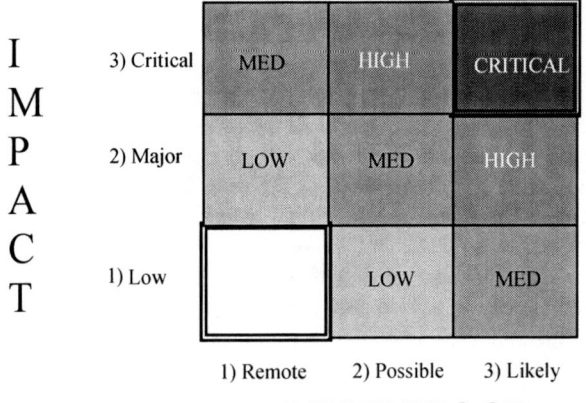

Fig. 9.1 Risk matrix

The cells in the matrix are shaded to illustrate the level of risk. Low risk is characterised by the light grey shading in Fig. 9.1 and refers to risks which are low impact to the company and where the likelihood of the risk occurring is remote. Medium risks are shown as medium grey and critical risks are shaded dark grey in the figure. This enables personnel to respond to all categories of risks to the business at a glance and plan contingencies. Risk management is an iterative process at M&S and is implemented at the beginning and half way through a project and is a continuous process, because some risks will reduce, others will increase and new ones will emerge.

To improve their risk strategy M&S aimed to develop intelligent risk takers in the company and in their supply bases, so that they could become more strategic in risk taking. M&S's risk management team actively aimed to engender, encourage and support the risk assessment/management process throughout the company. A risk strategy team was created to ensure that the risk management process was extended through the organisation and embedded in the M&S culture by making sure that M&S personnel and suppliers developed a risk management ethos in their processes and in the things that they were required to do. To become more risk aware required managers to be proactively thinking about risk, before a decision, during it and again after it. Hence, employees and suppliers were better able to identify, evaluate and assign ownership to risks, leading to a more robust risk strategy.

9.6.2 Consolidation of the M&S Supply Chain

M&S continued to consolidate its supply chain by increasingly disengaging with UK manufacturers and seeking new partnerships offshore that could offer value for money, quality, new technology, flexibility and innovation. M&S believed that the route to competitive advantage in the supply chain was through having supplier relationships which were mutually beneficial. Supplier relationships moved away from a purely dyadic relationship to a supply network, in which numerous suppliers were engaged throughout the product development process. M&S has devised a method of sourcing and procuring goods from around the world that will give them flexibility but most importantly reduce the likelihood of risk to their supply chain. M&S has identified contingency sources in case of unpredictable risks such as natural disasters or political risks that could pose a threat to M&S. As a result M&S has achieved greater flexibility as most of their material suppliers have their own spinning base, hence, this shortens the supply chain, reduces the likelihood of risks and makes the supply chain more responsive and quicker to market with M&S products.

As previously noted, M&S have moved away from their traditional sourcing patterns towards direct sourcing. This had a major impact on supplier relationships and the entire supply network changed from a large concentration of garment suppliers and a small number of raw material suppliers to a large concentration of raw material suppliers and a small number of garment suppliers. This enabled

M&S to achieve better margins as buying direct from the source of manufacture has obvious cost benefits than buying via a garment supplier that takes its share of the margin. The risk of direct sourcing is that M&S needs to understand and be aware of the manufacturing processes more than ever before. By going direct to source of manufacture also means that M&S needs to understand the performance and quality of particular fabrics, fibres and constructions and how these technicalities impact the overall style of the garment that they intend to produce. So acquiring the appropriate technical knowledge and competences is critical if M&S was to see the benefits of direct sourcing and risk management. It is perhaps too soon to say whether direct sourcing has achieved its long-term objective of mitigating risks in the supply network. However, internal M&S opinion suggests that M&S have been successful in mitigating some of the risks in their supply chain by direct sourcing.

The new global sourcing strategy required M&S to conduct risk assessment on the possible ramifications of sourcing abroad. From the risk assessment, M&S established that in order to compete through lower costs and to achieve long term stability, the risks of outsourcing appeared less than the risks of continuing to manufacture in the UK. Hence, M&S continued to extend their supply base around the world, constantly seeking new manufacturing locations that could offer quality, value and timely products. M&S strategically located some of their sources 'close to home' which were able to supply finished goods to stores in the UK in a matter of days. These supply bases would be utilised for M&S's high fashion ranges such as Per Una and Autograph, where timing dictates the success of the brand. In this way M&S is able to incorporate designer and catwalk trends into their products and bring them to the market at the most appropriate time. This strategy gave M&S much more opportunity to postpone the final design changes of products based on their sales performance, enabling M&S to benefit from both an agile and lean supply network. M&S has created a strong portfolio of relationships with suppliers, and through their strategic relationships M&S has mitigated the risks associated with off-shore manufacturing.

9.6.3 Internalising the Design Process to Manage Supply Chain Risk

A major part of M&S's recovery strategy was to re-orientate its design input, thus M&S made a huge investment in the design and product development process by developing a large design studio at their London head office and recruiting specialist design skills. M&S now have full control of the design process and are the main drivers in the product development process. There is an impetus in the business to start the design procurement process closer to the season than was previously the case. This is to avoid the costs of storing finished products for an indeterminate time and to be able to make last minute changes to products when they receive point of sale information from their stores. Therefore, to mitigate the risk of holding too much inventory, product introduction is postponed and the products are launched into the market as late as possible, so are driven by demand. This means that each season is phased, so that the final configuration of products can be delayed in manufacturing as late as possible. Based on the sales results of products launched first into the season, designers and manufacturers work together to modify the designs and tailor them to what the market demands. This process repeats throughout the season and avoids holding too much inventory and enables M&S to be more responsive to customer needs. Therefore products are changing quicker in the stores within a season and store space is utilised with good sellers and stock that does not sell is not replaced.

By investing in design and direct sourcing M&S has shifted risks in the supply network. It is now up to the supplier to produce exactly what M&S requires to very prescriptive briefs and to the quality and cost that M&S want. This has enabled M&S to mitigate the risk of miss-communication, poor performing products and incoherent designs, but it means that suppliers are pushed into holding more inventory to be responsive to M&S's new buying models. This change has a significant impact on suppliers and they find the product development process a more difficult and risky process especially when there are no long term guaranteed contracts between themselves and M&S. Examining the implications of the case in terms of design and product development, it is clear that M&S have capitalised on the design function. Good sales results indicate that M&S have mitigated some of the risks to the product development process through better segmentation and design of ranges, indicating that design is a strategic tool for managing risk in the fashion supply chain. M&S have built up a number of specific propositions aimed at increasing their relevance and attractiveness to a more contemporary and typically younger customer.

The shift to offshore manufacturing exposed the business to new risks, and despite the new design culture embraced by M&S, the company found that off shoring significantly lengthened its product lead times and reduced flexibility in its manufacturing processes. In order to respond to these risks in the supply chain, M&S internalised the product design process and sourced some of its fashion ranges from sources closer to the UK. By identifying design as a core competence to manage risk in their supply network, M&S has harnessed a method for managing risks of global sourcing; hence design could be perceived as a strategic tool for managing risk in M&S's supply network.

9.7 The Hidden Risks of Off Shore Sourcing

The M&S case study highlights that offshore sourcing appears to offer huge cost benefits initially but if sourcing and procurement processes are not managed effectively this can expose the company to many risks. There is evidence (Fitzgerald 2005) that many companies have been naïve in their decisions to offshore and outsource and it appears that these decisions have been made in haste and often to follow the general trend. It is acknowledged that many industries are reaping the rewards of the cost savings from off shoring. However, in unpredictable markets such as fashion where consumers demand quick response and short lead times, off shoring presents more uncertainty and risk than ever before.

Furthermore the globalisation of supply chains has exposed companies to increased global risks such as terrorist threats, natural disasters and political upheaval in certain parts of the world which must all be accounted for when outsourcing and manufacturing offshore. These risks if not managed effectively could cause severe disruptions in the supply chain with the risk of increased obsolescence and market failure which could result in significant losses.

9.8 Strategies for Managing Risk in a Global Market

Several strategies can be proposed to manage supply chain risks in a global economy, and emerging from the case study two strategies in particular can be identified for mitigating risks in the textiles and clothing sector:

1) *The integration of the design functions in to supply chain management.* The case study of M&S has shown the importance of investment in design to mitigate risks in the supply chain and to retain a competitive advantage. Hence, design must refer to much more than a process which brings style, aesthetic or ergonomic changes, it must be defined as an organisation's strategic tool which impacts the entire supply chain with a significant impact on market success and failure. Therefore design must be perceived as a prerequisite for identifying and managing risks early in the supply chain, where uncertainties are high.

2) *Adopting agile strategies across the supply network.* The M&S case study has shown that there is less flexibility with offshore manufacturing and changes cannot be made as fast as if manufacturing in the UK or closer to the UK, because of the time differences and language differences, which slow down processes significantly. Hence, companies must become more agile – a strategy which calls for increasing speed, responsiveness and flexibility and consequently determines the approach to design, in terms of products reaching the market quicker and responding to product changes through flexible manufacturing. As the M&S case identifies, having high fashion products made closer to home where changes can be implemented quicker mitigates supply chain risk.

Design and agility offer the benefits of greater product diversity, quicker new product development cycles and faster reaction to consumer demands, hence, it can be argued that design and agility are inter-related strategies for firms seeking competitive advantage. The M&S case indicates that companies need a robust approach to managing supply chain risk through the adoption of agile practices and the strategic repositioning of design in their extended supply network in order to support dynamic and competitive manufacturing in the UK.

9.9 Conclusions

The globalisation of supply chains and the trends to outsourcing and offshore manufacturing in the search for low labour costs have paradoxically exposed businesses to new risks. For example the abolition of the Multi Fibre Agreement which has removed quota restrictions on low labour cost economies, has enabled those countries to increase their penetration of the market with cheaper imports thus increasing competition for UK manufacturers.

This chapter indicates that companies are more exposed to risk and are increasingly faced with the challenges of dealing with unpredictable and uncertain events both internally and externally to the firm. The potential instability arising from outsourcing to achieve cost savings can actually create risks which may result in losses many times greater than the potential savings. The risks are greater in innovative and quick response markets such as textiles and clothing where failure to meet customer demand could have severe consequences. It is of paramount importance that companies from such a competitive and volatile industrial sector, which are exposed to high levels of risk, identify and manage risks in the supply chain and develop risk-reducing strategies to avoid adverse effects.

Competitive advantage has become more difficult to achieve in this dynamic environment, but the industry strives to remain competitive by means of higher productivity and through competitive strengths such as innovation, quality, creativity and design. The industry has also been adapting to new technologies at a fast pace with regard to information and communication technologies and new production techniques to remain competitive. Product designs can change dramatically and retailers and manufacturers are forced to adapt to these changes instantly or risk market failure. This raises the issue of how companies should manage their design capability in a global supply network, with the increased lead times that offshore manufacturing often generates. Consequently, the question is how can companies manage risks in a design driven supply chain and maintain agility in a global market?

The fact that many companies continue in their pursuit of offshore cost savings without a clear risk management strategy or contingency plans highlights the need to give attention to risk management in the supply chain. Risk management is not a new idea and companies have long identified, analysed, and mitigated risk. Furthermore, risk is not static and it must be recognised that as supply chains, networks and procedures evolve so do the risks and thus the need for their subsequent management. Despite the current interest in supply chain risk management research, strategies and tools to manage risk based on empirical research remain at an incipient stage.

References

Barry, J (2004) "Supply Chain Risk in an Uncertain Global Supply Chain Environment," *International Journal of Physical Distribution & Logistics Management*, vol. 34, no. 9, pp. 695–697.

Bhagwati, J (2004) *In Defence of Globalisation*. Oxford University Press.

Braithwaite, A. (2003) The Supply Chain Risks of Global Sourcing. *LCP Consulting Supply Chain Strategy and Trends – Globalisation*.LCP Consulting Ltd. UK.

Brown, D and Wilson, S. (2005) *The Black Book of Outsourcing*. John Wiley & Sons: New Jersey.

Cho, J. and Kang, J. (2001) "Benefits and Challenges of Global Sourcing: Perceptions of US Apparel Retail Firms". *International Marketing Review*, vol. 18, no. 5, pp. 542–561.

Christopher, M. and Lee, H. (2004) "Mitigating Supply Chain Risk Through Improved Confidence". *International Journal of Physical Distribution & Logistics Management*, vol. 34, no. 5, pp. 388–396.

Christopher, M. Lowson, R. and Peck, H 2004) "Creating Agile Supply Chains in the Fashion Industry". *International Journal of Retail and Distribution Management*, vol. 32, no. 8/9, pp. 367–377.

Elliott, L (2005) "Bra Wars: Europe Strikes Back." *The Guardian*, 26 August (http://www.guardian.co.uk).

Embleton P.R and Wright, P.C. (1998) "A Practical Guide to Successful Outsourcing". *Empowerment in Organisations*, vol. 6, no. 3, pp. 94–106.

Ettlie, J.E. and Sethuraman, K. (2002) "Locus of Supply and Global Manufacturing". *International Journal of Operations & Production Management*, vol. 22, no. 3, pp. 349–379.

Fagan, M.L. (1991) "A Guide to Global Sourcing". *The Journal of Business Strategy*, vol. 12, no. 2, pp. 21.

Fitzgerald, K. (2005) "Big Savings but Lots of Risks". *Supply Chain Management Review*, December.

Flynn, B.B, Sakakibara, S, Scroeder, R.G; Bates, K.A and Flynn, E.J. (1990) 'Empirical Research Methods in Operations Management'. *Journal of Operations Management*, Vol. 9, no. 2, pp. 250–284.

Frear, C, Metcalf, L and Alguire, M. (1992) "Offshore Sourcing: Its Nature and Scope" *International Journal of Purchasing and Materials Management*, vol. 28, no. 3, pp. 2–11.

Hendricks K.B and Singhal V.R. (2003) "The Effect of Supply Chain Glitches on Shareholder Wealth". *Journal of Operations Management*, vol. 21, pp. 501–522.

Howell R.A and Soucy S.R. (1991) "Determining the Real Cost of Doing Business in a Global Market". *National Productivity Review*, vol. 10, no. 2, pp. 157–165.

James, B. (1990) "Reducing the Risks of Globalization". *Long Range Planning*, vol. 23, no. 1, pp. 80–88.

Jüttner, U, Peck, H. and Christopher, M. (2003) "Supply Chain Risk Management: Outlining an Agenda for future Research", *International Journal of Logistics: Research and Applications*, vol. 6, no. 5, pp. 197–210.

Khan, O (2005) *"Risk and the Textile Industry: The Case of Marks and Spencer"*. PhD Thesis, Manchester Business School.

Lowson, R. (2001) "Offshore Sourcing: An Optimal Operations Strategy?" *Business Horizons*, vol. 44, no. 6, p. 61.

Ritter R.C and Sternfells R.A. (2004) "When Offshore Manufacturing Doesn't Make Sense". *The McKinsey Quarterly* **4**.

Samli, A.C., Browning, J.M. and Busbia, C. (1998) "The Status Of Global Sourcing As A Critical Tool Of Strategic Planning: Opportunistic Versus Strategic Dichotomy". *Journal of Business Research,* vol. 43, no. 3, pp. 177–187.

Trent, R.J. and Monczka, R.M. (2003) "International Purchasing and Global Sourcing – What Are the Differences?" *Journal of Supply Chain Management,* vol. 39, no. 4, pp. 26–37.

Trent, R.J. and Monczka, R.M. (2003) "Understanding Integrated Global Sourcing". *International Journal of Physical Distribution & Logistics Management,* vol. 33, no. 7, pp. 607–629.

Trent, R.J. and Monczka, R.M. (2005) "Achieving Excellence in Global Sourcing". *MIT Sloan Management Review* (Fall 2005), vol 47, no. 1, pp. 24–32.

Voss, C, Tsikriktsis, N and Frohlich, M. (2002) "Case Research in Operations Management". *International Journal of Operations and Production Management,* vol. 22, no. 2, pp. 195–219.

Walsh, V, Potter, S. Roy, Capon, C.H., Bruce, M. and Lewis, J. (1991) "The Benefits and Costs of Investment in Design: Using Professional Design Expertise in Product, Engineering and Graphics Projects". *Open University/Umist Design Innovation Group, Report* DIG-03, July.

Warburton R.D.H and Stratton, R. (2002) "Questioning the Relentless Shift to Offshore Manufacturing". *Supply Chain Management: An International Journal,* vol. 7, no. 2, pp. 101–108.

Zeng, A. Z. (2000) "A Synthetic Study of Sourcing Strategies". *Industrial Management + Data Systems,* vol. 100, no. 5, pp. 219–226.

Chapter 10: How Much Flexibility Does It Take to Mitigate Supply Chain Risks?

Christopher Tang* and Brian Tomlin

*Corresponding author: UCLA Anderson School, UCLA, Los Angeles, CA 90095, USA

10.1 Introduction

In light of the number of severe, and well publicized, supply disruptions over the past decade, it is not surprising that firms are instituting risk assessment programs to gauge their vulnerability. Using both formal quantitative methods and informal qualitative ones, risk assessment programs attempt to systematically uncover and estimate supply chain risks. What is surprising, perhaps, is that there has not been a concomitant investment in risk reduction programs. While the exact reasons for this are not known, a number of researchers, e.g., Rice and Caniato (2004), Zsidisin et al. (2001) and Zsidisin et al. (2004), have offered the following as potential explanations for why risk reduction efforts are less widespread: (1) Some firms are not familiar with the different approaches for managing supply chain risks; (2) Lacking credible estimates for the probability of a major disruption, many firms cannot perform the formal cost/benefit or return on investment (ROI) analyses to justify risk reduction investments.

Given the inherent challenge of performing a rigorous ROI analysis, Tang (2006) argued that risk reduction programs must provide strategic value to the firms regardless of the occurrence of major disruptions. Indeed, disruption risk is only one of a number of risk categories firms need to account for in their risk programs. Routine risks, e.g., those frequently-occurring problems that cause mismatches in supply and demand or higher-than-expected procurement costs, also need to be considered. In their empirical study across a wide range of industries, Hendricks and Singhal (2005) found that supply chain glitches (resulting in a supply-demand imbalance) had a considerable, and long-lasting, negative impact on a firm's operating performance. Risk reduction efforts directed toward these routine risks may also have the side benefit of making a firm less vulnerable to rare-but-severe supply disruptions.

Risks are often measured on two dimensions – the likelihood of occurrence and the impact if the event occurs. Risk reduction strategies can, therefore, be categorized according to whether they tackle the likelihood or the impact. In the first category (likelihood), some strategies, such as the Poka-Yoke system, aim to prevent a risk from occurring, while others, attempt to reduce the likelihood. For instance, Lee and Wolfe (2003) illustrate how certain technologies, say, biometric systems for positive identification of personnel and smart container systems for monitoring internal temperature and pressure of each container, can be used to prevent containers being tampered with during the shipping process. Applying TQM and Six Sigma principles, it is possible for firms to reduce the likelihood of certain supply-chain risks. For example, the Container Security Initiative (CSI) is based on the inspection-at-source principle of TQM. Under this initiative, all containers are to be pre-screened at the port of departure before they arrive at U.S. ports so as to reduce the likelihood of a terrorist attack in the U.S. The second risk-reduction category (impact) refers to those strategies that focus on reducing the negative impact of a risk, and that is the focus of this chapter. In particular, we focus on the power of flexibility to moderate the impact of many different risk types.

Li and Fung (www.lifung.com), the largest trading company in Hong Kong for durable goods such as textiles and toys, has a supply network of over 4,000 suppliers throughout Asia and this network provides it with a degree of flexibility to absorb risks by quickly adapting to market conditions. In 1997, the Indonesian Rupiah devalued by more than 50% and many Indonesian suppliers were unable to deliver their orders to their U.S. customers as they were unable to pay for imported materials. Li and Fung reacted to the situation quickly by (1) shifting some production to other suppliers in Asia, and (2) providing financial assistance to those affected Indonesian suppliers to ensure business continuity. (The reader is referred to St. George (1998) for further details.) The flexibility to shift production amongst its suppliers enabled Li and Fung to mitigate the impact of the currency crisis. Flexibility, in its many forms, is a critical strategy for reducing the negative impact of supply chain risks. Viewing flexibility through the prism of the "Triple-A" principles (Alignment, Adaptability and Agility) espoused by Lee (2004), flexibility enhances a firm's adaptability and agility.

Tang (2006) highlighted the strategic value of 9 different risk reduction programs that each call for an increase in a different type of supply chain flexibility. While it is clear that flexibility provides strategic value to a firm and it enhances the supply chain resiliency, it is unclear how much flexibility is needed to mitigate supply chain risk. Without a clear understanding of the benefit associated with different levels of flexibility, firms are reluctant to invest in risk-reducing flexibility strategies, especially if a lack of precise risk data prevents a detailed ROI analysis. In this chapter, we present a framework for examining the benefits of flexibility. Based on our analysis, it appears that firms can obtain significant strategic value by implementing a risk reduction program that calls for a relatively low level of flexibility. Our findings highlight the power of flexibility, and provide convincing arguments for deploying flexibility to mitigate supply chain risks.

The rest of the chapter is organized as follows. In the next section, we discuss some key supply chain risks and the role that flexibility can play in mitigating the risks. In Sect. 10.3, we introduce a flexibility measure and review some stylized models intended to illustrate the power of flexibility. Based on our models, we show that only a small amount of flexibility is required to mitigate risk. Section 10.4 concludes this chapter. We note that this chapter is based on research presented in Tang and Tomlin (2007).

10.2 Supply Chain Risks and the Role of Flexibility

In this section, we categorize and discuss a variety of supply chain risks and how flexibility can be deployed to mitigate the negative impact of these risks.

10.2.1 Supply Risks

In addition to the risk of severe-but-rare disruptions (discussed as a separate category later), firms face a number of more routine, but still-important, supply risks.

- *Supply Cost Risk.* The effective per-unit price that a firm pays can fluctuate over time due to variability in raw material prices and exchange rates, among other things. For example, Intercon Japan's connector manufacturer sourced a special type of bronze from a single metal supplier (Asahi Metal). This resulted in Intercon Japan having very little control over the raw material cost, and it, therefore, bore a significant risk of uncertain connector costs. (The reader is referred to Tang (1999) and references therein for details.)
- *Supply Commitment Risk.* Under a partnership agreement between Canon and Hewlett-Packard (HP), Canon has been the exclusive supplier of engines for the HP LaserJet printers. To keep supply costs down, the agreement dictated that HP place its order 6 months in advance and, furthermore, HP was not allowed to change the order quantity once the order was placed. This arrangement gave rise to a commitment risk for HP as it could not react to changes in demand by revising previously-placed orders. (The reader is referred to Lee (2004) for details.)
- *Supply Continuity Risk.* There are a number of risks (quality, labor unrest, etc.) that can interrupt supply for a short period of time. In April 2007, Ford temporarily closed 5 assembly plants in response to faulty transmission parts provided by a supplier but, according to a Ford spokesperson, they "certainly [did] not expect it to be a long period of time." (See Reuters 2006 for further details.) Rare-but-severe risks that cause lengthy interruptions are discussed separately in Sect. 10.2.4.

Firms can deploy a number of flexibility strategies to mitigate the negative impact of these supply risks. In particular, they can deploy one of the following:

- *Flexible Supply Strategy via Multiple Suppliers.* As discussed above, Intercon Japan faced a supply cost risk as a result of the single-sourcing strategy of its connector supplier. Firms that maintain an active set of pre-qualified suppliers for a given component can shift order quantities across these suppliers in response to variations in supplier costs. As discussed in the introduction, Li and Fung availed of its multiple-supplier strategy to mitigate supply cost risks by ordering from suppliers with the lowest costs. Clearly a firm has more supply flexibility as the number of suppliers increases but does it need a lot of suppliers, or just a few, to effectively mitigate supply cost risk? This is one of the questions addressed in this chapter. We note that a multiple-supplier strategy can mitigate routine supply continuity risk by enabling the firm to increase orders placed at other suppliers if one supplier suffers a short-term interruption.

- *Flexible Supply Strategy via Flexible Supply Contracts.* As discussed above, HP faced a supply commitment risk because they were not allowed to revise their order quantity once submitted to Canon. To reduce HPs supply commitment risk, Canon agreed to offer HP some adjustment flexibility, that is, they allowed HP to adjust their order quantity upward or downward, but limited the adjustment to be no more than a few percent. This type of supply contract, one that specifies an upward/downward adjustment limit, is called a Quantity Flexible (QF) contract. QF contracts enable firms to mitigate their supply-commitment risk by shifting their order quantities across time. Clearly a firm has more flexibility as the adjustment limit in a QF contract increases but does it need a large limit or will a small one suffice to effectively mitigate the firm's commitment risk? This is one of the questions addressed in this chapter.

10.2.2 Process Risks

To improve internal quality and capabilities, firms have invested heavily over the past decade in programs such as Total Quality Management (TQM), Lean Manufacturing and Six Sigma. However, internal operations (including inbound and outbound logistics) are still susceptible to issues that can cause fluctuations in effective capacity and quality. For example, in 2004 IBM announced that yield problems at its plant in East Fishkill, New York contributed to the $150 million first-quarter loss by its microelectronics division (Krazit 2004). The lower-than-expected yields reduced the plant's effective capacity and limited IBM's ability to meet customer demand.

Consider a simple case in which a firm produces two products, each in a plant dedicated to that product. If one plant happens to be suffering from a lower-than-expected capacity, then the firm may be unable to meet demand for the associated product. Firms can mitigate process risk, e.g., fluctuating-capacity, by pursuing the following strategy.

- *Flexible Process Strategy via Flexible Manufacturing Processes.* In this strategy a firm configures plants to be able to produce a range of products. Again, consider the simple two-product, two-plant example but now imagine that the firm has configured both plants to be capable of producing both products. In this case, the firm can shift some production of the affected product to the unaffected plant, thereby reducing the risk of not being able to meet demand. (We note that, analogous to the flexible process strategy, a firm could institute a flexible supplier strategy to mitigate some of the supply risks discussed earlier.)

For firms that produce many different products, it may be prohibitively expensive to configure plants to be able to produce every product. While this "total flexibility" would offer the greatest protection against process risk, is this level of flexibility necessary for effective mitigation. This is one of the questions addressed in this chapter.

10.2.3 Demand Risks

Customer demand for a product fluctuates over time, and, therefore, uncertainty about the level of demand for a product is a routine risk faced by all companies. However, volume uncertainty is not the only demand risk firms need to manage. To increase revenue, many firms sell their product in more than one country and may need to localize their product for each country. For example, to satisfy certain country-specific requirements such as power supply and language driver, HP has to develop multiple versions for each model of their DeskJet printers. Each version serves a particular geographical region (Asia-Pacific, Europe, or Americas). Due to uncertain demand in each region, HP faces the problem of overstocking certain printers in one region and under stocking certain printers in other regions. (The reader is referred to Kopczak and Lee (1993) for details.) This example reflects a risk facing companies that sell multiple products: not only is the demand volume unpredictable but so is the demand mix, e.g., the demand for each of the product variants. Demand risks therefore encompass uncertainties in both volume and mix.

Firms can deploy a number of flexibility strategies to mitigate the negative impact of these demand risks. In particular, they can deploy one of the following:

- *Flexible Product Strategy via Postponement.* To reduce the overstocking and under stocking costs associated with its demand-mix risk, HP redesigned its DeskJet printers by delaying the point of product differentiation. Specifically, HP first manufactures and ships generic printers to the distribution centers in different regions. These generic printers are then customized for different country-specific markets at each distribution center. The generic printers are produced according to a make-to-stock system, while the country-specific printers are customized in a make-to-order manner. This postponement strategy has enabled HP to respond to demand changes quickly and effectively. (The reader is referred to Lee and Tang (1997) for a detailed description of various mechanisms for delayed product differentiation such as modular design, standardization, commonality, etc., (see Feitzinger and Lee 1997) for a detailed description of successful implementations of various postponement strategies at HP.) As the generic printers are completely flexible, delaying the point of product differentiation until the last stage of the process would offer HP the highest level of product flexibility for mitigating demand risks. However, the cost of postponing the point of differentiation until the last stage of the process can be excessive and may not be necessary to effectively mitigate demand risk. Later in this chapter, we address the question of whether limited flexibility, e.g., placing the point of differentiation early in the production process, can effectively mitigate demand risk.

- *Flexible pricing strategy via Responsive Pricing.* While flexibility is often associated with operations, certain companies have developed important flexibilities in sales and marketing. Dell is well known for their ability to shape consumer demand in response to supply availability. If there is a mismatch in demand and supply, Dell can rapidly adjust the price of computer configurations to better match demand to the available supply. The combination of their online ordering system and their supply-chain visibility gives them the flexibility to shift demand across products in response to supply-demand mismatches. We note that Dell is also able to influence demand not only through prices, but also through the use of the radio buttons on the site that recommend certain configurations. As with operational flexibility, there is the question of how much marketing flexibility is necessary to effectively mitigate demand risk. Focusing on pricing flexibility, we investigate this question later in the chapter.

10.2.4 Rare-but-Severe Disruption Risks

According to a study conducted by Computer Sciences Corporation in 2004, 60% of the firms reported that their supply chains are vulnerable to disruptions. While severe disruptions might be rare for any particular company, there have been a number of significant disruptions in a variety of industries over the past decade. When Chiron, the flu-vaccine maker for the US market, halted production due to a bacterial contamination problem at Chiron's plant in Liverpool in 2004, Chiron announced that they would not be able to deliver 48 million doses of vaccine for the U.S. market. This shortfall amounted to 50% of the total estimated demand and, so, the shortage threatened the health, and, indeed, life of many senior citizens. (The reader is referred to Brown 2004 for details.) Other notable disruptions include the following: in 2006, due to a fire hazard, Dell recalled 4 million laptop computer batteries made by Sony; Ericsson lost 400 million Euros after their supplier's semiconductor plant caught on fire in 2000; Land Rover laid off 1400 workers after their supplier became insolvent in 2001; Dole suffered a large revenue decline after their banana plantations were destroyed after Hurricane

Mitch hit South America in 1998; and Ford closed 5 plants for several days after all air traffic was suspended after September 11 in 2001. (The reader is referred to Christopher 2004; Martha and Subbakrishna 2002; Monahan et al. 2003; and Chopra and Sodhi 2004 for more details on some of these disruptions.)

Many of the flexibility strategies discussed above, e.g., multiple suppliers, quantity-flexible contracts, flexible processes, and responsive pricing, can also be used to mitigate the negative impact of a severe disruption, e.g., Sheffi (2001) and Sheffi (2005). For example, as the supply of certain components from Taiwan was affected by an earthquake, Dell's response was to lower the price of certain products so as to entice their online customers to "shift" their demands to other Dell computers that utilized components from other countries. The capability to influence customer choice enabled Dell to improve its earnings in 1999 by 41% even during a supply crunch (Martha and Subbakrishna 2002). To avoid the influenza-vaccine crisis from occurring in the future, the U.S. government could consider offering certain economic incentives to entice more suppliers, instead of the current two, to re-enter the flu vaccine market. With more potential suppliers, the U.S. government would have the flexibility to shift their orders to different suppliers when faced with a severe disruption to one supplier.

As noted in the introduction, firms may be reluctant to invest in flexibility to reduce the impact of rare-but-severe disruptions because a lack of precise likelihood estimates prevents them from conducting accurate cost-benefit analyses. However, the fact that the types of flexibility that mitigate the more routine supply, process and demand risks also mitigate rare-but-severe disruption risks should encourage managers to invest in flexibility.

10.2.5 Other Risks

While we focus in this chapter on the role that flexibility can play in mitigating supply, process and demand risks, it is important to recognize that these are not the only supply chain risks. Firms should also include the following in their risk management programs:

- *Intellectual property risks.* While outsourcing or off-shoring can result in lower manufacturing costs, it makes it difficult to protect intellectual property (IP). For example, even though the reform of the Intellectual Property protection law has made some good progress after China's WTO entry in 2001 (http://www.chinaiprlaw.com/english/news/news5.htm), some unfortunate incidents can still occur in China. For instance, multinational firms are not necessarily protected legally when their Chinese suppliers start producing unauthorized products using virtually identical design and materials. To elaborate, when the relationship between New Balance shoes and one of their Chinese suppliers went sour, this Chinese supplier started producing different types of shoes using a logo that resembles the New Balance's block "N" saddle design. New Balance filed a lawsuit in China without success and the saga continues. The reader is referred to Chandler and Fung (2006) for more details. As such, it is still difficult to protect IP and to eliminate the risk of counterfeits when a multinational firm outsources their manufacturing operations to their Chinese suppliers under certain licensing or contractual agreements.

- *Behavioral risks.* As the number of partners increases in a global supply chain, the level of visibility and control can be reduced significantly. For instance, according to a study conducted by AMR Research in 2006, if supply chain visibility is relatively low: few companies have demand/inventory information from downstream partners and over 56% of the companies take more than 2 weeks to sense changes in actual demand. The low visibility level and the low control level reduce the "confidence" of each supply chain partner regarding the following information: the replenishment lead time/order status quoted by upstream partners, demand forecasts provided by downstream partners, etc. Christopher and Lee (2004) argue that a low confidence level may induce damaging behavior such that the entire supply chain enters a "risk spiral" that can be described as follows. Each supply chain partner either "inflates" their order or "disguises" their on-hand inventory because of the lack of confidence in the replenishment lead time, demand forecasts, etc. The confidence level deteriorates further as every partner starts gaming the system, and hence, the "risk spiral" continues. To break this vicious cycle, supply chain visibility, timely communication, and coordinated corrective actions are needed to restore the confidence level of each supply chain partner.

- *Political risks.* A global supply chain can fall victim to political upheaval or political interference. Airbus offers an example of the latter risk. Airbus, a four-nation consortium, is facing an opportunity loss of 4.8 billion Euros due to a 2-year delay in launching the super-jumbo A380. In addition to technical problems associated with the wiring system, political battles may be a key reason for the delay because of the political pressure to "balance" the interests of 4 different European countries. As reported in Gumbel (2006), Airbus' parent, EADS is struggling to develop a restructuring plan to replace political bargaining with industrial logic.

- *Social risks.* Firms are not immune from changing views as to what constitutes socially-acceptable labor practices. Because Nike sourced their athletic shoes in various developing countries such as China, Pakistan, Thailand, Vietnam, etc., many citizens raised their concerns over the issue of Child Labor/working children. According to the International Labor Organization, over 250 million children between the ages of 5–14 are working and 61% of the working children are found in Asia. As some concerned citizens became watchdogs and launched websites, such as www.saigon.com/~nike, boycotting Nike products, Nike came under pressure to develop various social responsibility programs to respond to public opinion (www.nike.com/nikebiz).

10.2.6 The Role of Flexibility in Mitigating Risks

In this chapter we focus on the 5 types of flexibility described above and summarized in Table 10.1. Since our focus is on the benefit of flexibility, we do not consider the cost for implementing flexibility in our models. One can combine the cost and the benefit associated with different levels of flexibility to determine the optimal level of flexibility. However, the determination of the optimal level of flexibility is beyond the scope of this chapter.

Table 10.1 Flexibility strategies for reducing supply chain risks

Supply chain risk	Flexible strategy reducing the negative impact of the risk	Underlying mechanism
Supply	Flexible supply via multiple suppliers. Flexible supply via flexible supply contracts.	Shift orders quantities across suppliers. Shift order quantities across time.
Process	Flexible process via flexible manufacturing processes.	Shift production quantities across internal resources (plants or machines).
Demand	Flexible product via postponement flexible pricing via responsive pricing.	Shift production quantities across different products. Shift demands across different products.

Although these five types of flexible strategies have been described separately, firms can, of course, implement some of these strategies jointly. Here are some examples.

A firm can combine the multiple-supplier strategy and the flexible-supply-contract strategy by implementing different flexible supply contracts with multiple suppliers. Hence, a firm can establish a "portfolio" of suppliers, say, a long-term inflexible supply contract with one supplier at a lower supply cost, and a more flexible supply contract with another supplier at a higher supply cost. This portfolio approach has enabled many firms to mitigate their supply chain risks. Specifically, Zara, the most profitable fashion company in Europe, sources their stable items from China at low cost with very little flexibility in changing order quantity and makes their fashion items at their own plant in Coruna. The reader is referred to Ferdows et al. (2004) for details. In addition, Billington (2002) highlighted how this portfolio approach mimics the financial portfolio theory and how it enabled HP to reduce the average and the standard deviation of the procurement cost.

By combining the multiple-suppliers and the flexible-manufacturing-process strategies, a firm can have multiple plants with flexible manufacturing processes in multiple countries so that the firm can shift the production volume of a portfolio of products from one plant to a different plant quickly. This combined strategy offers the "operational flexibility" that would allow a firm such as Li and Fung to reduce supply risks including currency exchange risks. The use of operational flexibility to exploit uncertain exchange rates is called "operational hedging" and it has been examined by various researchers including Huchzermeier and Cohen (1996) and Kouvelis et al. (2006). The reader is referred to Boyabatli and Tokay (2004) for a review of recent research in the area of operational hedging.

For the sake of completeness, we note that, in addition to these five types of flexible strategies, there are other strategies that can enhance the overall flexibility of the entire supply chain. For example, to reduce the exposure to various types of supply chain risks, one can shorten the overall lead time by redesigning the supply chain network. For instance, Liz Claiborne launched a campus in China by bringing all stages of the textile supply chain to the campus. This campus concept enabled Liz Claiborne to reduce the lead time from concept to retail store from the existing 10–50 weeks to fewer than 60 days (Tang 2006a). Sun Microsystems implemented the "one touch" supply chain strategy by shifting most of the manufacturing operations to its contract manufacturers. With fewer steps in the supply chain process, this "one touch" strategy has enabled Sun Microsystems to reduce lead time, reduce cost, and increase supply chain visibility (Gary 2005). In this chapter, however, we restrict our attention to the 5 types of flexibility described above.

10.3 The Power of Flexibility: How Much Flexibility Do You Need?

In the last section, we described and highlighted the benefits of 5 types of flexibility strategies. We now examine a fundamental question: How much flexibility does it take to mitigate supply chain risks? To help answer this question, we first introduce a general flexibility measure that can be used for each of the flexibility types described in the last section. Let f denote the level of flexibility for a particular strategy such that a higher f refers to a more flexible supply chain. For example, in the multiple-supplier strategy, f would refer to the number of suppliers, whereas, in the process-flexibility strategy it would refer to the number of plants each product could be produced in. Each of the five flexibility strategies has a minimum and maximum level of possible flexibility. The minimum level, denoted by f_{min}, corresponds to a supply chain with no flexibility, e.g., $f = 1$ in the multiple-supplier strategy. The maximum level, denoted by f_{max}, corresponds to a supply chain with the highest level of flexibility theoretically possible, e.g., $f = \infty$ in the multiple-supplier strategy.

Let $P(f)$ be a performance metric for a supply chain with flexibility level f. Depending on the context, the performance metric $P(f)$ might be measured in terms of cost or profit. For example, in the case of the multiple-suppler strategy that aims to mitigate the impact of uncertain supplier costs, $P(f)$ might be the expected per-unit cost. We can measure the "relative" benefit of flexibility by using the following:

$$V(f) = \frac{\frac{P(f) - P(f_{\min})}{P(f_{\min})}}{\frac{P(f_{\max}) - P(f_{\min})}{P(f_{\min})}} = \frac{P(f) - P(f_{\min})}{P(f_{\max}) - P(f_{\min})} \quad (10.1)$$

Notice that $V(f)$ measures the percentage of benefit obtained by a supply chain with flexibility level f as compared to one with the maximum possible level of flexibility. Specifically, $V(f_{\min}) = 0\%$ and $V(f_{\max}) = 100\%$. Given the performance metric $V(f)$ associated with a flexibility level f, we can evaluate the impact of flexibility associated with each of the 5 flexibility strategies.

Clearly, a more flexible supply chain performs better than a less flexible supply chain and, therefore, the measure $V(f)$ is increasing in f. However, what is less clear is whether $V(f)$ is concave or convex in f. (See Fig. 10.1) If $V(f)$ is concave, then significant benefits associated with a flexibility strategy can be obtained with a low level of flexibility; e.g., when f is small. On the other hand, if $V(f)$ is convex, then a firm needs to invest in a high level of flexibility in order to obtain significant benefit. In the remainder of this section, we analytically examine the flexibility measure $V(f)$ for each of the 5 flexibility strategies. This examination is based on the analysis of different stylized models as reported in Tang and Tomlin (2007)

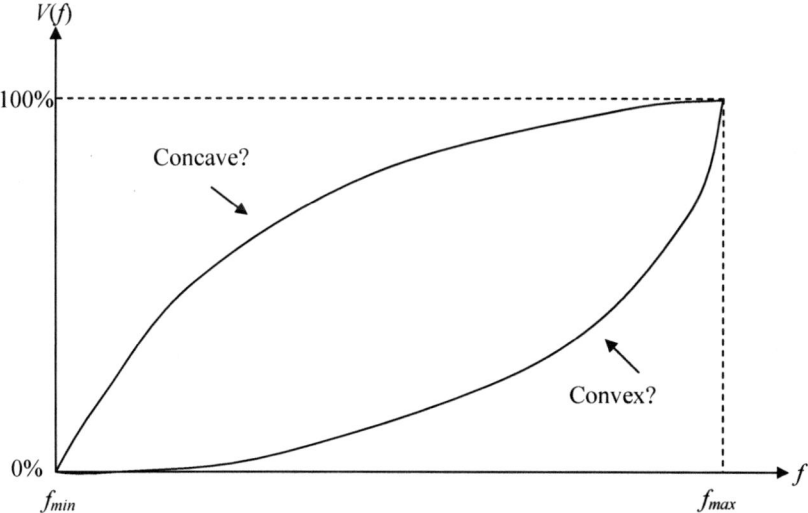

Fig. 10.1 The relative benefit of flexibility

As we will see in the following pages, $V(f)$ is in many instances concave. In other words, firms can significantly reduce their supply chain risk using only limited flexibility. This is of great practical importance. The higher the degree of flexibility required the more costly the investment and, therefore, the more likely it is that a precise ROI analysis will be required to justify the investment. The fact that a relatively low degree of flexibility is often sufficient may enable managers to justify flexibility investments more readily, even if precise estimates of costs, impacts, and likelihoods are not available.

10.3.1 Supply-Cost Risk: The Benefit of Flexibility via Multiple Suppliers

Firms faced with uncertain supplier costs may choose to maintain an active set of suppliers so that, at any given time, it can place orders with those suppliers who currently offer the lowest cost. Consider the following stylized example in which a manufacturer has an unlimited number of pre-qualified suppliers with uncertain supply costs. Let the unit cost of supplier $j = 1,2,..., \infty$, denoted by C_j, be \$5, \$10, or \$15 with equal probability 1/3. To satisfy the demand in each period, we assume that the manufacturer always orders from the supplier who offers the lowest unit cost. In this case, the flexibility level f can be defined in terms of the number of active suppliers and the performance metric $P(f)$ can be defined as the expected unit cost associated with sourcing from f suppliers.

Suppose that the manufacturer is committed to sourcing from one exclusive supplier, e.g., it chooses an inflexible sourcing strategy. Then the expected unit cost, denoted by $P(f_{min}) = P(1)$, is given as: $P(1) = 1/3 (5 + 10 + 15) = \$10$. Next, consider the case in which the manufacturer can source from 2 suppliers, and so it has some flexibility. Because the manufacturer selects the supplier with a lower unit cost, the corresponding expected unit cost associated with sourcing from 2 potential suppliers, denoted by $P(2)$, can be expressed as $P(2) = E(Min\{C_1, C_2\})$, e.g., the expected value of the minimum of the two supplier costs. By enumerating all possible scenarios, it can be shown that $P(2) = \$7.8$. Similarly, one can show that $P(3) = \$6.6$, $P(4) = \$5.9$, $P(5) = \$5.6$, and so on. Finally, if the manufacturer sources from $f_{max} = \infty$ suppliers, then $P(f_{max}) = \$5$. In this case, it is easy to check that $V(2) = 44\%$, $V(3) = 68\%$, $V(4) = 82\%$, and $V(5) = 88\%$. Therefore, 44% of the benefit associated with an infinite number of suppliers can be achieved when a firm orders from just 2 suppliers. As shown in Tang and Tomlin (2007), the underlying finding illustrated by this example, that $V(f)$ is concave, holds true regardless of the specific costs and probabilities used. Therefore, limited flexibility is very effective at managing supply-cost risk.

10.3.2 Supply-Commitment Risk: The Benefit of Flexibility via a Flexible Supply Contract

In many supply chains, contracts with suppliers limit the ability of a manufacturer to alter a previously placed order. A contract might specify an upper bound on the percentage by which the manufacturer can revise, upwards or downwards, a previous order. In this case, the flexibility level f can be defined in terms of the percentage bound placed on quantity revisions. Consider the following stylized supply chain comprising a supplier, a manufacturer and a retailer.[1] The supply cost is $c per unit, the wholesale price is $p per unit, and all unsold units have $0 salvage value. We consider a 2-period model in which the retailer places his order only at the end of period 1. However, due to the supply lead time, the manufacturer needs to place an order with the supplier at the beginning of period 1, which occurs prior to the actual order to be placed by the retailer. The ordering process, as described here, is similar to the process as depicted in the Sport Obermeyer case prepared by Hammond and Raman (1995).

At the beginning of period 1, the manufacturer estimates that the retailer will order a quantity $D = a + \varepsilon$ at the end of period 1, where ε corresponds to the uncertain market condition to be realized in period 1. Based on the information about c, p, and D, the manufacturer orders x units at the beginning of period 1. Under a flexible supply contract, the manufacturer is allowed to modify this order from x units to y units after receiving the actual order from the retailer at the end of period 1. Consider the case when the retailer orders $d = a + e$ at the end of period 1, where e is the realized value of ε. Under the f-flexible contract, the modified order y must satisfy: $x(1-f) < y \leq x(1+f)$, where $f \geq 0$ represents the allowable percentage adjustment as specified in the contract. Let $P(f)$ be the manufacturer's expected profit under the f-flexible contract based on the optimal initial order x^* and the optimal adjusted order y^*. When ε is uniformly distributed, Tang and Tomlin (2007) showed that the benefit associated with the f-flexible supply contract is increasing and concave in f. Therefore, significant benefits associated with the f-flexible contract can be obtained when f is relatively small, say 5%. Again, we find that limited flexibility is very effective at managing supply-cost risk.

10.3.3 Process Risk: The Benefit of Flexibility via Flexible Manufacturing Processes

Process risks, resulting from yield or quality issues for example, cause fluctuations in the effective capacity of plants. Firms that produce multiple products can mitigate this capacity variability by building plants that have the ability to produce more than one product. Consider the following stylized example in which a firm sells 4 different products (1, 2, 3, and 4), each with a demand of $D_1 = D_2 = D_3 = D_4 = 100$ units. The firm owns 4 different plants; the capacity of each plant j = 1, 2, 3, 4, denoted by C_j, is equal to 50, 100, or 150 units with equal probability 1/3. In this setting, there is no redundant capacity in the sense that the average total aggregate capacity of all 4 plants is 400 units, which is equal to the total aggregate demand of all 4 products. To illustrate the benefit of process

---

[1] We note that Tsay and Lovejoy (1999) analyzed QF contracts previously. However, due to the multi-period nature of their model, an analytical characterization of the value of flexibility is not feasible.

flexibility, we focus on the following system configurations: a system is considered to possess "f-flexibility" when each plant has the capability of producing exactly f products and when the system is configured as illustrated in Fig. 10.2 which depicts the f-flexibility² system for f = 1, 2, 3, 4. When f = 1, each plant j is capable of producing product j only, where j = 1, 2, 3, 4. Hence, 1-flexibility system corresponds to the system with no flexibility, and so f_{min} = 1. The 4-flexibility system corresponds to a system with total flexibility, and so f_{max} = 4.

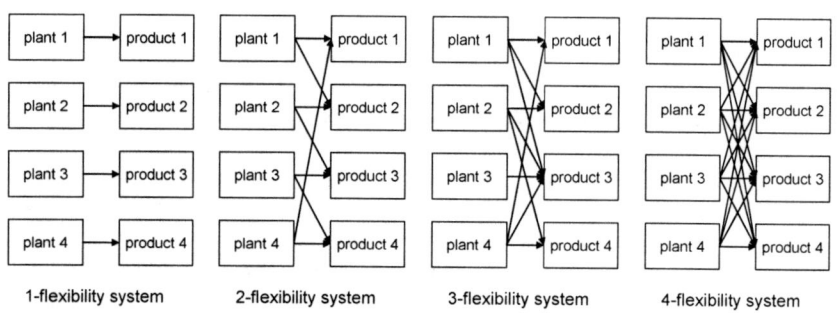

Fig. 10.2 f-flexibility manufacturing systems

Since each plant has 3 capacity scenarios, there are 81 possible plant-capacity scenarios for each of the f-flexibility manufacturing systems. By considering the probability of each of the 81 possible plant-capacity scenarios, Tang and Tomlin (2007) showed that the expected sales associated with the f-flexibility system, denoted by $P(f)$, is given as follows: $P(1)$ = 333.33, $P(2)$ =367.9, $P(3)$ = 367.9, $P(4)$ = 367.9. By noting that $V(2)$ = 100%, we can conclude that significant benefits associated with process flexibility can be obtained with limited flexibility, e.g., the 2-flexibility system. (We refer the reader to Tang and Tomlin (2007) for a more general treatment of managing process risks with limited process flexibility.) Therefore, to reduce process risks, it is sufficient to deploy a manufacturing system with limited flexibility.

10.3.4 Demand Risk: The Power of Flexibility via Postponement

Postponement, or delayed differentiation, is an increasingly popular strategy for managing demand risk. By postponing the point of differentiation, a firm has

---

² To simplify our exposition, we restrict attention to this particular type of system configurations, which Jordan and Graves have referred to as chain configurations. We note that our usage of f for the level of flexibility corresponds to the parameter h in Jordan and Graves (1995). The reader is referred to Jordan and Graves (1995) for an in-depth analysis of a model in which different plants are capable of producing different numbers of products.

increased flexibility in matching its production mix to the demand mix. It can, therefore, reduce the amount of inventory required to provide a high customer service. The following description is a simplified version of the postponement model presented in Lee and Whang (1998). A firm produces 2 end-products by using a 2-stage production process. The firm adopts an "f-postponement" strategy when it takes f time periods to produce a generic semi-finished product at the first stage and $(T-f)$ time periods to customize these generic products into two different end-products. Figure 10.3 depicts a process under the f-postponement strategy. Since the generic product is flexible, the production process is more flexible as f increases. We note that $f_{min} = 0$ and $f_{max} = T$. For any f-postponement strategy, define the performance metric $P(f)$ to be the optimal average inventory level of the two end-products.

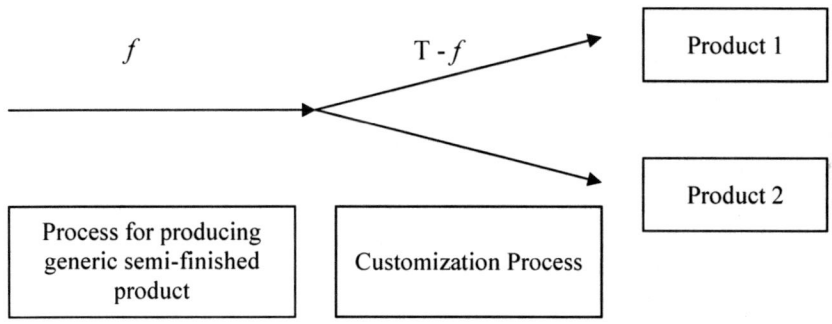

Fig. 10.3 An illustration of the f-postponement strategy

Let $D_i(t)$ denote the demand for product i to be realized t periods in the future, where i = 1, 2. Let the demand follow a Random Walk (RW) model; e.g., $D_i(t) = \mu_i + \varepsilon_{i1} + \varepsilon_{i2} + \ldots + \varepsilon_{i,t-1} + \varepsilon_{it}$, where i = 1, 2, $t = 1, \ldots, T$, and the ε_{it} are independently and identically distributed normal random variables with mean 0 and standard deviation σ. Lee and Whang (1998) proved the following result: $V(f)$ is increasing and concave in f. Therefore, significant benefits associated with postponement can be obtained even if the point of differentiation is placed at an early stage of the production process, e.g., when f is small. Again, we find that limited flexibility delivers much of the benefits.

10.3.5 Demand Risk: The Power of Flexibility via Responsive Pricing

To this point, we have focused on operational flexibilities such as maintaining multiple suppliers or configuring plants to be capable of processing multiple products. We now turn to a marketing flexibility, namely the flexibility of delaying the time at which prices must be set. While van Mieghem and Dada (1999) study the benefit of complete price postponement, we want to investigate the value of limited flexibility, or in other words, partial price postponement.

Consider a stylized model in which a manufacturer sells 2 substitutable products (1 and 2) through a retailer over a selling season that starts after period T. At the beginning of period 1, the manufacturer estimates that the total demand for product i over the selling season is given by $D_i(1)$, where $D_i(1) = a_i + S_{i1} + S_{i2} + \ldots + S_{i,T-1} + S_{iT} - b\ p_i + \delta(p_j - p_i)$, i, j = 1, 2, j ≠ i. In this case, a_i represents the primary demand of product i, S_{it} represents the "shock" to the primary demand of product i occurring in period t, b measures price sensitivity, and δ measures product substitutability. In our model, we assume that the shocks S_{it} follows an auto-regressive process of order one, e.g., AR(1), so that $S_{it} = \rho_i S_{i,t-1} + \varepsilon_{it}$ for i = 1,2, and t = 1, 2, ..., T, where $0 \leq \rho_i < 1$, and ε_{i1}, ε_{i2}, ..., $\varepsilon_{i,T-1}$, $\varepsilon_{i,T}$ are independently and identically distributed normal random variables with mean 0 and standard deviation σ_i. Without loss of generality, we set $S_{i0} = 0$ for i = 1,2.

Consider the case in which the manufacturer and the retailer are both owned and controlled by a single firm. We also assume that the manufacturer has the capacity to meet the actual demand of each product over the selling season that starts after period T. In this integrated supply chain, the unit cost of each product i is given as c and we only need to decide on p_i; e.g., the retail price of each product i. (The wholesale price is determined internally between the manufacturer and the retailer.) Suppose that the firm has the flexibility to set and announce the retail price of each product i at the end of period f, where f = 1, ..., T. Once the retail price is announced, we assume that the firm is committed to sell each product at p_i during the selling season that starts after period T. This implies that the firm must announce the actual retail price no later than the end of period T.

Clearly, the firm would benefit from the flexibility of postponing its pricing decision because it would allow the firm to gain more accurate information about the market demand before setting the actual retail price. To formalize this thinking, we say that the firm employs an "f-timing" strategy when the actual retail price is determined at the end of period f. Thus, timing flexibility increases as f increases, and $f_{min} = 1$ and $f_{max} = T$. Define the performance index $P(f)$ as the optimal expected profit associated with an f-timing strategy. Tang and Tomlin (2007) proved that $P(f)$ is concave increasing for $0 < \rho_i < 1$ and linear increasing for $\rho_i = 0$. Since $P(f_{min}) = P(1)$ and $P(f_{max}) = P(T)$ are independent of f, one can conclude that $V(f)$ is also increasing and concave in f for $0 < \rho_i < 1$. As with the operational-flexibility cases, we find that many of the benefits associated with price postponement can be obtained for low values of flexibility, e.g., when f is small.

10.4 Concluding Remarks

Throughout this chapter, we have focused our attention on "defensive" flexibility strategies, that is, strategies that mitigate the negative impact of undesirable events. This focus should not be allowed to obscure the fact that flexibility can also be used as a "proactive" mechanism that enables firms to compete more effectively in the marketplace. Let us consider three successful examples of firms taking advantage of flexibility to compete.

- Seven-Eleven Japan, in order to reduce the process risks arising from variable traffic conditions in Japan, has implemented a flexible delivery strategy that utilizes trucks, motorcycles, boats, and even helicopters to ship their products from various distribution centers to their stores throughout Japan. This flexible delivery strategy has allowed Seven-Eleven Japan to ensure a Just-In-Time delivery of fresh products to its stores. This capability has helped Seven-Eleven Japan to become the most profitable convenience store in Japan. In addition, this multi-mode delivery system earned the respect of many Japanese earthquake victims in Kobe when Seven-Eleven Japan was the first company to deliver 64,000 rice balls in Kobe within 6 h by using 7 helicopters and 125 motorcycles. The reader is referred to Lee (2004) for details.
- Honda, in response to Yamaha's development, in the late 1980s, of low cost and high quality motorcycles, improved its process flexibility so that it could introduce new models of motorcycles more frequently. This flexibility strategy allowed Honda to gain significant market share from Yamaha. (The reader is referred to Stalk and Hout [1990] for details.) More recently, Zara, the Spanish fashion company, has earned its reputation as the "Fast Fashion" company. Specifically, Zara used a flexible strategy to speed up both the design and the production process so that it can change its complete fashion collection within 2–3 weeks. Consequently, Zara has become Europe's most profitable fashion company with double digit growth rate annually for the last 10 years. The reader is referred to Ferdows et al. (2004) for details.
- The airline industry was revolutionized in the 1990s by the implementation of flexible pricing strategies via dynamic pricing. When selling limited seats on an airplane with uncertain demand, airlines can adjust their ticket price dynamically so as to meet uncertain demand with limited supply. Cook (1998) reported that dynamic pricing has generated "almost $1 billion of incremental annual revenue" at American Airlines. In the context of e-tailing, dynamic pricing can increase online traffic. For instance, Lands' End's Overstock site (http://www.landsend.com) generated additional traffic after they introduced the "on the counter" event, whereby, every Saturday, Lands' End puts a new group of products for sale at a reduced price. The price of each item is then reduced by 25% if it is not sold by Monday, 50% by Wednesday, and 75% by Friday. Pre-announcing the markdown price schedule encourages many online shoppers to monitor the sales of these items so as to time their purchase accordingly. As online traffic increases, the total sales can increase as well.

In this chapter, we have examined the benefits of different flexibility strategies in the context of supply chain risk management. By considering 5 different flexibility strategies, and reviewing the stylized models presented in Tang and Tomlin (2007), we have shown that a firm does not need to invest in a high degree of flexibility to mitigate supply, process and demand risks; most of the benefits are obtained at low levels of flexibility.

In addition to these 5 different flexibility strategies, there are other flexibility strategies worth considering. One other common flexibility strategy is "inventory pooling." When Toyota introduced its first hybrid car, the Prius, in the U.S. market in 2000, demand was highly uncertain because it was very unclear if U.S. consumers would embrace the hybrid concept. To encourage Toyota dealers to sell Prius without incurring the risk of overstocking, Toyota decided to own and stock all Prius at a central location and to take dealers' orders via the internet (Lee et al. 2005). The success of the Prius inventory pooling concept was key to convincing the dealers to share Scion's inventories among dealers when Scion was launched in the U.S. in 2003. Along with inventory pooling, creating a flexible workforce via cross-trained teams is another promising flexibility strategy. So et al. (2003) reported the benefits (in terms of productivity) associated with flexible cross-trained teams at the Federal Reserve Bank in Los Angeles.

We hope that the findings presented in this chapter and the arguments presented in Tang (2006a) provide a convincing argument for implementing flexibility strategies. In many real-life settings, exact cost-benefit analyses of flexibility investments are not feasible due to limitations of data availability. However, the robustness of the insight that only limited flexibility is needed to mitigate risk should encourage firms to build flexibility into their supply chains. Of course, when implementing a particular strategy in a particular context, a firm needs to establish a structured evaluation process that includes risk identification, risk assessment, decision analysis, mitigation and contingency planning. The reader is referred to Zsidisin et al. (2001) and Zsidisin et al. (2004) for a discussion of structured approaches.

References

AMR (2006) AMR Research Report on Managing Supply Chain Risk. AMR Research, Inc.
Billington C (2002) HP Cuts Risk with Portfolio Approach. Purchasing Magazine Online, February 21.
Brown D (2004) How U.S. Got Down to Two Makers of Flu Vaccine. Washington Post, October 16.
Boyabatli O, Tokay B (2004) Operational Hedging: A Review with Discussion. INSEAD working paper.
Chandler C, Fung A (2006) Not Exactly Counterfeit. Fortune, May 1, pp. 108–116.
Chopra S, Sodhi M (2004) Avoiding Supply Chain Breakdown. Sloan Management Review, vol. 46, pp. 53–62.
Christopher M, Lee H (2004) Mitigating Supply Chain Risk Through Improved Confidence. Cranfield School of Management working paper.
Feitzinger E, Lee H (1997) Mass Customization at Hewlett Packard: The Power of Postponement. Harvard Business Review, vol. 75, pp. 116–121.
Ferdows, K, Lewis M, Machuca J (2004) Rapid Fire Fulfillment. Harvard Business Review, vol. 82, pp. 104–117.

Gumbel P (2006) Trying to Untangle Wires. Time Europe Magazine, October, (http://www.time.com/time/europe/magazine/printout/0,13155,1543879,00.html).

Hammond J, Raman A (1995) Sport Obermeyer, Ltd. Harvard Business School Case # 5-696-012.

Hendricks K, Singhal V (2005) Association Between Supply Chain Glitches and Operating Performance. Management Science, vol. 51, pp. 695–711.

Huchzermeier A, Cohen M (1996) Valuing Operational Flexibility Under Exchange Rate Risk. Operations Research, vol. 44, pp. 100–113.

Jordan W, Graves SC (1995) Principles on the Benefits of Manufacturing Process Flexibility. Management Science, vol. 41, pp. 577–594.

Kopczak L, Lee H (1993) Hewlett-Packard: DeskJet Printer Supply Chain. Stanford Graduate School of Business Case.

Kouvelis P, Dong L, Su P (2006) Operational Hedging Strategies and Competitive Exposure to Exchange Rates. Olin School of Business, (Washington University) working paper.

Krazit T (2004) Trouble in East Fishkill? IBM chip group struggles. InfoWorld.com, April 21.

Lee H (2004) The Triple-A Supply Chain. Harvard Business Review, vol. 19, pp. 102–112.

Lee H, Tang CS (1997) Modeling the Costs and Benefits of Delayed Product Differentiation. Management Science, vol. 43, pp. 40–53.

Lee H, Whang S (1998) Value of Postponement. In Produce Variety Management: Research Advances, eds. T.H. Ho and C.S. Tang, Kluwer Publishers.

Lee H, Wolfe M (2003) Supply Chain Security Without Tears. Supply Chain Management Review, January/February issue, pp.12–20.

Lee H, Peleg B, Whang S (2005) Toyota: Demand Chain Management. Stanford Business School Case #GS42.

Martha J, Subbakrishna S (2002) Targeting a Just-in-Case Supply Chain for the Inevitable Next Disaster. Supply Chain Management Review, September issue.

Reuters (2006) Ford Shuts Down Truck Plants for Part Problem. April 27. Reuters.com.

Rice B, Caniato F (2004) Supply Chain Response to Terrorism: Creating Resilient and Secure Supply Chains. Supply Chain Response to Terrorism Project Interim Report, MIT Center for Transportation and Logistics, MIT, Massachusetts.

Sheffi Y (2001) Supply Chain Management under the Threat of International Terrorism. International Journal of Logistics Management, vol. 12, pp. 1–11.

Sheffi Y (2005) Building a Resilient Supply Chain. Harvard Business Review, vol. 10, pp. 3–5.

So K, Tang CS, Zavala R (2003) Model for Improving Team Productivity at the Federal Reserve Bank. Interfaces, vol. 33, pp. 25–36.

St. George A (1998) Li and Fung: Beyond 'Filling in the Mosaic'. Harvard Business School Case # 9-398-092.

Stalk G, Hout T (1990) Competing Against Time. The Free Press, New York.

Tang CS (1999) Supplier Relationship Map. International Journal of Logistics: Research and Applications, vol. 2, pp. 39–56.

Tang CS (2006a) Robust Strategies for Mitigating Supply Chain Disruptions. International Journal of Logistics: Research and Applications, vol. 9, pp. 33–45, 2006a.

Tang CS (2006b) Perspectives in Supply Chain Risk Management. International Journal of Production Economics, vol. 103, pp. 451–488.

Tang CS, Tomlin B (2007) The Power of Flexibility for Mitigating Supply Chain Risks. Anderson School (UCLA) working paper.

Tsay AA, Lovejoy WS (1999) Quantity Flexible Contracts and Supply Chain Performance. Manufacturing & Service Operations Management, vol. 1, pp.89–111.

van Mieghem J, Dada M (1999) Price Versus Production Postponement: Capacity and Competition. Management Science, vol. 45, pp. 1631–1649.

Zsidisin G, Panelli A, Upton R (2001) Purchasing Organization Involvement in Risk Assessments, Contingency Plans, and Risk Management: An Exploratory Study. Supply Chain Management: An International Journal, pp. 187–197.

Zsidisin G, Ellram L, Carter J, Cavinato J (2004) An Analysis of Supply Risk Assessment Techniques. International Journal of Physical Distribution and Logistics Management, vol. 34, pp. 397–413.

SECTION THREE - SUPPLY CHAIN RISK MANAGEMENT

Chapter 11: Enterprise and Supply Risk Management

Michael Henke

Chair for Financial Supply Management at the European Business School (EBS) International University Schloss Reichartshausen, Oestrich-Winkel; Research Director Financial Supply Management at the Supply Management Institute SMI™; 65201 Wiesbaden, Germany

11.1 Introduction: Enterprise Risk Management for Supply Management

The Committee of Sponsoring Organizations of the Treadway Commission (COSO) introduced the "Enterprise Risk Management (ERM) Framework" in 2004. COSO is an independent private sector initiative that is dedicated to improving the quality of financial reporting through business ethics, effective internal controls, and corporate governance. Since the publishing of the COSO report in 1992, its recommendations have become a guideline for the evaluation of internal control systems. During the past decade several companies such as Worldcom, Enron, and Parmalat have experienced significant financial breakdowns. In response, COSO codified the close relationship between monitoring and risk management and further developed the COSO report with the ERM framework. The reliability of reporting was therefore expanded from merely financial reporting to all internal and external company reports in order to improve monitoring. "Business reporting" thus replaces "financial reporting" to better supply shareholders and stakeholders with the information they need. This strategic orientation has been added to the framework as a target category (first dimension of the ERM model, please see Fig. 11.1). Furthermore, the framework now includes the necessary components for risk management (second dimension of the ERM model). As a result, the ERM model brings together the topics of both monitoring and risk management systems.

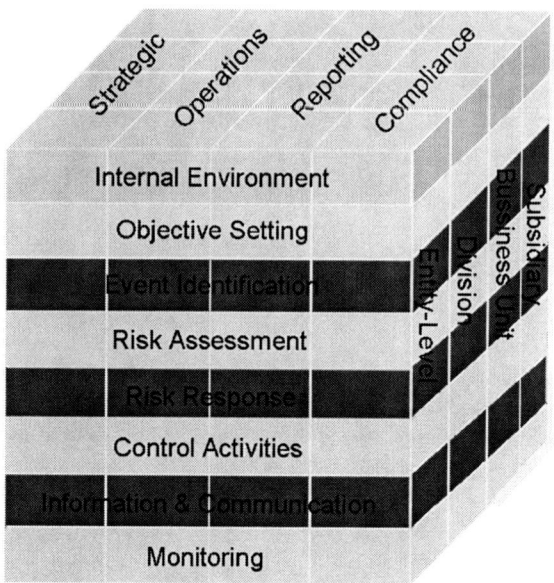

Fig. 11.1 Enterprise Risk Management (ERM) – the COSO model

The third dimension contains all organizational entities of the company; the first and second dimensions are to be applied to the third dimension.

Enterprise risk management is a process, effected by an entity's board of directors, management and other personnel, applied in a strategy setting and across the enterprise, designed to identify potential events that may affect the entity, and manage risk to be within its risk appetite, to provide reasonable assurance regarding the achievement of entity objectives. (The Committee of Sponsoring Organizations of the Treadway Commission 2004, p. 16).

ERM was deliberately given a broad definition for two reasons. The first was to establish a consistent awareness for monitoring and risk management throughout the entire company. Second, this general model on monitoring and risk management was developed for all companies regardless of size or structure (Eichler and Bungartz 2004, pp. 108 and 113).

Based on its universal application, ERM opens up new possibilities to strive toward optimizing internal control systems, risk management systems and thus the entire corporate governance systems (Eichler and Bungartz 2004, pp. 108 and 112–113). Its importance is further exemplified by the "Sarbanes-Oxley Act of 2002" (SOX), where "COSO's framework may become an important tool for implementing the directives set forth in the Sarbanes-Oxley Act of 2002" (Hermanson 2003, p. 3).

From an organizational perspective, risk management is often structured at specific organizational levels (Institut der Wirtschaftsprüfer in Deutschland e.V.

(IDW) 2004, p. 51). Supply management inherently has links to several cross-divisional functions and organizational areas. The application of target categories and components of the ERM model for Supply Management takes many organizational entities into account. Therefore Supply Management seems well-oriented for adopting ERM as a holistic risk management tool in the form of Supply Risk Management.

The purpose of this chapter is to introduce and describe the development of an ERM-compliant Supply Risk Management approach and how the respective processes can help companies better manage supply (chain) risks. This is also intended to help answer the question if ERM is a good point to start the further development of an integrated and process-oriented Supply Risk Management approach.

11.2 ERM Starting Points

Improvement is particularly needed in recognising Supply Risk Management as a senior level activity in Supply Management itself. The tools available to Supply Management professionals must be adapted to current and future challenges of the business world and further developed to comprehensively manage supply risk (Zsidisin et al. 2005).

The following practical example will demonstrate the necessity for the development of new solutions for Supply Risk Management. Several years ago DaimlerChrysler had to stop its production for diesel vehicles for 18 days in the Sindelfingen plant due to a defective diesel fuel injector that had been delivered by its 1st-tier automotive supplier Bosch. As a result DaimlerChrysler experienced a significant sales decline because vehicles were not being delivered to customers. Daimler claimed that Bosch, as the supplier for diesel fuel injectors, was responsible for the losses associated with stopping production. Bosch stated that the company did not commit any mistakes in the production process and made its American supplier Federal Mogul responsible for delivering sockets with defective coatings for the fuel injector. Federal Mogul also denied responsibility because it identified granulates for the coating of the sockets from its supplier DuPont as the source of the defective fuel injectors (Büschemann 2005, p. 19).

This short example shows the importance of inbound supply dependency and the related risks for companies. No matter which company in the value chain is identified as the source of failure, every single participant is obliged to manage risk to avoid financial losses or a detrimental effect to its supply chain constituents.

Companies should develop and implement a comprehensive Supply Risk Management system that can be integrated into corporate or Enterprise Risk Management (ERM). The process model (Matzenbacher 2003, p. 109) for Supply Risk Management (Henke and Jahns 2005, pp. 229–230) represents an important step in the right direction towards proactive and regular handling of supply risks (please see the centre of Fig. 11.2).

A process-oriented approach deals primarily with the combination of individual tasks to a logical chain of activities (Labbé and Langen 2004, p. 720). Therefore, the process model of Supply Risk Management needs to proactively influence the (potential) supply risks after formulating and reworking the *supply risk strategy* in the next process phases:

Identifying, analyzing and assessing supply risks do not mean solely focusing on the functional area of procurement and purchasing. This approach would lead to failure (Matzenbacher 2003) since today modern and strategic Supply Management stretches well beyond the more operative areas of procurement and purchasing. Risk management is moreover one of the support processes necessary for the integral supply process. Therefore the procurement and purchasing processes must be analyzed in connection with the other sub-processes of Supply Management such as procurement market research. A process-oriented Supply Risk Management approach can help organizations coalesce interdependencies between individual processes.

Supply risk regulation itself has the following challenge: the optional regulation measures are often only described in unspecific and rudimentary terms in both the management literature and corporate practice. A solution to this problem and thus a general improvement of risk regulation is inevitably tied to a specification of the "four principle actions" (prevention, reduction, transfer and compensation) for addressing supply risk. Still the question remains, if and also how, ERM can substantially reduce the shortcomings in specification and operationalization of monitoring and risk management for Supply Management.

ERM may offer a logical approach for the further development of Supply Risk Management in recourse to the known and systematic method of the COSO model, since the process model for Supply Risk Management already includes specialized central ERM objectives and ERM components. The integrated Supply Risk Management process, which should be run by the Supply Managers as operational risk owners, incorporates both the ERM components "internal environment" and "objective setting" in the process phase of formulating and revising the supply risk strategy. The subsequent business process phases are identical, almost equivalent in terminology, to the ERM components "event identification," "risk assessment" and "risk response."[2] Furthermore, three ERM target categories can be accounted for in the process model: "strategic" (supply risk strategy), "operations" (integral supply process) and "compliance" (e.g., SOX; The Committee of Sponsoring Organizations of the Treadway Commission 2004, pp. 5–6).

In order to account for the ERM objectives and ERM components for Supply Management not yet included in the process model, the central process phases of Supply Risk Management must be incorporated by other functions and institutions. This includes internal monitoring and external monitoring, among others.

11.3 ERM Monitoring and Managing of Supply Risks

Supply Controlling involves the relocation and specification of management accounting for Supply Management. Along with Supply Management, Supply Controlling is constantly becoming a more independent field of management-oriented business administration (Jahns 2005, pp. 349–358). Supply Controlling can functionally be defined as a goal-oriented coordination of planning, information supply, control and regulation in Supply Management (Lück 1998, p. 1). Therefore, by definition, the Supply Controlling function already includes two ERM components that are not reflected in the process model of Supply Risk Management. These components are "control activities" and "information and communication," which in turn are both connected to the ERM target category "reporting" (The Committee of Sponsoring Organizations of the Treadway Commission 2004, pp. 5–6). If a process can be dependent on "monitoring" and thus needs to be regulated, then the assignment of this same ERM component to risk-oriented Supply Controlling activities is equally justified. This assignment is supported by the fact (Eichler and Bungartz 2004, p. 110) that monitoring supply risk is conducted in the framework of normal Supply Management tasks.

Monitoring supply risks is a multi-faceted information and decision process that stretches beyond Supply (Risk-) Controlling to internal and external monitoring. From the agency theory perspective controlling as well as internal and external monitoring can be considered "governance mechanisms." They especially fulfil information and control functions that seem fitting as a solution for agency problems between the principal and its agents (Ebers and Gotsch 2002, pp. 214–215). Extensive cooperation along with comprehensive information exchange between Supply Controlling and internal and external auditing are essential based on common ground and mutual interest in securing an efficient and goal-oriented internal and external monitoring of supply risks.

The approach to comprehensive handling and monitoring of supply risks in terms of ERM is as follows:

- Methodical support of risk management in Supply Management and Supply Controlling through Supply Risk-Controlling (for risk-controlling please see Lück and Henke 2003, p. 291).
- Evaluation and improvement of effectiveness of Supply Controlling and Risk Management through internal auditing (e.g., The Institute of Internal Auditors: Standards for the Professional Practice of Internal Auditing, 2004, pp. 12–13).
- Appraisal of applicability of form and results of Supply Controlling and Risk Management by an external auditor, a certified public accountant (see generally in Germany § 317 Abs. 4 HGB).

The relationship between Supply Controlling and Supply Risk-Controlling can best be illustrated by the "House of Supply Controlling and Risk Management." This contains the process model of Supply Risk Management at its core and can be supplemented by the more institutional Supply Controlling as well as the more functional Supply Risk-Controlling. The "House of Supply Controlling and Risk Management" uses the ERM framework as a basis to show the action steps for supply managers and controllers. It also needs to be monitored internally and externally in terms of monitoring supply risks and their management. This concept represents an ERM-compliant option to specify monitoring and risk management for Supply Management.

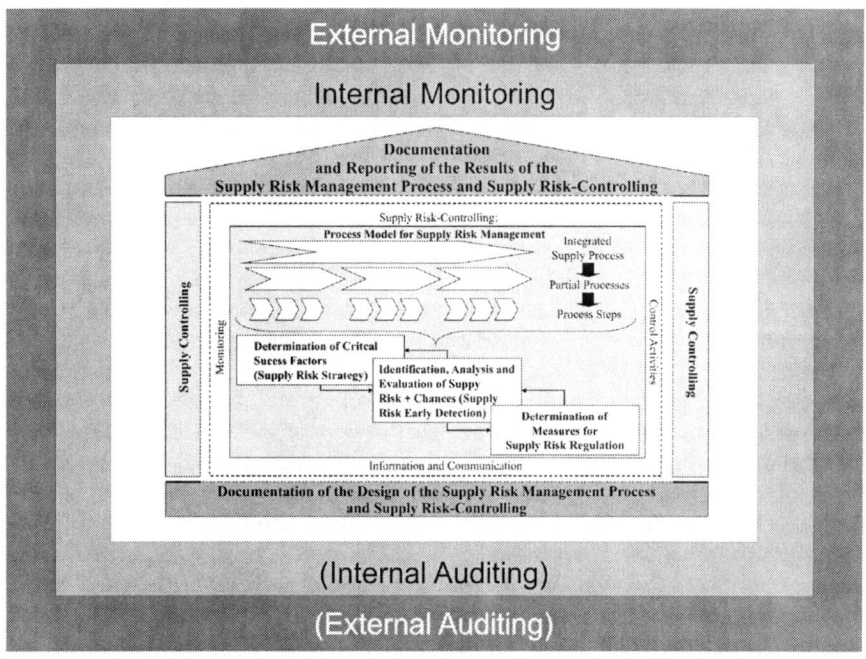

Fig. 11.2 Internal and external monitoring of the House of Supply Controlling and Risk Management

The House of Supply Controlling and Risk Management can serve as an action-oriented basis for the risk-oriented analysis of Supply Management by the certified public accountant. The accountant can thus determine his/her desired values for Supply Controlling and Risk Management since the law generally does not issue any regulations for the desired values for internal control and risk management systems. In the framework of these audit-oriented consulting services in terms of supply risks, the employment status of certified public accountants can at the same time lead to internally monitored "House of Supply Controlling and Risk Management" becoming a "state-of-the-art" specialization of ERM in the audited companies.

11.4 Advances in Operationalizing Tasks for Supply Controlling and Risk Management

It is important to note that the ERM framework recognizes that each of the management components includes decisions at the strategic and operational levels (Institut der Wirtschaftsprüfer in Deutschland e.V. (IDW) 2004, p. 59). Two tasks need to be worked on in the Supply Risk Management process and its controlling and monitoring. Firstly, strategic considerations of the ERM framework need to be specified and secondly, decisions need to be operationalized for their implementation in corporate practice. To address these issues an empirical investigation of monitoring and managing risk based on the ERM model was conducted. This study amongst supply managers and controllers as well as internal and external auditors examined the influence of individual occupation groups on the relevance of Supply Risk Management. The study results lead to conclusions on future action steps to operationalize an integrated Supply Risk Management approach in terms of the ERM.

First, Supply Risk Management clusters were identified and characterized based partially on assessments of current and future management of supply risks in participating companies ("Advanced," "Awareness" and "Beginners"). The selected classification of all responding companies was then transferred to the individuals and occupation groups polled. Thus the survey results were operationalized (i.e., two sub-groups: 1. supply managers and controllers, 2. internal and external auditors). This way both the differentiation of the survey criteria and the various target groups can be taken into account in respect to the survey goal.

The most important empirical research results can be summarized as follows:

- The majority of supply managers and controllers consider Supply Controlling & Risk Management of great importance. Nonetheless, there still was a noticeable implementation gap in 2005. This gap should be significantly reduced by the year 2010.
- The current and future tasks to reduce this implementation gap for Supply Controlling & Risk Management still need to be consolidated into an integrated approach by 2010. Supply Risk Management that supply managers and controllers can actually use needs to be conceptualized in such a way that it can satisfy the requirements of being a top management job, having considerable influence on the value of the company.
- The advances in operationalizing tasks for Supply Controlling and Risk Management emphasize the need to professionalize all Supply Risk Management clusters.
- According to the majority of internal and external auditors, Supply Controlling and Risk Management are currently not of great importance. But in order to help reduce the implementation gap for Supply Controlling and Risk Management over the next years (until 2010), internal and external auditors need not only to expand their audit services, but also their consulting services on Supply Risk Management. Then, the process-independent services of

internal and external auditors could actually keep pace with the changes in Supply Management process and in process-dependent controlling.
- The current and future tasks for monitoring Supply Controlling and Risk Management suggest that internal and external auditors will attribute Supply Risk Management much greater significance in the future. Therefore, increased internal and external monitoring is inevitably associated with an integrated approach.

The advances in operationalizing tasks for the monitoring of Supply Controlling and Risk Management emphasize that nearly every other internal and external auditor still needs to be convinced of the economic need for Supply Controlling and Risk Management as well as its internal and external monitoring.

The empirical survey results comparing the two sub-groups, taking into account all requests of the individuals and employee groups from the Supply Risk Management clusters, illustrate the need to establish and respect Supply Controlling & Risk Management as a field of activity with increasing importance for supply managers and controllers in the next years. The (process-independent) internal and external monitoring of Supply Controlling and Risk Management must occur at the same time frame and match the increasingly important field of activity of internal auditors and certified public accountants.

11.5 Conclusions

Universities and corporations should work together toward developing and analyzing theoretical and empirical approaches for managing supply risks. The ERM-compliant specification and operationalization of monitoring and risk management are intended to serve as an integrative approach for designing and implementing Supply Risk Management for firms today and in the future.

However, an ERM-compliant Supply Risk Management concept needs to be tested first. Then, theory-based approaches for implementing ERM can be deduced from corporate practice. If and how these fields of activity of Supply Risk Management within the ERM can be transferred to other corporate areas remains to be seen and can only then be ultimately resolved.

In summary, it is important to remember that ERM provides an action-oriented basis for the specialization and operationalization of monitoring and risk management. In the case of Supply Management this was clarified in the form of an ERM-compliant Supply Risk Management process.

The performed analysis contributes to closing the research gap in Supply Risk Management. This topic will continue to develop in corporate governance research and corporate governance application. Proactive management of supply risks will become a competitive advantage. If the success of Supply Controlling and Risk Management and its internal and external monitoring can be quantified by academe in the future, then a sustained ripple effect can be expected from this successful example. In sum, Enterprise Risk Management and Supply Risk

Management are neither separately nor in combination merely theoretical problems, but rather obligatory practical necessities.

References

Büschemann, K.-H., 2005. Eine Dieselpumpe bringt Autohersteller in Verlegenheit. Zulieferer Bosch auf Fehlersuche/Kosten bei BMW etwa 20 Millionen Euro/ "Individuelle Werkstatt-Termine" bei Mercedes. Süddeutsche Zeitung from February 4, 2005, 19.

Ebers, M., Gotsch, W., 2002. 7. Instiutionenökonomische Theorien der Organisation. In: Kieser, A. (Ed.), Organisationstheorien. 5th Edition. Stuttgart, 199–251.

Eichler, H., Bungartz, O., 2004. Enterprise Risk Management—aktuelle Entwicklungen im Bereich unternehmensinterner Risiko- und Überwachungssysteme. Zeitschrift Interne Revision 39 (3), 108–114.

Henke, M., Jahns, Chr., 2005. The Importance of Supplier Performance Measurement and Key Performance Indicators (KPIs) for the Systematic Management of Supply Risks. In: Proceedings. 16th Annual North American Research Symposium on Purchasing and Supply Management. Tempe, Arizona. March 17–19, 2005, 219–239.

Hermanson, H. M., 2003. COSO: More relevant than ever. Internal Auditing 18 (4), 3–6.

Institut der Wirtschaftsprüfer in Deutschland e.V. (IDW), 2004. Comment Letter on the Enterprise Risk Management Framework. IDW-Fachnachrichten, 49–60.

Jahns, C., 2005. Supply Controlling. Diskussionen über den Zustand einer Disziplin. Controlling 17 (6), 349–358.

Labbé, M., Langen, D., 2004. General Management: Eine prozessorientierte Perspektive. Der Betrieb, 720–723.

Lück, W., 1998. Controlling. Ergebnisse einer empirischen Untersuchung zum Controlling in der Brauwirtschaft. Krefeld.

Lück, W., Henke, M., 2003. Risiko-Controlling in Wachstumsunternehmen. In: Achleitner, A.-K., Bassen, A. (Ed.), Controlling von jungen Unternehmen. Stuttgart, 281–298.

Matzenbacher, J., 2003. Risikominimierung bei der Beschaffung von Maschinen und Neuanlagen. In: Biedermann, H. (Ed.), Risikominimierung im Anlagenmanagement—Risiken beim Planen, Errichten und Betreiben von Anlagen. 17. Instandhaltungs-Forum. Reihe Praxiswissen für Ingenieure—Instandhaltung. Köln, 105–122.

The Committee of Sponsoring Organizations of the Treadway Commission, 2004. Enterprise Risk Management—Integrated Framework. Jersey City.

The Institute of Internal Auditors, 2004. Standards for the Professional Practice of Internal Auditing. Performance Standard 2110—Risk Management. Altamonte Springs. Effective January 1, 2004.

Wildemann, H., 2006. Risikomanagement und Rating. München.

Zsidisin, G. A., Ragatz, G. L., Melnyk, S. A., 2005. The DARK SIDE of Supply Chain Management. Supply Chain Management Review 9 (2), 46–52.

Chapter 12: Pre-Contract Risk in International PFI Projects

Simon A. Burtonshaw-Gunn

Salford Business School, University of Salford, UK and, Risktec Solutions Limited, UK

12.1 Introduction

In life there are risks: in driving a car, crossing the road or playing various sports. So too in business although in many cases such risks are associated with financial consequences. Firms are subject to market volatility, and hence the ability to realistically provide expectations based upon a risk versus reward trade off. This chapter discusses the relationship between risk management and the use of Prime Contracting which is becoming one of the popular choices of procurement systems for large scale construction projects, as encouraged by two principle Government supported industry reviews reported in the publications 'Constructing the Team' (Latham 1994) and 'Rethinking Construction' (Egan 1998). These two review publications have supported an almost universal client requirement to achieve the benefits of increased value in infrastructure projects from the facilities management and ergonomic perspectives in both the private and public sectors. As a first step many UK clients have moved away from competitive tendering to favour the use of contracts with closer supply chain management through advances in both project and long-term strategic partnering arrangements between themselves and their facilities providers, constructors, designers and occasionally with some of their second-tier supply chain members. Although there are a number of examples showing an increased use in 'partnering', the actual extent of this commitment has been found from UK studies to vary in practice (Galliford 2000; Burtonshaw-Gunn 2001).

The choices available by which companies may trade are often described by the term 'contract strategy', which covers the legal contractual arrangements for the execution of a specific project. In addition to establishing the defined legal undertakings, the decisions taken during the development of the contract strategy also have an impact on the responsibilities of the client, the contractor and the supply chain and hence the co-ordination of these groups in the project. In parallel with the 'partnering' initiatives the UK has also witnessed Government support in the development of a contract strategy to construction procurement based on supply chain integration known as 'Prime Contracting'. This development has been aided by the United Kingdom Treasury and the National Audit Office as their recommended procurement method for central government construction projects through the Private Finance Initiative (PFI) and Public Private Partnership (PPP). Both these procurement initiatives bring together the infrastructure project management of design and construction with external private sector funding arrangements through bank loans, equity provision/exchange or investment through a number of innovative funding arrangements. Use of such PFI arrangements are growing in number around the world. Typically PFI projects occur when traditional client funding is either unavailable or when other priorities have stronger demands on a government department's limited finances. Irrespective of the project location, under such funding arrangements the private sector builds, finances and operates the public infrastructure facility such as a road, school, hospital, and rail link, and recovers the cost through service provision charges (sometimes referred to by the term 'power by the hour'). These types of PFI projects necessitate a very long-term commitment between the facility provider/manager, the user and the ultimate client who will not acquire ownership of the facility until the expiry of a significant contracted period when the user revenues have repaid the original project costs, covered the facilities operational costs and provided the facility providers with a profit on their investment.

The selection of the most appropriate contract strategy prior to commencement of the project plays an important role in defining the policies appropriate for the management of project risk and performance implications including the extent of control which may be transferred between the client, the contractor and the supply chain. In many ways the key decisions for the client organisation considering the selection of a contract strategy is based not on the physical attributes of the facility but upon the following four key parameters:

- The *project characteristics* which comprises the primary objectives of time, cost and functional performance together with identification of project constraints, secondary objectives and the identification of the project risks. The project management organisation can then determine who is best placed to manage these risks.
- The *project organisational system* which covers the size and scope of tender work packages, the roles of designers, contractors and their supply chains and sponsors and/or investors.
- *Contract selection* to determine the most appropriate type of contract with respect to payment choices and timescales together with the tendering process with respect to the choice of appointment of consultants, suppliers etc., whether pre-qualification and a competitive tendering process will be required and some early analysis to determine suitable conditions of contract.

- The fourth parameter is that of *risk management* which impacts on the type of contract that would be agreeable to the contractor and client organisation in considering the commercial undertakings. Indeed understanding the risks associated with the proposed project will have a strong influence on the selection of an appropriate and acceptable contract strategy to all parties within the supply chain.

Whilst the successful management of construction projects presents a challenge in any environment, the topic of this chapter is to present an understanding of the early risks which need to be assessed in such PFI infrastructure projects particularly those at international locations.

12.2 Construction Prime Contracting and the Management of Risk

The term 'Prime Contracting' has been defined and interpreted in several different ways by constructors, clients, and project financiers. The UK Government, who sponsored the 'Building Down Barriers' report defines Prime Contracting as *'a systematic approach to the procurement and management of buildings, based on the role of a Prime Contractor in integrating all the activities of a pre-assembled supply chain. The approach also draws together a number of best practices, including through-life costing, value engineering and risk management, to achieve significant efficiency of the completed building'*. (Defence Estates Organisation/ Tavistock Institute 1999, p. 1). In addition the UK's National Audit office publication 'PFI: Construction Performance' (2003, p. 21) suggests the following definition: *'A contract involving a main-supplier, the Prime Contractor, which has a well established supply chain of reliable suppliers of quality products to encourage increased quality and value for money resulting from an element of consistency and standardisation'*. A third definition which comes from the Chartered Institute of Building states that *'The prime contractor will be expected to have a well-established relationship with a supply chain of reliable suppliers. The prime contractor co-ordinates and manages throughout the design and construction period to provide a facility, which is fit for purpose and meets its predicted through life costs'*. (2002, p. 30).

From these three definitions a number of common attributes emerge. The first is focusing on having a single point of responsibility for the design, building, operation and, sometimes, maintenance of the facilities until the ultimate delivery of the project to the client. Another common feature is seen to be the role of the Prime Contractor in managing the whole supply chain. It is proposed that the areas of management which Prime Contracting is likely to cover are shown in Table 12.1. below.

Table 12.1 Proposed Prime Contracting Management Areas

Main areas for Prime Contractor	Detail areas to be addressed
Commercial Management	Risk management
	Project financing
	Through life costing
	Contractual Management
Design Management	Detail design
	Concept design
	Value Engineering
	Value Management
Project Management	Organisation structure
	Cost and Price issues
	Customer expectations
	Management of cost, time and quality
Construction Management	Planning
	Site Safety
	On site construction
Supply Chain Management	Competition
	Relationships
	Selection Process
Total Quality Management	Organisational procedures
	Continuous improvement
	Performance measurement
Human Resources Management	Attitudes
	Staffing
	Social/behaviour
Operations and Facilities Management	Operation
	Maintenance
	Transfer

Although there is no universal form of a documented 'Prime Contract', those contracts that are used to commercially formalise the arrangement have a number of similarities in their intent to provide opportunities for improved efficiencies and savings throughout the contract period. These savings are typically achieved through a higher level of co-ordination and integration of the activities of the supply chain members (sub-contractors) aimed to meet the overall specification, on-time delivery and efficient operation of the facility.

With an increasing reliance on the use of private sector funded development and long-term operation of major projects around the world, the Prime Contractor with single point of accountability will need to consider the through-life costs to a far greater extent than when merely providing a new facility for a client without the added responsibility of its day-to-day operation and longer-term maintenance costs. As such, just like the concept of Partnering, the adoption of Prime Contracting represents a major advance in terms of improving investment returns rather than past traditional procurement method based on competitive tenders and lowest cost. This will typically include early involvement of the total supply chain from facility design through to its eventual operation.

It is widely accepted within the construction industry that a number of client-customer interfaces and the imprecise allocation of risks have long-since hampered traditional procurement. The use of a Prime Contracting procurement strategy provides a degree of confidence with respect to managing project uncertainty together with a reduction in the amount of required contingency provision typically witnessed with traditional procurement methods. Furthermore the use of Prime Contracting encourages and cultivates the development of non-adversarial collaborative working witnessed by construction partnering. This fosters a more systematic approach to value engineering, value management and risk management by employing multi-party workshops as key activities in assessing options to achieve the required fitness for purpose and quality standard. Although these quantitative performance factors are widely recognised as being important, as in all types of collaboration, these are best achieved if the 'softer' factors of appropriate leadership style, facilitation, training and a shared commitment to continual improvement are also present.

The likely magnitude of Prime Contracting projects in both financial scale and complexity requires clear identification and assessment of the project risks throughout the collaborative supply chain. As such Prime Contracting is benefited by a supply chain approach to the management process of allocating project risks to where they can be most appropriately addressed. This could be the contractor, an appropriate member of the supply chain or indeed be retained by the client organisation.

Unlike the term Prime Contracting discussed earlier, there is a widespread agreement of what is meant by the term Risk Management within the construction industry. The Association of Project Management defines risk management as *'the process whereby responses to the risks are formulated, justified, planned, initiated, progressed, monitored, measured for success, reviewed, adjusted and (hopefully) closed'* (1997, p. 16). Similarly the British Standard Institute's 'Guide to Project Management' offers its definition as *'the process whereby decisions are made to accept a known or assessed risk and/or the implementation of actions to reduce the consequences or probability of occurrence'* (BS 6079:1996, p. 3). A final common definition comes from outside of the construction industry: The Royal Society. As the independent scientific academy of the UK charged with promoting excellence in science, it proposes that risk management is *'the process whereby decisions are made to accept a known or assessed risk and/or the implementation of actions to reduce the consequences or probability of occurrence'* (1992, p. 5).

Within the construction industry a typical risk process covers Risk Identification to capture all of the potential risks which could arise within the project. This is followed by Risk Classification, where the identified risks are grouped into internal risks (which reside within the company or organisation) and external risks (those factors that condition the environment or are conditional on the environment in which the organisation has to operate) which are outside its direct control. The next step would often be to undertake a Risk Analysis to quantify and evaluate the risk on the project. The final stage is Risk Response, which addresses how the risk will be managed. This final activity typically covers a range of actions including risk reduction, risk avoidance, risk transfer and risk retention. To assist in this task there are a number of proprietary project planning tools which can be supplemented by add-on or stand-alone risk management programmes providing risk simulation, scenario and sensitivity analysis covering the individual project stages with accumulation over the whole project.

12.3 Risks in International PFI Projects

Having examined the role of the Prime Contractor and the topic of construction risk management in a general way, the focus of this chapter now moves to examining the risks associated with undertaking major project work in an international environment. Whether in mature, developing, or underdeveloped regions one common requirement is the necessity for countries to construct, repair, refurbish, and modernise their infrastructure – the 'built environment'. An increasing number of governments *inter alia* Africa, Indonesia, China, India and Europe is increasingly using Prime Contracting in conjunction with private funded infrastructure project development and operations. Indeed, as international interest has grown, a number of different forms of PFI arrangements to accommodate foreign direct investment, long-term leasing and private funded ownership of public infrastructure facilities have evolved. From the client perspective the main difference in these arrangements centres on the timing of when ownership is transferred back to the public sector corporation or government department. The projects which are most often seen as suitable for private funding and operation before transfer typically range from roads, bridges, water systems, airports, ports and public service buildings such as museums, prisons, hospitals and schools. In considering a PFI funded project there are a number of arrangements which can be used in conjunction with a Prime Contracting procurement approach. These would provide private developers with a range of options on the amount of their involvement in key project activities including design, finance, construction, and operations.

As previously mentioned, the duration of the project arrangement can be very lengthy with the return of the facilities to the community or Government not occurring until the end of an agreed payback period. These agreements are always contractually fixed and usually range between 25 and 40 years. From a Government point of view, one of the reasons that PFI schemes have become so popular is because it relies on the private sector to build the project with no or little public money involvement and then at the end of the contract/concession period the facility becomes a State asset, be this a road, railway or a more capital intensive project such as an airport.

Whilst construction project risks can typically be divided into three phases: Planning and Pre-construction risks; Construction and Commissioning risks; and finally Operational risks, it is the general, pre-contract risks that are suggested to require careful assessment and management if an international PFI project is to have the best chance of success. These general areas of international risks which need to be considered, understood and addressed at the pre-contract stage are listed in Table 12.2. below.

Table 12.2 Pre-contract Risk Considerations

Risk area	Pre-contract considerations
Technical	Evolution and maturity of design
	Site investigations
	Source and availability of materials
Employment	Productivity of resources
	New or different methods of construction or operation
	Safety and security of employees and equipment
	Health, Safety and Environmental legislation
	Working patterns – hours, holidays
Financial	Inflation
	Fluctuation of foreign exchange
	Payment delays
	Local taxes
	Advisors fees
Political	Stability in terms of war or revolution,
	Constraints on availability or employment of expatriate staff
	The use of local companies and suppliers
Logistical	Availability of resources
	Customs procedures
	Import duties
	Embargo
Geographic and Social	Weather and seasonal implications
	Prohibitive weather patterns – typhoon, monsoon etc
	Cultural understanding including work practices season and religious beliefs

In considering an international PFI facilities project, it is suggested that well before any potential Prime Contractor formally submits an expression of interest or holds discussions about a project – let alone produces a fully-costed proposal – an outline assessment of its indigenous political and economic landscape is needed to understand the risks that either foster or dampen the project's attractiveness and result in performance implications.

Such an assessment is important as it will provide an initial 'snapshot' of the risks of undertaking general business in the identified environment. This can be further developed to examine specific risks associated with political stability and any potential change; the laws and regulations associated with commercial activities in the project location, and any special technical standards and environmental issues which may adversely impact on the project to be undertaken. In addition, at this early stage, a further investigation will also need to be undertaken to examine the national employment legislation particularly those associated with the use of local and overseas employees. This may constrain the involvement of the Prime Contractor's already established supply chain if local suppliers are expected to be significantly involved. These early risk assessment activities are important as they will influence not just the construction programme but also the ease of attracting the required project funding either through banks, private investors or Governments. In the case of many of the PFI arrangements, understanding the forecast revenue opportunities, charge basis and profit potential needs to be fully explored at the earliest stage to allow an assessment of any project constraints and performance implications affecting revenue generation capabilities. This knowledge can reduce the occurrence of a conflict of interest which is often seen in the differing demands of project facility operators and those of the project's financiers, where the latter will be looking for the earliest possible financial return on their investment.

The effective management of a supply chain potentially comprising local and foreign stakeholders during the design, construction, commissioning and operational phases of the project will also need to be established. Given then that Prime Contracting is seen to be appropriate there will in addition need to be some level of confidence by the public or private client with respect to the acceptance of the supply chain relationships with raw material suppliers, main sub-contractors, specialist service providers and any nominated preferred suppliers by the client organisation. Thus the pre-contract risk assessment and risk management approach phase is clearly important in the provision of current information on which to base any strategic decision to undertake the potential work proposed. Whilst there will be the financial costs to the business in undertaking this investigation, it will provide an informed basis on which it may then base any future decision as to whether to pursue the opportunity based not just on its expected fees and profit in isolation but the inherent risks involved. Another benefit of conducting a due diligence investigation is its use in assisting the process of raising project financing from banks or other lenders. It has to be recognised, however, that time and effort spent on such pre-bid, inception, investigations are no guarantee of appointment for such projects. Indeed should the client organisation select another company for the Prime Contractor role then the cost of this exploratory work has to be borne by the 'non-successful' Prime Contacting organisation.

For the successful bidder, the length of the concession period must be sufficient for the recovery of investments for all of the funding parties, and determined through the contractual agreement with the client. Menheere and Pollalis (1995) comment that the status of the economy is important to the success of the PFI project. If the economy inhibits investment from private companies then it will be impossible for the project to be developed. Whilst risk and reward have to be balanced, risks are often assessed to be higher in a weak economy. Funding parties too will need to be convinced that there is a need for the project/facility and only after a full market analysis that is able to justify the need will private investors be willing to participate in providing financing and becoming involved in the project.

International projects often result in complex contractual finance mechanisms with the establishment of a 'Special Purpose Vehicle' (SPV) arrangement to allow operating contracts with key participants to be agreed. One of the features of an SPV *'which nearly always comprise several companies often including a construction company and a facilities management provider'* (National Audit Office 2003, p. 21) will be the risk allocation and apportionment between the client and the Prime Contractor, sub-contractors, financiers, and insurance companies. An almost inevitable requirement on such international projects will be the provision of a guarantee to the lenders to cover their investment in the project prior to completion. Looking at the contract for construction of the facility, this is usually to a fixed price with cost penalties for late completion as the project financiers will be seeking an immediate use of the facility to allow an early return on their investment.

The Prime Contractor's offering as the single point of responsibility increases its level of risk exposure, which may lie outside the governance, and the level of risk aversion of its organisation. These risks may arise from construction risks, particularly those around increased costs and project time extensions which itself may attract a cost penalty, and from operational risks where long-term facilities operation is required before transfer to the ultimate client. As first point of contact the Prime Contractor is the *de facto* main focus of liability for claims by and amongst the supply chain members and the client organisation – be this private or public. Previously mentioned was the acceptability of foreign supply chain members to adopt collaborative working agreements. This may require additional Prime Contractor project management efforts or investments in training to support the required collaborative *modus operandi*.

The final phase of the project will be covered by an operation contract covering the financial charges of the facility. PFI project risks are thus better identified and addressed when the inter-related financial, planning, construction, commissioning and operational risks assessments are brought together with a common Risk Management process under the control of the Prime Contractor.

On completion of the PFI contract period the facility is transferred cost free to the public sector in accordance with the financial contract and operating agreement. In the event that the State wishes to take possession of the facility earlier than the contracted transfer date, then some financial compensation will need to be agreed. Ideally all project costs will have been fully recovered by the time of its transfer, in which case the State may wish not to charge users anymore. This is more likely to be the case with PFI road projects and bridges than with a more technological facility such as an airport which will continue to require on-going high cost plant and equipment maintenance and have its own investment program to allow it to conform to international safety requirements for its continued operation.

12.4 Conclusion

This chapter began by noting that significant changes had taken place in the UK construction industry over the last 10 years from the initiatives of Latham (1994) and Egan (1998) with closer collaborative working relationships and a move towards single point responsibility offered through the use of Prime Contracting.

Such moves have been supported by the UK Treasury and National Audit Office in the procurement of central government construction projects through the Private Finance Initiative (PFI) and Public Private Partnership (PPP). Those project types have provided a number of construction companies with the experience to look for work in the international environment offering to cover the capital cost funding, design, construction, commissioning and operations.

From the client perspective there are clear advantages to offer projects using a Private Finance Initiate approach including the opportunity for third party funding and risk sharing or even full transfer. Importantly client organisations (usually Governments) can preserve their own funds for other uses whilst still meeting their social requirements through a PFI contract procurement route where such assets can return to State ownership after an agreed contract period. In looking at PFI projects, Prime Contractors will have to give serious consideration to determine if the project attractiveness is acceptable to them when considering the risks of the venture. Indeed the risk assessment also has to take into account their wish to do business in the particular region or country; the projected costs and anticipated return on investment; the marketing value in the country or with respect to other potential projects and the amount that it can utilise its own products and services or those of its known supply chain partners. At the pre-contract stage the identified Prime Contractor Management Areas shown in Table 12.1 will each have a number risks which will need to be considered and addressed. The areas of risk for these Prime Contractor Management areas are shown in Table 12.3 below:

An additional risk area is that of the external environment which will need to be assessed and managed. This will include the selection of project location - country, area, and regional development, as well as the political stability and support for third-party operation of a 'state' asset. Furthermore, the influence of other governmental areas and development schemes and changes in State legislation can also affect the project. As a greater number of PFI projects come to the market there is a view that many contractors offer a very conservative approach to design and construction in order to reduce both project risks and costs. Furthermore, innovative ideas are only encouraged when they will make the infrastructure facility more profitable in the longer term through enhanced performance achieved by better cost-effective maintenance, for example. Additionally, if contractors assess that the risks are too high and find it difficult to secure investors with an interest in the project, then these projects may only be available for development on a traditional financial basis requiring funding from the client organisation.

Some would argue that establishing a PFI project is a complicated process due to the number of different contracts, SPVs, and the number of organisations involved. However, this will not necessarily result in a more expensive project when efficient design, construction and operation techniques are used together with the management of a mature supply chain and a proven process for risk assessment and management to provide improved performance during the facilities project lifetime. This is especially true when the project is part of a series of similar facilities such as schools and prisons, and undertaken by an established Prime Contracting/investment consortium. Whether such projects are offered as part of a series of projects or they are effectively 'one-off' projects, as in the case of airport developments, the use of this form of procurement with benefits of supply chain integration and effective risk management continues to grow in its popularity and use.

Table 12.3 Prime Contracting Management Areas of Risk Assessment and Management

Main areas	Areas for pre-contract risk assessment
Commercial Management	The financial viability of the project which will need to be assessed and built on a sound business plan together with agreeing on the contractual framework. There will be a management need by the Prime Contractor to have a role in the position of 'influencer' as a major shareholder.
Design Management	The limitations of the PFI Consortium will need to be assessed and managed in the event that it may not be able to influence design which may already have been done with the risk of incorrect or inappropriate design features impacting performance.
Project Management	With high profile projects, good performance of meeting time, cost and quality will need to be demonstrated as failure to do so will impact on other opportunities in the country or may extend to adversely influence other PFI projects for the Prime Contractor or SPV consortium in other locations for different clients.
Construction Management	Any changes in environmental legislation may increase the risk to the operating assumptions made at the start of the project
Supply Chain Management	An agreement that the Prime Contractor's supply chain partners are acceptable to the rest of the Consortium and/or local Government representatives. In addition management of the supply chain will need to be seen as a risk as its poor performance will have an impact on the projected financial performance.
Total Quality Management	The risk that the traffic/passenger forecast which represents the income potential on which design is based may be flawed or over optimistic.
Human Resources Management	There may be pressure to use local suppliers proposed by Government. In addition the Financing Consortium also may wish to appoint other suppliers than those preferred by the Prime Contractor. Such changes may affect financial performance and even the overall reputation of the Prime Contractor.

References

British Standards Institute (1996) Guide to Project Management, BS 6079, British Standards Institute, London.
Burtonshaw-Gunn, S. A. (2001) Strategic Supply Chain Management: Critical Success Factors for Partnering Relationships in the UK Construction Industry PhD thesis, Manchester Metropolitan University, Manchester, UK.

Chartered Institute of Building (2002) Code of Practice for Project Management for Construction and Development, Blackwell Publishing, UK.

Defence Estates Organisation and the Tavistock Institute (1999) 'Building Down Barriers Programme'—Prime Contractor Handbook of Supply Chain Management Sections I and II May 1999.

Egan, Sir John (1998) Rethinking Construction. The Report of the Construction Task Force to the Deputy Prime Minister, on the Scope for Improving the Quality and Efficiency of UK Construction. July 1998.

Galliford (2000) Partnering Survey.

Latham, Sir Michael (1994) Constructing the Team, Final Report of the Government/Industry Review of Procurement and Contractual Arrangements in the UK Construction Industry, HMSO, London, July 1994.

National Audit Office (2003) PFI: Construction Performance. Report by the Comptroller and Auditor General HC371 Session 2002–2003: 5 February 2003. The Stationery Office, London.

Menheere, S. C. M. and Pollalis, S. N. (1996) Case Studies on Build Operate Transfer, Delf University of Technology, The Netherlands.

The Royal Society (1992) Risk: Analysis, Perception and Management. The Royal Society, London, UK.

The Association of Project Management (1997) Project Risk Analysis and Management guide, Simon, Hillson and Newland (eds.).

Chapter 13: Supply Chain Risk Management for Small and Medium-Sized Businesses

Uta Jüttner* and Arne Ziegenbein

*Corresponding author: Cranfield School of Management, Cranfield University, Bedford, UK

13.1 Introduction

Over the last ten years, supply chain vulnerability (SCV) and its managerial counterpart supply chain risk management (SCRM) have received considerable attention by practitioners as well as academics (see for reviews Jüttner 2005; Peck 2006). Disruptions in global supply chains caused by a sequence of large scale environmental, political and company-driven events provide vivid illustrations of the entwined global marketplace that characterize today's supply chains. At the same time, they have demonstrated the limitations of conventional risk management approaches emerging from a single company context. However, supply chain risk management approaches must have a broader scope than that of a single organization in order to capture the risks caused by the linkages among multiple supply chain parties and their subsequent ripple effects. Thus, based on the consensus in the literature that a supply chain at its simplest degree of complexity comprises at least three entities: a company, a supplier and a customer (Mentzer et al. 2001), it has been suggested that any approach to managing risks in the supply chain should adopt the same cross-company, supply chain orientation (Jüttner et al. 2003; Ziegenbein 2007). This condition increases the complexity of the managerial approaches sought as well as the resource requirements for their successful implementation. As a consequence, it appears that mainly large companies that typically have substantial control over supply chain activities and possibly act as the 'channel captain' are in position to effectively manage supply chain risk. The few existing contributions reporting on companies' experiences with the implementation of SCRM seem to support this view. They seem to focus on large, international enterprises with abundant know how and financial resources such as the telecommunication provider Ericcson (Norrman and Jansson 2004). For small and medium sized businesses (SMEs), which are increasingly trading globally and therefore exposed to similar global supply chain risks as their large international firm counterparts, it is more difficult to manage risks since they are missing the necessary resources, structures and processes (Ritchie and Brindley 2000). This is even more concerning when considering the fact that

SMEs are often affected disproportionately by supply chain risks. As second or third tier suppliers, in many supply chains they have to shoulder a significant size of the risk burden which is pushed upstream in the supply chain by the other parties. In a large project on SCV conducted in England in the time between 2000 and 2003, interviews with managing directors of 15 SMEs revealed that the requirements of SMEs for SCRM approaches differ substantially from those of larger businesses (Cranfield University 2003).

This chapter addresses the gap identified in the current body of knowledge on SCRM by developing a practical, IT-supported SCRM methodology targeting the specific requirements of SMEs. The research reported in the chapter is part of a Swiss government-funded programme of research into the application of SCRM in Swiss companies. It has been carried out jointly by the University of Applied Sciences in Lucerne and the Institute of Technology in Zürich. The chapter is structured into three main parts. In the next section, the details about the research project objectives and methodology are outlined. This is followed by two sections contrasting the specific SCRM requirements of the companies involved in the project with the approaches discussed in the academic and practitioner-orientated supply chain literature. Finally, in the third section, our own practical SCRM approach for SMEs is introduced.

13.2 The Research Project

The objective of the research project that this chapter is based upon is to develop a structured approach for the identification, assessment, and mitigation of supply chain risks based on the specific requirements of SMEs and the existing literature. The project team is composed of the two academic institutions named above, four Swiss companies from a range of different industries and the Swiss Sourcing Industry association with ~800 member companies. Three out of the four companies involved are small to medium sized businesses, namely, a cabling supplier (Cable Co[1]), a contract electronic manufacturing services provider (Electronic Co) and a wood/timber wholesaler (Timber Co). The fourth industry partner is a global supplier of speciality chemical products with 750 employees in a Swiss subsidiary and around 8,500 employees worldwide (Chemical Co). While they certainly do not qualify as a SME, the company still has an important role to play in the project. Firstly, their approach to managing risks in the supply chain is advanced and, secondly, for the researchers, they serve as a surrogate 'control group' to help to reveal the main differences in the SCRM approaches between large and small to medium sized businesses.

The methodology applied in this study is classified as action research (see Argyries 1994; Reason and Bradbury 2001) because the participating companies

---

[1] We will disguise the names of the companies throughout the paper and refer to these synonyms.

and industry association had initially defined the scope of the project as well as serve as active participants in the study. Still, two specifications need to be made in order to position the approach taken within the wide and rather diverse field of action research. Firstly, the research objective and project plan has been driven by the researcher's agenda rather than by the participating company representatives. Secondly, the project plan has been motivated primarily by the development of a generic tool for managing supply chain risks and not by the aim of transforming the individual organization's supply chain practices.

13.2.1 Researcher-Driven Objective

Both researching institutions have actively contributed to the development of knowledge in the field of SCRM and have been aware of the gap of practical approaches for SMEs. In addition, their intimate knowledge of Swiss SMEs, which account for more than eighty percent of all companies in Switzerland, led them to suggest that there should be a cross-industry interest in the research project. Due to the high labour costs in Switzerland, the vast majority of SMEs follow a differentiating strategy with high quality, availability and lead time as 'order winners' (Aitken et al. 2005). Still, in the last ten years, low cost global competition has forced them to add a cost focus to their supply chains. This, for example, has triggered a trend to global sourcing which is not congruent with the rather risk averse and quality conscious Swiss (industry) mentality. Interestingly however, recent studies confirm that many Swiss companies have abandoned global sourcing in the case of negative experiences. In a cross-industry survey carried out in Switzerland with more than 150 responding companies, it was found that 44% of all Original Equipment Manufacturers (OEM) and 33% of all suppliers have outsourced either parts of their production and/or supply base from low cost countries (Waser and Hanisch 2005). However, in 2005, 9% of the companies have moved their production and/or sourcing back to Western European countries. Among the reasons reported were firstly, risks related to quality targets (65%), secondly, to lead time targets (54%) and thirdly, difficulties related to flexible capacity planning (19%) (Waser and Hanisch 2005). Overall, these findings have encouraged the researchers to undertake the research project in the chosen industry context.

13.2.2 Project Plan

The project design follows the requirements of the government funding institution. They demand that the knowledge is initially developed in a close relationship between research institutions and selected sponsoring companies but disseminated widely across industries at a later stage. As a consequence, the project serves two different target groups: the four partnering companies in the project team and all further Swiss SME companies interested in applying the methodology (see Fig. 13.1 for an overview).

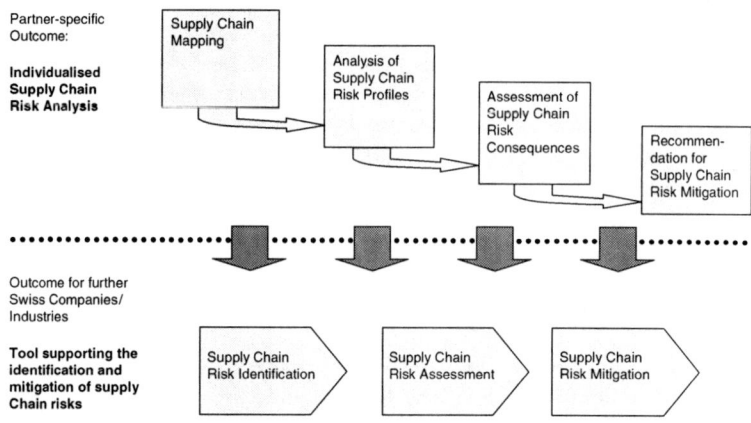

Fig. 13.1 Project plan and outcome

The output supports the partnering companies in their individual analysis of their supply chain risks, gives them a useful evaluation of existing SCRM practices and offers recommendations for improving their SCRM. Furthermore, the developed structured methodology will also be valuable for other companies, especially small-medium businesses, by providing them with a structured approach to SCRM.

The data collected throughout the project is primarily based on the four longitudinal case studies within the partnering supply chains. However, at various milestones we have had the interim findings externally validated through further SME representatives, both within Switzerland and Europe.

13.3 Requirements for Managing Supply Chain Risks in SMEs

In the initial supply chain risk analysis phase, four selected supply chains of the partnering companies have been closely investigated and potential improvements were discussed. At the same time, the requirements for a SCRM methodology have been derived from the case analyses.

13.3.1 The Four Partnering Companies

Interestingly, although the four partnering companies are from different industries and appear to show few similarities at first sight, they also share some fundamental commonalities regarding their supply chain strategies and overarching business models. Specifically, all four companies:

- have concentrated on core competencies and outsource non-competence related activities and are therefore dependent on their suppliers;
- serve business customers and not end consumers (business-to-business markets);
- work with suppliers from low cost countries (e.g., Asia and Eastern Europe);
- follow a high quality differentiating strategy with lead time and availability as order winning factors;
- have optimized their supply chain processes over the last years, yet, primarily with a focus on efficiency (cost) and not risk.

Cable Co is a third-generation family-owned cabling supplier with 500 employees and 150 million Euro turnover. More than 70% of the revenue is export-based and a particular challenge to their supply chain is the huge number of customers in four strategic business areas served through a variety of sales channels. The company structures their supply chains along technologies and products and the one selected for the initial analysis was based on a new product with a high revenue potential.

Electronic Co has also ~500 employees and offers contract electronic manufacturing services (EMS) primarily to large and powerful OEM customers. Their customer value proposition covers the complete value chain, from engineering through manufacturing, after sales services and product lifecycle management of electronic components and products. Their main supply chain management challenge stems from the specific role and position they have within the supply chain. As an outsourcing partner they are heavily dependent on their customers' supply chain management strategy. For example, many of the OEMs are both customers and suppliers to Electronic Co. Since the customers are primarily looking for cost savings when working with Electronic Co, it is difficult to make them aware of their own influence on supply chain effectiveness and efficiency. The supply chain chosen for the initial analysis carried a structural component customized for a specific customer.

Timber Co is the only trading and non-manufacturing company within the project team. With 350 employees and an annual revenue of 120 million Euro they are the biggest Swiss wholesaler for timber. In contrast to Cable Co they only serve the Swiss market but deal with around 500 foreign suppliers. Similar to most trading companies, their main supply chain challenge is the availability of overall 28,000 SKUs without having to carry vast amounts of inventory. In addition, Timber Co stresses that compared with manufacturing companies they lack the ability to absorb risks through flexible production planning. The supply chain selected for the initial analysis was a specific board with high yet unstable demand which has experienced cost as well as availability problems in the past.

Chemical Co, a non SME partnering company is a leading global supplier of specialty chemical products and industrial materials. They supply processing materials for sealing, bonding, damping, reinforcing and protecting load-bearing structures in construction and industry and generate 1.8 billion Euro turnover worldwide. Interestingly, although their supply chain management approach is clearly highly advanced with a global and national sourcing structure (e.g., national sourcing departments and international lead buyers for key components/ raw material) and specialized sales and operations planning (SOP), they also experienced great difficulties in finding the relevant information for the initial supply chain risk analysis. The supply chain selected was an after glassing repair adhesive for the automotive industry. This is a product classified as 'AY' which means it is of high value and has high demand variability.

13.3.2 Findings from the Initial Supply Chain Risk Analyses

In order to understand the specific risk situation within the four focused supply chains, between 8 and 16 interviews have been conducted in each partnering company. We have applied a process-orientated approach based on the original SCOR model (Supply Chain Operations Reference) (Supply Chain Council 2006) but extended the perspective to include supplier and customer-driven risks (see Fig. 13.2). The interview partners have either been responsible for the entire supply chain or for selected processes. Supplier or customer representatives have not been involved. While this might appear as a weakness, it accurately reflects the situation in many SMEs which have little or no influence over their external supply chain partners and will find it difficult to involve them in a joint SCRM approach.

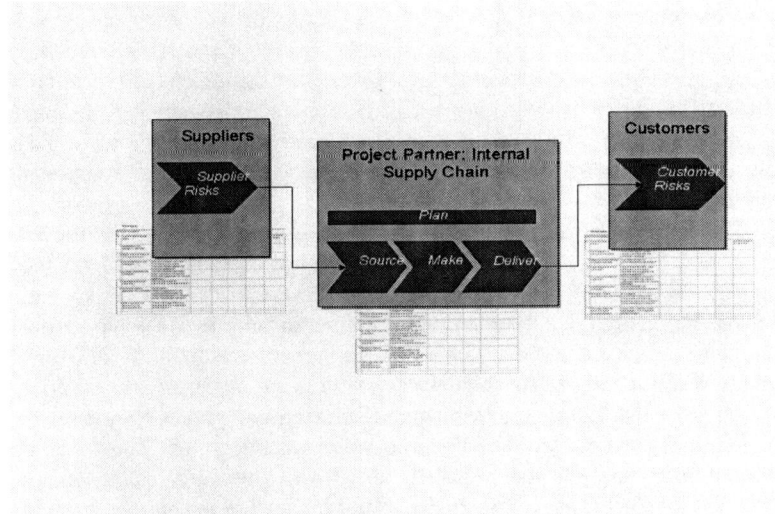

Fig. 13.2 A process-driven approach to supply chain risk analysis

The risks within each of the six core processes have been identified and assessed, leading to combined 'risk catalogues' for all focused supply chains. Table 13.1 shows an excerpt from the risk catalogue of the supply chain analysed for Cable Co.

Table 13.1 Supply chain risk catalogue Cable & Co

Suppliers	S1	Structural component supplier – delayed supply
	S2	Band supplier – delayed supply
Source	So1	Delayed incoming goods
Make	M1	Breakdown of flow machine
Customers	C1	Uncertain demand of key customer
Deliver	D1	Transport problems
	D2	Flawed consignments
Plan	P1	Conflicting supply chain objectives
	P2	Flawed forecasts
	P3	Communication problems between Cable Co and suppliers
	P4	Lack of visibility of inventory levels

When comparing the supply chain risk profiles identified, a range of interesting similarities are apparent. A key risk in all supply chains is the lack of information from the market. These include the lack of access to point of sales data, inaccurate forecasts and short term sales planning. This finding has been of particular interest to the key contact people within the organizations who mainly represent purchasing and were initially reluctant to take a comprehensive supply chain perspective. Moreover, and in line with the literature, a lack of communication and/or cooperation in the internal as well as external supply chain is a key risk which appears to drive high levels of inventory (e.g., Christopher and Lee 2004). On the supplier side, a relatively high supplier dependence seems to be caused by the predominant quality-focused market strategies. Even for purchased materials with low value added such as packaging, the specifications imply that supplier changes are costly and time consuming. A final commonality in the risk profiles identified is that none of the companies are exposed to risks emerging from Just-in-time relationships or other lean supply chain management approaches (see Svensson 2002). In order to mitigate existing supply chain risks, all three SMEs use their flexibility as a primary means of risk absorption. For example, short communication lines and the ability to make decisions quickly, help to react more quickly to unforeseen supply chain incidents. Also, the industry partners are currently undertaking various initiatives to improve the information flow both within the internal as well as external supply chain. The methodology to be developed in the project is seen as an important means of communication. Other strategies such as establishing contracts that address the risk consequences rather than its sources, which appear to be widespread in large companies (see e.g., Jüttner 2005), are not a viable option for the case companies.

13.3.3 Requirements for the Practical SCRM Methodology

The supply chain risk profiles identified by the researchers have been validated in a workshop with the case company representatives. In addition, the requirements for the methodology have been derived. All four partners agree that any practical and useful approach to managing risks in their supply chains must be designed to meet their specific needs. These risk requirements can be classified into those related to the scope of the approach, its objectives and implementation. Table 13.2 summarises the complete list of requirements.

Table 13.2 SCRM requirements for SMEs

Scope of analysis	.. take an internal as well as external supply chain perspective
	.. support the identification of critical supply chains to be analysed in greater depth
Objectives	.. enable the *efficient* identification of supply chain risks
	.. enable pragmatic not scientific risk assessments
	.. guide SMEs in their decision about mitigation measures
Implementation	.. support a cross-functional, systematic exchange of 'intuitive, personalised knowledge' within the organization
	.. be designed for an annual or bi-annual application
	.. should be supported by a user-friendly, simple IT tool

13.3.4 Evaluating Existing Supply Chain Risk Management Approaches

The methodology to be developed should not only meet the requirements from the practitioners but, at the same time, build on the knowledge within the field. We have therefore compared and contrasted our own requirements with existing approaches in the literature.

13.3.4.1 Scope of Analysis

A first requirement of our methodology is the need to take an internal as well as an external supply chain perspective. A range of contributions in the literature taking either an inbound supply risk perspective (e.g., Zsidisin 2003; Wu et al. 2006) or an external supply chain risk perspective only (e.g., Johnson 2001; Svensson 2002) are therefore not suitable. In SMEs in particular, the dependence on the other parties in the supply chain both on the customer as well as supplier side suggests that internal and external supply chain processes are inextricably linked. These linkages need to be captured by a SCRM methodology. Our process-driven approach applied in the initial supply chain risk analysis meets the requirement and has been adopted in the methodology. The approach distinguishes between four internal supply chain processes (source, make, deliver, plan) and two external processes (supplier and customer risks).

The second requirement is that a SCRM methodology should help to identify the supply chain for analysis. This has been addressed in the literature before. For example, Christopher (2005) argues that for complex supply chains it is not practical to map the entire supply chain. In his proposed seven step framework he suggests to include two steps in which companies need to understand the supply chain and identify the critical paths. Similarly, Kiser and Cantrell (2006) state that the supply chains need to be mapped and that the strategic materials need to be identified. A strategic material is any material that is of strategic importance to the business, whether this is caused by high spend or the fact that the material is purchased from critical suppliers. In our supply chain risk analyses it has been evident that firstly, the case companies structure their supply chains differently (e.g., by components, products or even customers) and secondly, that despite the linkages between supply chains in the overall network they refer to them as distinct 'planning units'. In the methodology we have considered these facts by including an initial step in which the supply chains are compared and contrasted and those requiring closer analysis are identified.

13.3.4.2 Objectives

Overall, the requirements related to the objectives stress the need for a methodology which, as one of the industry partners pointed out prominently, *'enables us to identify and handle 80% of the most prevailing supply chain risks with 20% of the potential costs and time'*. Of the large number of approaches for risk classification, identification, assessment and mitigation in the literature, many had to be excluded because despite their rigour, they would stretch the resources of SMEs too far.

Looking at risk classification and identification first, a methodological challenge emerges from the definition of risk itself. Typically, risk is defined as 'variation in the distribution of possible outcomes, their likelihood, and their subjective values' (March and Shapira 1987). The definition illustrates the vast number of risks potentially affecting a supply chain. In order to avoid omitting any risks, comprehensive classifications and brainstorming sessions are suggested in the literature for the identification of risks (e.g., Harland et al. 2003; Norrman and Jansson 2004). Still, these highly company specific and almost unlimited risk catalogues do not meet the requirements of our SME partners. We have decided to adopt the six internal and external supply chain risk processes as a rigid risk classification and, moreover, to predefine for each process the three most salient potential risks: risks related to the costs, the quality or the lead time/availability. Lead time further distinguishes between a delay (e.g., delayed supply, delay in the production processes) and a complete failure to meet the time targets (e.g., failure to supply because of a supplier going bankrupt or failure to produce because of a machine breakdown). These key risks are derived from the main supply chain objectives or outcomes (Lambert and Pohlen 2001) and are in line with the definition of risk. In order to avoid the approach becoming too rigid and restrictive, companies have to further elaborate the main causes that these cost, quality and lead time risks in the key supply chain processes are driving.

Looking at the risk assessment next, in line with most of the literature a risk evaluation based on the Failure Mode and Effect Analysis (FMEA) is proposed (e.g., Christopher 2005; Zsidisin et al. 2004). In the systematic yet easy to apply technique each risk is assessed on a one to five or one to ten point scale according to the criteria impact (potential damage), probability (possibility of occurrence) and likelihood of detection (predictability of an event). In our approach, 'likelihood of detection' has been replaced by the 'degree of current mitigation actions/measures'. This is explained by the fact that SMEs in particular, are often forced to implement intensive risk mitigation measures which in turn can trigger further risks. For example, SMEs often hold high levels of inventory firstly because their downstream supply chain partners request short lead times and are not prepared to hold inventory themselves and secondly, because they don't provide the necessary information which would enable the SMEs to substitute 'information for inventory' (Christopher and Lee 2004). High inventory levels are hence a risk mitigating measure but can at the same time cause risks for the SMEs' supply chain cost targets. To summarise, evaluating the 'appropriateness' of current actions is an important dimension of the risk assessment in SMEs.

Finally, regarding risk mitigation strategies, a range of generic classifications is suggested in the literature. For example, Jüttner et al. (2003) distinguish the four strategies of 'avoidance, control, co-operation and flexibility' and Chopra and Sodhi (2004) list eight options: 'adding capacity, inventory, having redundant suppliers, increasing responsiveness, flexibility and capability, aggregate or pool demand and have more customer accounts'. We adopted a rather coarse structure by prompting the SMEs to think about potential measures addressing either the risk consequences (e.g., business continuity plans) or the risk sources (e.g., improved forecasting). These are meant as 'search areas' but are deliberately not too specific so that the necessary imagination and creativity of those applying the approach is not jeopardized.

13.3.4.3 Implementation

In the literature, recommendations for SCRM implementation range from collaborative supply chain risk strategies (Harland et al. 2003) or the suggestion to establish a supply chain continuity team (Christopher 2005) to the proposed use of computer software as an enabling tool (Wu et al. 2006). One of the most detailed descriptions of a company implementing a SCRM approach and the necessary changes this caused for the organizational structure and processes is the case analysis of Ericsson (Norrman and Jansson 2004). Ericsson has implemented a SCRM matrix organization that spans the entire company, from the corporate and strategic level to the functional and finally process-oriented operational level.

Whereas our approach for SMEs should also be designed for cross-functional implementation, the project partners have stressed that it would be most beneficial if it was organised as an iterative process with joint workshops and discussions as well as information gathering tasks carried out by the separate functions involved.

Thereby, not only the efficiency targets would be met but it enables the company to tap and exchange the personalized and often intuitive knowledge within the organization. Again, such an approach is not feasible for large organizations unless it was implemented at a highly strategic level. However, the situation in our partnering company Chemical Co suggests that although desirable, this is not realistic unless a company has experienced major disruptions in their supply chains. Given that for small organizations, the same time constraints apply, an annual or bi-annual yet continuous application seems appropriate. A final requirement supporting the implementation of the approach is an enabling IT tool designed to lead the team involved through the sequential steps. Through the literature review as well as the validating discussions with external companies, a variety of standard and customized IT tools have been investigated by the researchers. The key disadvantages of the majority of tools available, is firstly, that they are designed for specific tasks and do not cover the entire comprehensive approach. For example, Ericsson applies a risk management evaluation tool referred to as 'ERMET' (Norrman and Jansson 2004). The tool analyses risks emerging from both internal as well as external suppliers but is restricted to an assessment of supplier related risks. As a consequence, it has to be combined with additional supporting tools such as contingency planning and other risk management actions. Secondly, some tools are based on highly sophisticated technology which is not suitable for a cross-functional application in SMEs. As an example, one of the project partners stated: '*as soon as our IT department has to get involved the tool is too complicated*'.

13.4 A Practical SCRM Approach for SME

Due to the fact that none of the existing SCRM approaches analysed, matches all of the requirements for SMEs, the researchers developed a systematic and practical methodology based on the three classical risk management phases of risk identification, risk assessment and risk mitigation (see Fig. 13.3).

The objective and output of the first phase is a list of all relevant risks for the supply chain selected for analysis. The output of the second, subsequent phase is a supply chain risk portfolio visualising the result of the individual risk assessments. Finally, having completed the third phase, the SME applying the methodology obtains an action plan with detailed measures and responsibilities for the mitigation actions agreed upon. In line with the requirements, the entire approach is supported by a simple and user-friendly Microsoft Excel based IT tool. Since it has been programmed in German, Fig. 4 serves only as an illustration from the tool which shows part of the identification phase. On the left hand side, the key phases of the methodology are listed as sequential push buttons, enabling a user friendly navigation at any time. On the top of each page, structured instructions

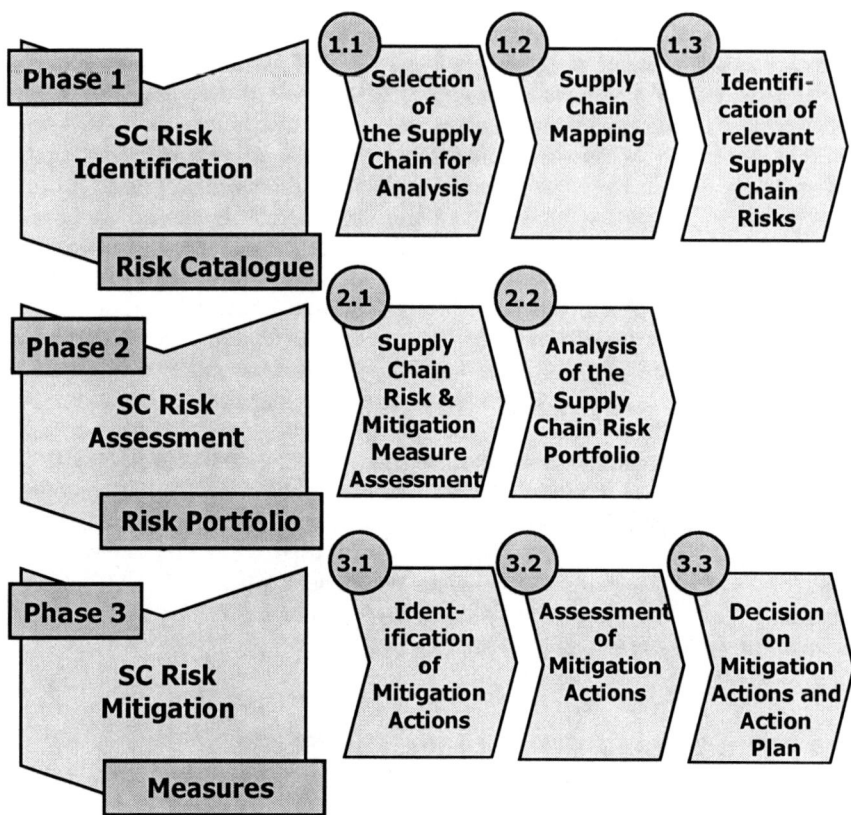

Fig. 13.3 A SCRM methodology for SMEs

guide the user. In the case of the selected risk identification step, a figure illustrating all six internal and external supply chain processes is also set up with push buttons. By clicking on the buttons, the user is transferred to a separate page with instructions to select the relevant risks within each process from a predefined list and to describe the main risk sources. All risks identified in the six processes are summarised as a risk catalogue at the bottom of Fig. 13.4.

In the remainder of this chapter, each of the steps in the three phases is described in greater detail.

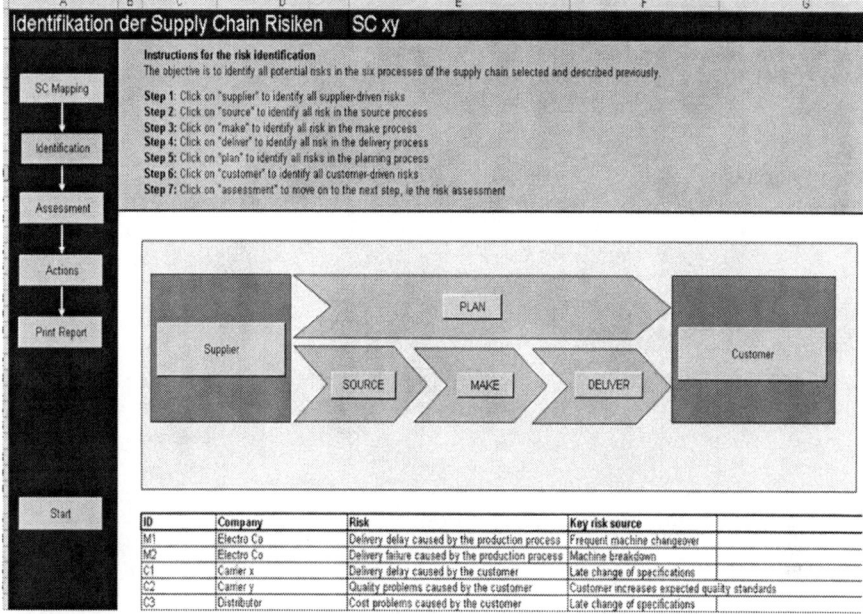

Fig. 13.4 The IT supported methodology

13.4.1 Phase 1 – Identification of Supply Chain Risks

13.4.1.1 Step 1.1 – Defining the Supply Chain for Analysis

Firstly, the supply chain to be analysed has to be defined. In the methodology, the SME can select a supply chain for the subsequent detailed risk analysis by positioning either all or an intuitive short list of supply chains in a portfolio with two dimensions: 'strategic importance' and 'vulnerability' (see Fig. 13.5).

For example, supply chains with a high revenue (strategic importance) as well as high perceived uncertainties or having experienced disruptions in the past (vulnerability) should be selected for closer investigation. Yet, even for a supply chain with reasonable (perceived) certainty which is of critical importance to the company, a closer risk investigation can be useful. It is obvious that this initial first step might not always be necessary. For example, critical external as well as internal events in a supply chain such as political unrest in a sourcing country or rumours about financial problems of an important supplier can trigger the need to investigate specific supply chains.

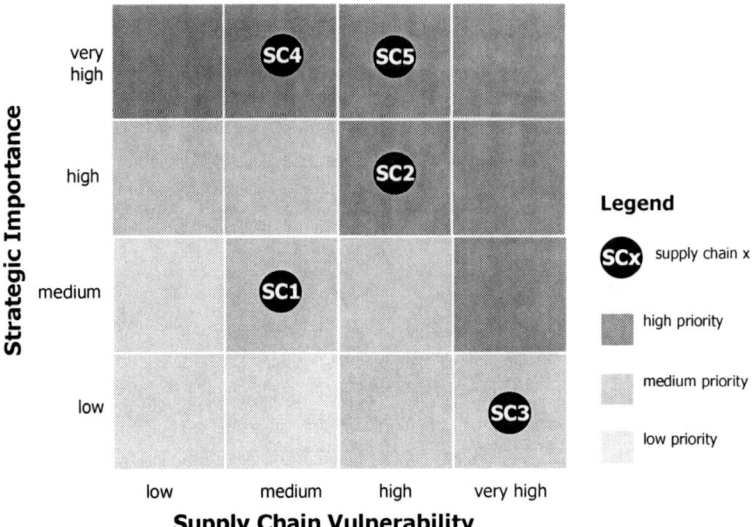

Fig. 13.5 Identifying critical supply chains

13.4.1.2 Step 1.2 – Supply Chain Mapping

In order to obtain a transparent overview on the related organizations and processes, the supply chain should be described and visualized. The methodology enables the SMEs to sketch the supply network to the third level both upstream and downstream. In addition, the most important information for each supply chain partner is captured, such as products/materials sourced or bought, costs/prices, quantities, replenishment lead time and whether or not it is a sole or single source or a key customer, respectively. Supply chain mapping is a typical example of a step within the methodology which will most likely start in joint workshops and then triggers individual task assignments, since the information is not easily accessible and needs to be obtained through further investigations or even market research.

13.4.1.3 Step 1.3 – Identification of Relevant Supply Chain Risks

Based on the supply chain map, the relevant risks for the supply chain can be efficiently identified by means of the six processes as well as the predefined key risk sources within each process. Whereas these predefined processes and risks are appreciated by the SMEs, care was still taken not to be too prescriptive. Therefore, for each of the relevant risks the company is asked to specify the main risk source. As will be seen in phase three, information on the risk sources is a precondition for any effective risk mitigation measure.

13.4.2 Phase 2 – Assessment of Supply Chain Risks

13.4.2.1 Step 2.1 – Assessment of Supply Chain Risks and Mitigation Measures

As outlined above, the risk assessment is in line with the literature and comprises firstly, the assessment of the probability of occurrence of each supply chain risk; secondly, the evaluation of the business impact; and thirdly, an assessment of the degree of existing mitigation actions. Looking at the probability of occurrence of a certain risk, the assessment can either be quantitative, using mathematical rules, or by a qualitative, experience-based score. For SMEs, a qualitative assessment with a scale from 1 (very low) to 10 (very high) is deemed to be sufficient. Secondly, the possible consequences and thus the business impact of each risk is also assessed with a qualitative score. Together with the partnering companies a scale has been developed ranging from 1 (low: little or no effect on the company's own costs) to 5 (medium: big impact on the company's own costs) and finally 10 (very high: loss of customer and/or market share). In the literature, the importance of financial expressions of risk consequences is stressed and suggestions for key indicators based on the multiplication of cost and recovery time figures can be found (e.g., Kiser and Cantrell 2006; Norrman and Jansson 2004). However, whereas such an approach might be suitable when assessing the impact of major supply chain disruptions, it seems impossible to get accurate numbers for disruptions caused by a large number of risks inherent in the six supply chain processes. Thirdly, for the evaluation of the degree of current mitigation measures, a simple three points scale with 'high' 'medium' or 'low' is used.

13.4.2.2 Step 2.2 – Analysis of the Supply Chain Portfolio

Once all risks have been assessed, a supply chain risk portfolio as well as a table ranking all risks according to their overall risk prioritisation index are composed (see Fig. 13.6 for the portfolio). Analysing the entire supply chain risk profile is an important first step for deriving mitigation measures. For example, the analysis can reveal that most of the risks are either within the internal or, alternatively, within the external supply chain processes. Similarly, a supply chain risk profile can be characterized by a predominance of risks on the supply side or, alternatively, on the demand side. Agreed interpretations of the supply chain risk profile are an important means of the SMEs' commitment to changes.

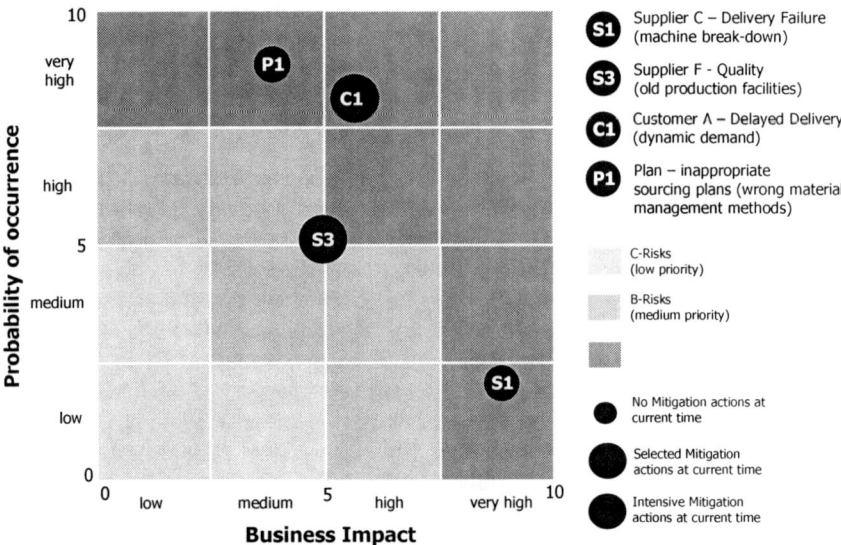

Fig. 13.6 Supply chain risk portfolio

13.4.3 Phase 3 – Supply Chain Risk Mitigation

13.4.3.1 Step 3.1 – Identification of Mitigation Actions

With a clear view of the supply chain's risk profile, a brainstorming session on potential risk mitigation measures should follow. As outlined above, it is ill-advised to provide a predefined, potentially incomplete list of measures to choose from. Instead, the decision making team is prompted to think about as many cause-oriented (e.g., eliminating unreliable suppliers to reduce supply risk) and impact-oriented measures (e.g., safety stock) as possible. Mitigation measures can be implemented at the strategic level (e.g., alternative suppliers), the tactical level (e.g., improved demand forecast) as well as at the operational level (e.g., business continuity plans). The result of this step is a list of different options to mitigate the most severe risks in the supply chain processes.

13.4.3.2 Step 3.2 – Assessment of Mitigation Actions

The mitigation options have to be compared and assessed. Stated simply, the foremost criteria when assessing a risk mitigation measure is whether or not it reduces the supply chain risks identified. In the methodology, for any mitigation measure the potential to reduce either the likelihood and/or the impact of each risk is evaluated. For example, the risk of quality problems caused by a supplier's lack of quality control which has been assessed as having a high likelihood (e.g., 8) and a low to medium impact (e.g., 4) is positively affected by a supplier development programme (mitigation measure). When assessing the mitigation measure, this specific risk would now have a lower likelihood (e.g., −4) whereas the impact is not affected. Once the mitigation measures have been assessed, a comparison of the supply chain risk profiles before and after the (successful) implementation of any combination of measures is illustrated in the IT tool (see Fig. 13.7 for an example). Finally, the costs for the implementation of the measure should be compared with the benefit of the measure. The potential benefits can go beyond risk reduction. For example, a supplier development programme may also lead to cost reductions, increased supplier capacity or reduced inventory levels.

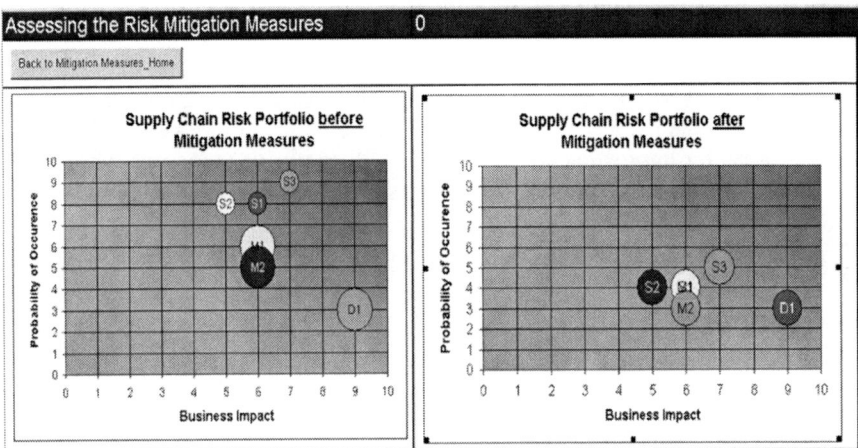

Fig. 13.7 Assessing risk

13.4.3.3 Step 3.3 – Decision on Mitigation Actions and Action Plans

Based on the detailed analysis of the mitigation measures, the company has to decide upon the measures. In most cases, these decisions will have to be made by the company board or owner. However, when interorganizational processes are addressed, even a commitment from suppliers and/or customers might be needed. For example, an effective mitigation measure can be the early exchange of forecast information from customers. In this case, the tool's systematic illustration of the positive effects of forecast information for the supply chain could convince customers to share the information. Similarly, higher inventory levels for components from a 'high risk supplier' can be necessary if the tool illustrates that the replacement time for the supplier is high and delivery failure causes severe ripple effects in the supply chain. We are encouraged that the systematic documentation of the supply chain risk analysis provided by the IT tool helps to 'make a strong case' for the necessary actions to be taken either internally or in cooperation with external supply chain parties. Finally, the firm must be committed to implementing measures, meeting deadlines, and tracking progress.

13.5 Conclusion

Due to increasing turbulences in today's global markets, supply chains are exposed to numerous risks. Companies are starting to realise that SCRM can not only protect against unforeseen and costly supply chain disruptions but can even become a competitive advantage (Sheffi 2005). However, in industry and research, there is a lack of techniques and measures which meet the specific SCRM requirements of small to medium sized businesses. The chapter strives to close this research gap and presents a SCRM approach for SMEs which has been developed jointly by a project team comprising two research institutions and four companies.

At the current stage of the research project the industry partners are implementing the approach within their supply chains. They have been advised to generate 'rich' SCRM case studies and, at the same time, validate the methodology and identify areas for improvements. The results will enable us to refine the approach and, more importantly, provide further insights into the SCRM practice and business needs of SMEs.

References

Aitken J, Christopher M and Towill D (2002) Understanding, implementing and exploiting agility and leanness, International Journal of Logistics: Research and Applications, Vol 5, No 1, pp 59–74

Argyries C (1994) Knowledge for Action, San Francisco, Cambridge University Press

Chopra S and Sodhi M (2004) Managing risk to avoid supply chain breakdown, MIT Sloan Management Review, 2004, Vol 46, No 1, pp 53–61

Cranfield School of Management (2003) Creating Resilient Supply Chains: A Practical Guide, Cranfield School of Management: Department of Transport

Christopher M (2005) Managing risk in the supply chain, Chapter in Logistics and Supply Chain Management: Creating Value-Adding Networks, 3rd ed, FT Prentice Hall, Harlow

Christopher M and Lee H (2004) Mitigating supply chain risk through improved confidence, International Journal of Physical Distribution & Logistics Management, Vol 34, No 5, pp 388–386

Harland C, Brenchley R and Walker H (2003) Risk in supply networks, Journal of Purchasing and Supply Management, Vol 9, No 2, pp 51–62

Johnson E (2001) Learning from toys: Lessons in managing supply chain risk from the toy industry, California Management Review, Vol 43, No 3, pp 106

Jüttner U (2005) Supply chain risk management – Understanding the business requirements from a practitioner perspective, International Journal of Logistics Management, Vol 16, No 1, pp 120

Jüttner U, Peck H and Christopher M (2003) Supply chain risk management: Outlining an agenda for future research, International Journal of Logistics, Vol 6, No 4, pp 197–210

Kiser J and Cantrell G (2006) 6 Steps to managing risk, Supply Chain Management Review, Vol 10, No 3, pp 12–17

Lambert DM and Pohlen TL (2001) Supply chain metrics, International Journal of Logistics Management, Vol 12, No 1, pp 1–19

March JG and Shapira Z (1987) Managerial perspectives on risk and risk taking, Management Science, Vol 33, No 11, pp 1404–1418

Mentzer J, DeWitt W, Keebler J and Min S (2001) Defining supply chain management, Journal of Business Logistics, Vol 22, No 2, pp 1–25

Norrman A and Jansson U (2004) Ericsson's proactive supply chain risk management approach after a serious sub-supplier accident, International Journal of Physical Distribution & Logistics Management, Vol 34, No 5, pp 434–456

Peck H (2006) Reconciling supply chain vulnerability, risk and supply chain management, International Journal of Logistics: Research and Applications, Vol 9, No 2, pp 127–142

Reason P and Bradbury H (2001) Handbook of Action Research, London, Sage

Ritchie B and Brindley C (2000) Disintermediation, disintegration and risk in the SME global supply chain, Management Decision, Vol 38, No 8, pp 575–583

Sheffi Y (2005) The Resilient Enterprise – Overcoming Vulnerability for Competitive Advantage, The MIT Press, Cambridge

Supply Chain Council (2006) Supply Chain Operations Reference (SCOR) Model, Version 8.0, Pittsburgh: Supply Chain Council

Svensson G (2002) A conceptual framework of vulnerability in firms' inbound and outbound logistics flows, International Journal of Physical Distribution & Logistics Management, Vol 32, No 2, pp 110–134

Waser B and Hanisch C (2005) Gestaltung der Wertschöpfungskette Schweizer Produktions-Unternehmen im internationalen Wettbewerb, Mitteilungen aus der PI-Erhebung 2003, Switzerland (in German)

Wu T, Blackhurst J and Chidambaram V (2006) A model for inbound supply risk analysis, Computers in Industry, Vol 57, pp 350–365

Ziegenbein A (2007) Supply Chain Risk – Identification, Assessment and Mitigation, Zurich: vdf Hochschulverlag ETH Zurich (in German)

Zsidisin G (2003) Managerial perceptions of supply risk, Journal of Supply Chain Management: A Global Review of Purchasing and Supply, Vol 39, pp 14–25

Zsidisin G, Ellram L, Carter J and Cavinato J (2004) An analysis of supply risk assessment techniques, International Journal of Physical Distribution & Logistics Management, 2004, Vol 34, No 5, pp 397–413

Chapter 14: Psychological Foundations of Supply Chain Risk Management

Michael E. Smith

Western Carolina University, Cullowhee, NC 28723

14.1 Introduction

Population statistics may be useful in overall characterization of risk, but ultimately, as Zuckerman (2007) notes, the way that risk is experienced is a matter of subjective assessment as represented in individual perception. In a managerial context, it is aggregation and interaction of these individual assessments in relatively small (at least as compared with society) organizational groups that drives business strategies and adjustments at the tactical level. Thus, organizational risk management responses are a function of coordination of subjective perceptions at a level sufficient to gain coherence in the responses from key employees. Of course, supply chain risk management often involves coordination of responses requiring such coherence at the level of inter-organizational responses. While the emerging literature in supply chain risk management focuses on types, magnitudes, and appropriate responses to risk at the organizational and inter-organizational level, this chapter will focus on the roots of perceived risk in the psychology of critical persons because detection and assessment of risk still requires thinking, judgment, and decision making by individuals.

14.2 Perception in Risk

If we take as our starting point the standard definition of risk, we combine two distinct perspectives on events that might occur. First, we have some event that, if it materialized, would be detrimental to organizational performance. Note that this event could be the occurrence of something negative or the failure of something good to happen. Secondly, we have some probability, less than one, or certainty that the event will occur. While there have been a number of other definitions proposed that include more terms, for example hazard, exposure, consequences and probability (Ropeik and Gray 2002), the formulation of risk in two terms (the magnitude of a negative future event and its probability of occurrence) is the most common approach (Adams 1995) and has become generally accepted in framing supply risk as the product of these two elements (Zsidisin and Ellram 2003).

Note that for many supply risks the magnitude of detriment to organizational performance can, at least in theory, be specified with substantial accuracy based upon the organization's financial information. However, this is not generally the case with the probability of an event, because only in extreme forms of risk, for example, death, are general statistical data maintained (Zuckerman 2007). Management in environments that present risk can often have consequences for the employment of the person classifying events and collecting data. This can be among many forms of trouble in better estimating the likelihood of adverse events (Adams 1995). Further, Smith and Buddress (2006) have found that for most businesses, organizational data related to the frequency of supply disruptions are not maintained, or are not collected in a form that would be useful in attempts to better estimate the probabilities of supply disruptions. Thus, we see that the discipline has resorted to managerial perceptions in attempts to assess and manage risk (e.g., Zsidisin 2003).

Essentially, we have resorted to perception because supply risk management involves making decisions under uncertainty. This uncertainty is a fundamental reality of the business environment, and is commonly seen as increasingly influencing our ability to manage the relationships within and between organizations. The challenges range from the dynamic complexity of business relationships (e.g., Roberts 1984; Senge 1990; Stacy 1992) to cognitive limitations (Reed 1982).

The challenges that the decision maker faces in risk detection and assessment lead to a substantial number of shortcuts, often referred to as heuristics, and the associated biases (Tversky and Kahneman 1974) can induce flawed decisions. Thus, it is important to understand the psychological roots of faulty perceptions in order to improve managerial effectiveness in assessing and preventing or mitigating supply risk.

14.3 The Challenged Decision Maker

A complex, chaotic business environment presents the supply management professional with a dizzying array of situations that tax and often exceed the limitations of the human mind, which after all evolved for problem solving in an environment vastly different from that imposed by modern society. Risk assessment tests the limits of human rationality (Zeckhauser and Viscusi 1990). Limitations that may impact effective supply chain risk management decision making include many topics in the field of cognitive psychology (Eysenck and Keane 1995). A multitude of constraints on rational thought can be found in structural limitations on perceptual processes and pattern recognition, limited attention, limited memory, limitations of mental representations including mental models and the impact of language on conscious processing, limitations in reasoning, and the impact of emotional responses. Such constraints drive simplistic decision making strategies characterized by rules outside of prescribed decision making constructs that would be considered best practices in the managerial context (Janis 1989; Stanovich 1999). As Stich (1990) has observed, many of the normative strategies that are seen as hallmarks of rational decision making would "require a brain the size of a blimp" (p. 27).

In fact, even with such a brain, supply management professionals might be unable to exercise complete rationality in anticipating and managing risk. Simon (1976) enumerated three perspectives to describe failures in objective rationality:

1) "Rationality requires a complete knowledge and anticipation of the consequences that will follow on each choice. In fact, knowledge of consequences is always fragmentary.
2) Since these consequences lie in the future, imagination must supply the lack of experienced feeling in attaching value to them. But values can be only imperfectly anticipated.
3) Rationality requires a choice among all possible alternative behaviors. In actual behavior, only a very few of all these possible alternatives ever come to mind" (p. 81).

A large amount of research has focused on how people deal with these limitations in thought processes aimed at solving problems. There is a general notion that such thought processes can be divided into two branches of cognitive operations associated with quick, intuitive thought, and slow, effortful reasoning. In a recent review of the research in this realm, Kahneman and Frederick (2005) propose the perspective that when faced with a problem, the first branch of thought processes initially addresses the problem with a quick intuitive answer, and the second branch monitors and adjusts the proposed answers. The limitations previously noted mean that there are substantial limitations to the extent that intuitive first reactions are evaluated and recalibrated. From this stance, our shortcuts in thinking can be seen to considerably influence and limit human behavior. The shortcuts that allow us to get on with life, bias what we do. Action may be valued over complete rationality when faced by a hungry tiger in the jungle wilds, but failures to adjust and override intuitive biases in solving the problems of modern society may be quite a different matter. Short-sightedness causes us to predictably mis-estimate risk when we understand that something may go wrong, and, as aptly pointed out by Tenner (1996), we may fail to see that a great solution to one problem can have the result of creating a new, potentially more substantial problem, such as those we often label unintended consequences.

14.4 The Shortcuts We Take

It is safe to say that none of the readers are suffering from such proportional challenges as to have a head large enough to hold the brain described earlier, and it is also the case that there are not any niches in the business environment that can be safely characterized as simple or stable. Therefore, it is safe to state that all of us take shortcuts in our decision making.

As you read of the list of limitations that we face, you may well have seen yourself in a number of the categories of limitations. You may even have begun to realize that the cognitive limitations and environmental challenges that are part of decision making give rise to an almost unlimited number of potential simplifications that could be part of the overall decision making landscape. At this point, you should readily recognize that no simple list of simplifications and biases is likely to describe the entire range of possibilities, so you might begin to question the value of any such list.

The simple truth is that it is not possible to make fully informed decisions. We must rely on models of reality in order to communicate (language requires the use of mental models), and thinking is also driven by our simplified images of a vastly more complicated reality (all mental images are simplifications of reality). In fact, to mistake our understanding of the elements of a problem as being the same as reality might reasonably be likened to going to a restaurant and taking a bite out of the menu because you mistook the representation for the meal (Bateson 1972).

It should now be clear to you that within any reasonably comprehensive list of simplifications you should be able to find approaches that you regularly employ. Further, exploration of such simplifications should suggest biases that you introduce into your decision making. Since the first step in addressing problems is awareness, this presentation will now turn to selected heuristics that have applicability to our ability to manage supply chain risk. Coverage of heuristics in this section is not intended to be all-inclusive, indeed such a listing would not be possible, but instead is directed toward uncovering positive steps that can be taken to improve supply risk management decision making.

One of the most common errors that we can make is to assume that one thing is like something else. This should come as no surprise, since we have already seen that our thinking progresses by means of models of what is observed. The next step is to recognize that in our search for means by which we can understand the world we inhabit, we look for patterns to help us understand and predict what we experience. When we use the extent that something seems like another thing to decide if we can let the first object concept or event stand in our thinking as a representation of the second object concept or event, we can end up with incorrect comparisons. This is the representativeness heuristic identified by Tversky and Kahneman (1974) and it can have important implications for estimation of probabilities associated with risk assessment. This shortcut allows substitution of some event about which we have knowledge for an event or events about which we have little or no knowledge. This can be seen to be useful in conditions under which our cognitive abilities would have evolved. For example, if while we are foraging for food, we happened upon a previously unknown large animal with sharp teeth, then it would probably be useful to assume that it might be as dangerous as another well known dangerous predator. Quick recognition of the potential for harm and rapid retreat are adaptive in such a situation. However, in most business settings such a quick response is probably neither necessary nor desirable.

Among the problems that have been observed based upon representativeness is misrepresentation of the role of chance in events. For example, in games of chance, we may well assume that someone who wins multiple times is cheating, but such strings do not necessarily indicate dishonesty. In this case, we see strings as representative of controlled processes, and assume that the person is controlling the process instead of recognizing that in independent events, a string of results in one direction is as valid as one that appears random. There are many coincidences in common experience that result from just such dynamics (Eastaway and Wyndham (1998) provide an interesting cataloguing and description of instances of this type of issue in everyday life).

As another example, performance extremes, whether positive or negative, are likely to be followed by performance more representative of the norm, quite independent of our interventions. This phenomenon is often referred to as regression to the mean, and is a property of chance processes. If we act on small samples, we may view something as representative of one thing (perhaps excellent performance as a supplier) when more data would better indicate what is actually happening (e.g., perhaps overall average performance). In this case, the decision to reward our apparently outstanding supplier based upon a small sample size, when coupled with the regression phenomenon can lead to disappointment when performance is seen as eroding. By the same token, when penalizing poor performance leads to improvement, we may want to congratulate ourselves on our sound supply management practices, only to find that over the long run, performance has not really improved.

Obviously, failures to attend to probabilities and chance have serious implications for risk assessment and management. A substantial challenge to our management in this realm comes when we substitute our sense of understanding based upon small samples and unreasonable comparisons for events that we do not fully understand.

Another set of challenges is presented in what comes to mind when we are faced with a problem. This has been described as the availability heuristic (Tversky and Kahneman 1974). Things that happen frequently are easy to recall. In the wilds, being prepared for those things that we are likely to face has real value. However, in modern society our sense of the frequency of things can be influenced by factors that have nothing to do with how often something actually happens. For example, information technology certainly makes it easier to find information about rare events. This capability, coupled with choices made by the media about newsworthiness (Sorenson et al. 1998) can present an image quite at odds with reality. For example, in an analysis of newspaper coverage of causes of death (Allman 1985) found that diseases that are frequent causes of death receive relatively little coverage compared with the coverage afforded accidents that are relatively rare. Perhaps the most telling figure was that the study found approximately three times as many stories about homicides as there were about disease-caused deaths, in spite of nearly 1,000 percent more deaths actually caused by disease when compared with homicide.

The frequency with which we are presented with information, and how situations are represented can have considerable influence on how we think about problems. Things that are subject to lots of news stories or are sensational when reported tend to be seen as more frequent and more serious than are things subject to less coverage and that do not stand out as clearly. For example, evidence seems to persistently suggest that people are not very good at estimating the probability that a given type of event will cause death, or at comparing the relative frequencies of death from various events as in the case of attempts at ranking major causes of death with respect to frequency (Slovic 1987; Slovic et al. 1982).

In the supply risk realm, major disasters are likely to color managerial perceptions of the frequency of major disruptions. Combining a misconception of how widespread major disruptions are with a tendency to focus on avoiding negative events to the exclusion of more positive events may drive actions where supply management professionals spend more on preventing rare events than is warranted, while at the same time paying less attention to comparatively common sources of risk than is appropriate.

Another heuristic is based on the search for a place to start in making a judgment in the face of uncertainty and a lack of information. Tversky and Kahneman referred to this as adjustment from an anchor (1974). The challenge is that we would obviously like to make informed decisions, and calibration is obviously critical to providing meaningful probability estimates, but in order to calibrate estimates, there has to be a starting point. Experimental evidence suggests that the starting point often significantly influences judgments about probability, as well as other quantitative estimates.

In the supply risk management realm, there are two areas that my experience suggests are particularly problematic from the perspective of anchoring effects. In one case, since we tend to survey groups of supply management professionals in our attempts to assess risk, it is challenging to gather these estimates in a manner that ensures proper calibration. In the second case, when assessing probabilities for events, it is often challenging to appropriately account for the relationships between events, and we confuse probabilities of something happening with the probability of things happening together (e.g., we confuse disjoint and conjoint probabilities, more on this later).

In addition to the specific examples indicated above, it may be useful to look at some broad categories of biases that have been observed as a result of the application of judgment heuristics. Overconfidence (Griffin and Tversky 1992) is typical in assessment of uncertain quantities. In the context of supply risk management such overconfidence is likely to arise when a particular manager who has substantial experience infers (e.g., makes an intuitive judgment) about the probability of some event, such as the probability of supply disruptions associated with various suppliers, without collecting and utilizing performance data. In such cases, the evidence suggests that the intuitive solution is likely not to be well calibrated, but unfortunately, the manager is prone to ardently believe and thus act on the false wisdom contained in the intuitive assessment.

There is good evidence that within the judgment and decision making realm, confidence is related to self evaluation and optimism (Wolfe and Grosch 1990). It is hardly novel to think that positive self evaluations might accompany professional advancement and accomplishments. That this would be linked to optimism in the face of uncertainty seems obvious as well. The challenge is that there is distinct benefit that can result from positive self evaluation, confidence, and optimism in that these help people to approach uncertain and difficult tasks that they might otherwise avoid. However, as noted by Griffin and Tversky (1992), there is a cost to decision making that is accomplished under excessive levels of these three positive traits in that this can mean that relatively risky courses of action are pursued. In supply risk management, it may be the case in some situations, based in part on past successes, that a manager may be excessively confident in his or her ability to mitigate risk with a given approach, and due to excessive optimism in an ultimate triumph, fails to see the warning signs that should have promoted steering a new course in managing the particular risk involved. Unfortunately, research has shown that unfounded optimism is not easy to reverse (Weinstein and Klein 2002). A particularly important concern in the face of this type of bias is that supply management professionals need to be particularly diligent in utilizing data to assess their effectiveness in risk management. In an uncertain and continually changing environment, adjustments based upon feedback represent one of few possible antidotes available to prevent being ruined by our past successes.

The net effect of the heuristics and associated biases is that overall it is common that we see overestimation of low risks, and underestimation of high risks. While the intuitive portion of arriving at judgments may seem obvious for decisions made by lay people without the training associated with becoming a professional, research has shown that experts are subject to the same shortcomings. Without conscious attention to how we are making decisions, our intuitive shortcuts have the potential to overcome even the best of training.

There has been substantial debate in the research literature about the number of heuristics, the major categories into which heuristics can be classified, and even the extent to which the experimental work that has uncovered heuristics should be taken as representative of decision tasks in the real world, such as that faced by supply management professionals (Gigerenzer (2004) presents a substantial review of these concerns). However, in the end, there is agreement that we use heuristics, that we can work with to recognize how we are making decisions, and that with careful attention, we can improve the decisions that we make. In the next section we will apply an understanding of psychological foundations to the re-examination of a common supply management concern as a way to illustrate limitations in how we address problems, and the value of understanding these limitations.

14.5 The Number of Suppliers Issue – Reconsidered

In the United States following the terrorist attacks of September 11, 2001, logistics networks were suddenly halted, and did not resume functioning for a number of days. The resulting disruptions in industrial activity had a predictable effect: businesses began to increase the number of their suppliers (Assaf et al. 2006). We have seen similar responses worldwide in reaction to well publicized disruptions resulting from large-scale events like terrorism, natural disasters, and labor strife.

As brief as this description is, a number of sources of biases in the thinking that leads to increasing the number of suppliers should be readily apparent. The first and perhaps most obvious is that these events are likely to evoke the availability heuristic and be subject to easy recall based upon several features. These events all impacted lots of people and businesses, caused major suffering, and were subject to a tremendous amount of news media reporting. All these features should make events easy to recall, and therefore, we tend to overestimate the probability of such events.

The magnitude of negative impact, one part of our assessment of risk, has also been shown to affect our assessments of the likelihood of an event. Thus, since the types of events listed all cause major supply disruptions, they tend to be very salient, standing out in our memories. Thus, unless we actually collect the data and calculate the frequencies, the odds are good that we will confound the two terms in our algebra of the supply risk associated with major events, since as a short cut to understanding the risk, we substitute the extent of damage for how representative such an event is of daily experience.

We are likely to believe that disaster is imminent following well-publicized business disruptions with their roots in public events that are experienced at least in part through supply failures. From this perspective we can see that following such events we are likely to see a wide-spread drive to do something to address supply risk.

The previously noted drive toward enlarging the supply base clearly indicates what many business organizations have chosen to do, but is this a wise course of action? How common are large-scale disruptions? While events such as September 11, 2001 in the U.S. get a great deal of press coverage, and while the magnitude of the disruption and harm is tremendous, such events are not a very common feature of the business environment in most locations. Such events are part of a set of problems where everything fails at the same time, which creates its own unique set of challenges to determining what to do, since we have little direct control with respect to triggering events. It is clear that taking reasonable steps is prudent. Given a lack of control, we generally seek to minimize the damage wrought as a result of such events, and so we search for what we can do in the face of events that seem so random.

Why do we add suppliers? The roots of this decision may be found in how we look at the probability of the risk. Imagine for a moment that you have ten coins, and that these coins are "fair" when tossed into the air and allowed to land on a surface (e.g., that in a long series of such tosses, the overall probability of landing on one side is the same as landing on the other side – we will ignore the trivial probability of standing the coin on its edge). Now we will take one side as "success" and the other as "failure." If we toss the first coin, it will land as either a success or a failure. What is the probability of failure?

If indeed the coin is fair, we should have a probability of failure of 0.50. If we are only playing with one coin out of the entirety of my potential "coin base" of ten coins, and that coin indicates failure, I have total failure on my hands with respect to this game. Now we all know what to do to avoid complete failure under such circumstances, right? If I have lots of coins at play in my coin base, it becomes quite unlikely that all of them will land on the failure side.

Now I enlarge my coin base to two coins. What is the probability of complete failure? Where previously the probability of complete failure was one out of two possibilities (0.5 probability of complete failure), with two coins, there are four possibilities, out of which one represents two failures or complete failure (0.25 probability of such complete, or in statistical terms, conjoint failure). With a third coin, there is one way out of eight potential unique coin arrangements that we can get complete failure (probability of 0.125). Note that the toss of one coin does not impact the alternatives for the next coin that is tossed, so the tosses are independent. Most people have an intuitive sense that with more independent events, the probability of everything going the same way becomes relatively unlikely, and of course formal study of statistics serves to confirm this.

Enlarging our supply base can reduce the chance of complete failure for a given commodity if there is independence between the suppliers with respect to causation of failure, or to the extent that such independence exists. In calculating the probability of complete or conjoint failure (we'll take any disruption as representing failure) in such cases, we simply apply the product rule, multiplying the individual probabilities of failure. In the case of our coin base analogy, with two coins, we multiply 0.5 by itself to obtain a 0.25 probability of conjoint failure, and so on.

While we have previously seen supply base reductions as an important way to better manage relationships with suppliers, in the face of substantial concern about total failure of our supply base for critical commodities, we might decide that the seemingly small risk of more relationship issues is more than offset by the benefits as we add more suppliers. Note that this is a risk versus benefit comparison.

By contrasting risk compared with benefits, we have framed the decision in one way. Notice, however, that this particular way of framing the problem may invoke the optimism bias. I can see benefits, and I perceptually minimize the risks associated with turning away from previous supply chain management practices. One of the major ways to overcome biases is to reframe problems to open new approaches to problem solution (Russo and Schoemaker 1989). In this case, the problem can be reframed by casting it not as a risk-versus-benefit problem, but a risk-versus-risk problem. Such a reframing can have the impact of significantly shifting the final decision because it shifts individual thought processes and the terms of debate in group decision-making processes.

What are the competing risks in this case? On one hand, the risk is associated with not enlarging the supply base, and appears in the form of the probability of conjoint failure, and on the other hand, the risk is associated with enlarging the supply base, and appears in the form of increased probability of disjoint failure (e.g., in the increased number of individual supplier disruptions). In this form, the problem is really about balancing the risk of one course of action with the risk of another course of action, a rather direct comparison that is not as likely to invoke biases of the same magnitude invoked by comparing positive outcomes with negative ones.

Let us look at the pattern of the reductions in risk probability for conjoint failure given the increases in the number of coins that we flipped. Note that there is a diminishing return on increasing the number of coins, such that as more coins are added, we see smaller reductions in the probability of conjoint failure. This is the general pattern that you will find for this type of situation. What this means is that there are very real limits on the value of increasing the size of the supply base.

Another concern is that we may significantly misrepresent the nature of catastrophic events when we postulate independence of suppliers. Certainly, all members of the potential supply base for a given commodity are from one industry, and many potentially disruptive events are common to entire industries. Another problem is that many of the large scale events that might trigger our concern to add suppliers have very large-scale effects, affecting regions, entire countries, and even having global impact. In the case of the terrorist attacks on the U.S., the reactions to the attacks crippled the entire logistic infrastructure of the U.S. While having relatively local suppliers may have been useful in the longer term, it did not really matter where your suppliers were based initially, because the disruption was nearly complete. For catastrophic events, there may be simply no way to prevent problems, but some spreading of your risk seems prudent, if the costs of such a course of action are not too great.

With the cost of adding suppliers in mind, we will now take a look at the risks associated with individual supplier failures as we add more suppliers. Going back to the example of the number of coins, with the first coin, the probability that some disruption will appear is one in two possibilities, or 0.5. When we add the second coin, we now have the appearance of failure on at least one of the coins for three out of the four possible configurations, or 0.75, and when we add the third coin, failure appears on at least one of the coins for seven out of eight possible configurations. Thus, we see that the disjoint probability of failure is increasing as we add suppliers, making it more likely that we will have to deal with disruptions. In essence, we have increased the marginal cost in terms of managerial effort in order to reduce the probability of complete failure. Note also that complete success, that is the situation where you do not have any disruptions among your suppliers for a given commodity, is also a conjoint event, and so, by adding more suppliers, you reduce the probability of complete success just as you reduce the probability of complete failure.

You probably would run rapidly away from a supplier that was causing you problems half of the time, and so your individual probabilities of failure would be much smaller than is the case for the coin example. In such cases, the amount of conjoint risk becomes fairly small after the second supplier, and diminishes rapidly beyond that. On the other hand, the frequency with which you would observe disruptions with individual suppliers increases quite rapidly (given that you would usually add suppliers with worse performance records than those of your preferred suppliers).

This example should show two things. First, we generally add suppliers to reduce risk, but we trade reductions of risk that is relatively rarely experienced (but catastrophic when it is observed) for risk that is relatively frequently experienced. Second, there is tremendous value to actually knowing how your suppliers are performing in the form of how frequently you experience disruptions that require intervention, because this is the only way that you can really understand what you are trading off. In general, it is prudent to have multiple suppliers, but the number is probably best kept reasonably small.

The type of example we addressed here would be difficult to address in the same way as you add complexity in real situations. Such complexity includes differences in performance and sourcing allocations that vary across the supply base (which obviously represents the appropriate supply–management response to differences in performance). This is where statistical knowledge can be useful. In particular, our concern would be in dealing with subjective probabilities, and in particular, in dealing with the value of information, such as the value of knowing the frequency and magnitude of disruptions that arise from individual suppliers. Bayesian statistics which deals with the probabilities of some event given knowledge about some indicator can be useful in addressing such problems.

A substantial body of research suggests that the use of conditional probabilities under a Bayesian framework supports better assessment of risk. For example, such an approach was found valuable in medical diagnoses, and accurate evaluation was indicated as a critical factor in applying computers to supporting decision making in medical treatment (Ledley and Lusted 1959). For society, the risks associated with our technologies benefit from establishing risk targets based upon Bayesian analysis (Starr and Whipple 1980). Indeed, from a social perspective, the challenge is to establish reasonable levels of risk. Generally, as is the case with our sourcing activities, the intent is not to eliminate all risk, but to determine how we can balance competing risks (Zechhauser and Viscusi 1990). Situations in which conditional probabilities provide valuable insight include the risk of cancer (Cornfield 1977), and the incidence of hurricanes (Simpson 1973).

The use of diagnostic tests in the practice of medicine illustrates important concepts for the future of risk management in purchasing and supply management. In the medical situation, the task is to determine when risk warrants intervention, and indeed, we seek to diagnose the supply system in similar fashion. In medical diagnosis, tests can be conducted, but the results of those tests are not entirely unambiguous. Conditional probabilities represent a way to determine medical treatment based upon limitations in medical tests. In medicine, many diagnostic tests are approaching a level of sophistication such that they very rarely provide a negative result in situations where a given illness is present, but this level of detection comes with an increased probability of false positive tests (e.g., Eddy 1982). In such cases, given a negative test, decisions based on Bayesian decision-making move away from additional tests or treatment as the probability of false negatives approaches zero (Bottom et al. 2002). The relative levels of false positives and false negatives impacts the extent to which test results suggest risk that illness is present.

Similar logic can be applied to looking at elements of supply risk. For example, inspection results have been applied to evaluating risks posed by quality failures (Brint 2000; Chun and Rinks 1998). Under acceptance sampling, Bayesian analysis has been demonstrated as an effective approach to quantifying risks for both the producer and the consumer (Chun and Rinks 1998), and such analyses are particularly important when sampling costs are high, allowing accurate balancing of failure rates and inspection costs (Brint 2000). Recent studies suggest that Bayesian approaches may represent a cost-effective means for determining the appropriate level of managerial concern and action where some finite level of risk aversion is present and where action depends upon a-priori assignments of probabilities (Adams 2004), conditions that seem to mirror real-world situations for supply managers relative to risk in many spheres (Zsidisin 2003).

14.6 Practical Steps for Supply Management Professionals

The first step in addressing any potential challenge is awareness, and this chapter should be regarded as a step in that direction for some of the psychological issues involved in accurately assessing and managing supply risk. However, awareness must be coupled with distinct actions to be meaningful. This section provides a summary of some steps that can be taken to improve decision making in the face of complexity and uncertainty.

Given the challenges of the modern business environment and the limitations of the human mind, a common approach is to assign problems to teams for solution. As noted by Russo and Shoemaker (1989), while many may think that this represents a solid step toward addressing cognitive limitations, it can only be meaningfully considered a potential solution in the case of well-functioning teams. Part of the value of teams is that they truly can look at a problem from multiple perspectives, thus avoiding many of the biases noted earlier in this chapter. However, to do so, the team must diverge in terms of the thinking applied by the members. All too often, teams converge on a particular line of thought in a process known as group think, and under such processes, they are likely to pursue risky actions (Janis 1989; Russo and Shoemaker 1989).

One source of the rapid convergence on a course of action characterized by group think is that many teams are not at all diverse in terms of their membership. Instead, teams are often composed of people who get along, and since we tend to like those that are similar to ourselves (Cialdini 1993); we tend to end up with teams with more homogeny than diversity. Even differences that are readily apparent, such as racial or gender differences, although they can serve to bring different backgrounds to the table that may present additional perspectives, do not guarantee that the team members truly think about problems in different ways. Effective teams must be composed of people with divergent backgrounds and divergent ways of thinking. Further, genuinely effective teams must be well facilitated so that problem identification is followed by a period of divergence in thinking about the problem. Such divergence requires that the different perspectives be encouraged and respected. Finally, once the various perspectives are on the table, the effective team turns to converging on the best course of action. The well-functioning team is appropriately deliberate in pulling the problem apart, while at the same time driven to address the problem (Russo and Shoemaker 1989).

We have already seen that statistical thinking and training can be valuable in overcoming the shortcuts that we can end up taking in the face of difficult decisions. In supply risk management it is important that the probability of potentially negative events be determined as accurately as reasonably possible, so that both the extent of the risks, and the potential for avoiding the negative consequences of the realized risk can be soundly assessed. Indeed, it is preferable to avoid negative impact as opposed to blunting its down side if this is cost effective. The only way to know if this is the case is to know how likely realization of the risk really is.

Statistical training has been shown to increase the likelihood that statistical thinking will be invoked in decision making (Nisbett et al. 2002). Thus, while experts are subject to the same biases as are those without expertise, the magnitude of deviation is lower for the experts, and statistical training coupled with knowledge of accepted risk modeling and the specific decision making environment associated with risk in the supply chain represents the best bet for well-calibrated supply risk management decision making (Koehler et al. 2002). In the future, we may find that in supply chain risk management, just as in the realm of medical diagnosis, work to develop solid actuarial procedures (Dawes et al. 2002), perhaps coupled with decision support systems, will reform how we approach many problems in supply chain management.

Finally, as was illustrated here, reframing represents a valuable approach to overcoming some of the biases associated with decision making under uncertainty. One major concern that should be addressed by such reframing is that in risk management decision making we should seek to compare like concerns. Although common, comparing risks with benefits is likely to result in biased decision making.

As shown in the example of reconsidering the number of suppliers, starting with a new frame can substantially impact perception of the problem. When coupled with some of the other actions discussed in this chapter, it is probable that reframing may lead to changes in many of the conventional notions of how we should manage supply risk.

References

Adams J (1995) Risk. UCL Press, London
Allman WF (1985) Staying alive in the 20th century. Science 85 6(8): 31–37
Assaf M, Bonincontro C, Johnsen S (2006) Global sourcing & purchasing post 9/11. J. Ross Publishing, Fort Lauderdale FL
Bateson G (1972) Steps to an ecology of mind: a revolutionary approach to man's understanding of himself. Ballantine, New York
Bottom WP, Ladha K, Miller GJ (2002) Propagation of individual bias through group judgment: error in the treatment of asymmetrically informative signals. Journal of Risk and Uncertainty 25(2): 147–159

Brint AT (2000) Sequential inspection sampling to avoid failure critical items being in an at risk condition. The Journal of the Operational Research Society 51(9): 1051–1059

Chun YH, Rinks DB (1998) Three types of producer's and consumer's risks in the single sampling plan. Journal of Quality Technology 30(3): 254–268

Cialdini RB (1993) Influence: the psychology of persuasion (2nd edn). Quill, New York

Cornfield J (1977) Carcinogenic risk assessment. Science 198(4318): 693–699

Dawes RM, Faust D, Meehl PE (2002) Clinical versus actuarial judgment. In: Gilovich T, Griffin D, Kahneman D (eds) Heuristics and biases: the psychology of intuitive judgment. Cambridge University Press, Cambridge, pp 716–729

Eastaway R, Wyndham J (1998) Why do buses come in threes? The hidden mathematics of everyday life. Barnes & Noble, New York

Eddy DM (1982) Probabilistic reasoning in clinical medicine. In: Kahneman D, Slovic P, Tversky A (eds) Judgment under uncertainty: heuristics and biases. Cambridge University Press, Cambridge, pp 249–267

Eysenck MW, Keane MT (1995) Cognitive psychology: a student's handbook (3rd edn). Lawrence Erlbaum, Hove UK

Gigerenzer G (2004) Fast and frugal heuristics: the tools of bounded rationality. In: Koehler DJ, Harvey N (eds) Blackwell handbook of judgment & decision making. Blackwell Publishing, Malden MA, pp 62–88

Griffin D, Tversky A (1992) The weighing of evidence and the determinants of confidence. Cognitive Psychology 24(3): 411–435

Janis IL (1989) Crucial decisions: leadership in policymaking and crisis management. Free Press, New York

Kahneman D, Frederick F (2005) A model of heuristic judgment. In: Holyoak KJ, Morrison RG (eds) The Cambridge handbook of thinking and reasoning. Cambridge University Press, Cambridge, pp 267–294

Koehler DJ, Brenner L, Griffin D (2002) The calibration of expert judgment: heuristics and biases beyond the laboratory. In: Gilovich T, Griffin D, Kahneman D (eds) Heuristics and biases: the psychology of intuitive judgment. Cambridge University Press, Cambridge, pp 686–715

Ledley RL, Lusted LB (1959) Reasoning foundations of medical diagnosis. Science 130(3366): 9–21

Nisbett RE, Krantz DH, Jepson C, Kunda Z (2002) The use of statistical heuristics in everyday inductive reasoning. In: Gilovich T, Griffin D, Kahneman D (eds) Heuristics and biases: the psychology of intuitive judgment. Cambridge University Press, Cambridge, pp 510–533

Reed, SK (1982) Cognition: theory and applications. Brooks/Cole Publishing, Monterey CA

Roberts EB (ed) (1984) Managerial applications of system dynamics. MIT Press, Cambridge MA

Ropeik D, Gray G (2002) Risk: a guide for deciding what's really safe and what's really dangerous in the world around you. Houghton Mifflin, Boston

Russo JE, Schoemaker PJH (1989) Decision traps: the ten barriers to brilliant decision-making & how to overcome them. Doubleday, New York

Senge PM (1990) The fifth discipline: the art and practice of the learning organization. Doubleday/Currency, New York

Simon HA (1976) Administrative behaviour: a study of decision-making processes in administrative organizations (3rd edn). Free Press, New York

Simpson RH (1973) Hurricane prediction: progress and problem areas. Science 181(4103): 899–907

Slovic P (1987) Perception of risk. Science 236(4799): 280–285

Slovic P, Fischoff B, Lichtenstein S (1982) Facts versus fears: understanding perceived risk. In: Kahneman D, Slovic P, Tversky A (eds) Judgment under uncertainty: heuristics and biases. Cambridge University Press, Cambridge, pp 463–492

Smith ME, Buddress L (2006) How many suppliers? A Bayesian perspective. 4th Worldwide Research Symposium on Purchasing and Supply Chain Management. Institute for Supply Management and IPSERA. San Diego, CA, April 8, 2006, proceedings available at http://www.ht2.org/conference/pdf/107.pdf

Sorenson SB, Manz JG, Peterson B, Berk RA (1998) News media coverage and the epidemiology of homicide. American Journal of Public Health 88(10): 1510–1514

Stanovich KE (1999) Who is rational? Studies of individual differences in reasoning. Lawrence Erlbaum, Mahwah NJ

Starr C, Whipple C (1980) Risk of risk decisions. Science 208(4448): 1114–1119

Stich SP (1990) The fragmentation of reason. MIT Press, Cambridge MA

Tenner E (1996) Why things bite back: technology and the revenge of unintended consequences. Alfred A. Knopf, New York

Tversky A, Kahneman D (1974) Judgment under uncertainty: heuristics and biases. Science 185(415): 1124–1131

Weinstein ND, Klein WM (2002) Resistance of personal risk perceptions to debasing interventions. In: Gilovich T, Griffin D, Kahneman D (eds) Heuristics and biases: the psychology of intuitive judgment. Cambridge University Press, Cambridge, pp 313–323

Wolfe RN, Grosch JW (1990) Personality correlates of confidence in one's decisions. Journal of Personality 58(3): 515–534

Zeckhauser RJ, Viscusi WK (1990) Risk within reason. Science 248(4955): 559–564

Zsidisin GA (2003) Managerial perceptions of supply risk. Journal of Supply Chain Management 39(1): 14–25

Zsidisin GA, Ellram LM (2003) An agency theory investigation of supply risk management. Journal of Supply Chain Management 39(3): 15–27

Zuckerman M (2007) Sensation seeking and risky behaviour. American Psychological Association, Washington DC

Chapter 15: Behavioural Risks in Supply Networks

M. Seiter

International Performance Research Institute, Rotebühlstr 121, 70178 Stuttgart, Germany

15.1 Introduction

Risks within supply networks are currently an intensively discussed topic (e.g., Brindley 2004; Gaudenzi and Borghesi 2006). A variety of different types of risk have been investigated, e.g., inventory risks, delay, quality risks and even terrorist attack. However, the dimension of behavioural risk has been largely neglected in previous studies. Yet, the relevance and significance of this type of risk has significant implications in all supply chain contexts. This chapter presents the results of an explorative study conducted by the author who demonstrates that behavioural risks occur frequently and cause high losses, e.g., resulting from supply networks interruptions. These findings are supported by other studies, (e.g., Hendricks and Singhal, 2005) who showed that supply networks interruptions not only cause short-term losses but long-term underperformance from a stock-market perspective. The exploratory study also facilitated the identification of different behavioural risk types, (e.g., opportunistic behaviour or conflicts between partners), which represent the main theme of the chapter and are explored in depth subsequently.

Several scholars have conducted studies into the risk of opportunistic behaviour (e.g., Das and Teng 1999; Joshi and Stump 1999; Das and Teng 2000; Jap and Anderson 2003; Rokkan and Buvik 2003; Hallikas and Virolainen 2004; Wuyts and Geyskens 2005). Williamson (1975, p. 6) defines opportunism as "self-interest seeking with guile" and as the breaking of formal contracts. Contracts are necessarily incomplete because not all future circumstances can be anticipated at the inception of the contract (Tirole 1999). Therefore, normally there are implicit agreements between the partners, (e.g., solidarity or flexibility, besides the formal contract). The breaking of the informal agreements, Williamson (1991 p. 273) calls "lawful opportunism". In this chapter opportunistic behaviour will be interpreted as the breaking of explicit and/or implicit contracts between the partners within the supply networks.

The literature review shows a gap in empirical evidence about the effects of instruments which are designed to prevent opportunistic behaviour in supply networks. This chapter contributes to addressing this gap in two ways. Firstly, the instruments which are used most frequently in practice to avoid opportunistic behaviour will be identified from the empirical study. This empirical approach is necessary at this stage as it is not yet clear whether the proposed instruments are really in use in practice. Subsequent to this first objective, the second objective is to test the effect of identified instruments on the prevention and management of opportunistic behaviour.

A model is developed initially based on the Principal-Agent Theory. The results of the empirical study are incorporated in the development of this model. The model is then employed to evaluate the effects of the instruments on opportunistic behaviour, both the direct and the indirect effects. The indirect effects are evaluated by introducing the construct "asymmetric information" as a mediating variable. The model is then tested against data from a questionnaire-based study. A structural equation modelling approach is employed as this is considered to be especially suited for this purpose.

The remainder of this chapter is divided into four parts. Initially, the model is developed and explained. The details of the methodological approach employed are articulated followed by the presentation and analysis of the empirical results. The fourth section discusses the managerial and theoretical implications including the limitations of the present study.

15.2 Conceptual Model

The Principal-Agent Theory serves as the theoretical basis for the model (Eisenhardt 1989). This theory helps to explain the occurrence of opportunistic behaviour between a principal and an agent as a result of asymmetric information between partners within the supply network. Actions to reduce asymmetric information within supply networks are proposed as an approach to coping with or managing the risks of opportunistic behaviour.

Pratt and Zeckhauser (1985, p. 2) suggest that a Principal-Agent-Relationship exists "whenever one individual depends on the action of another". Normally, relationships of delegation are understood as part of the Principal-Agent-Relationships. As a result, the agent agrees to undertake a duty, which is compensated by a reward. Hence, the behaviour of the agent affects the achievement of his own objectives (and rewards) and the achievement of the principal's goals.

This characterization of a Principal-Agent-Relationship allows the interpretation of a supply network as a set of Principal-Agent-Relationships (Fig. 15.1). Starting at the end of the supply chain, the companies with direct access to the customers are to be seen as principals ("P"). Companies at the first tier supplier level are agents ("A") of these principals. However, the same companies also face companies at the second tier level, taking on the role of principal to these second tier agents. Thus, they play a dual role of principal and agent, as shown in Fig. 15.1.

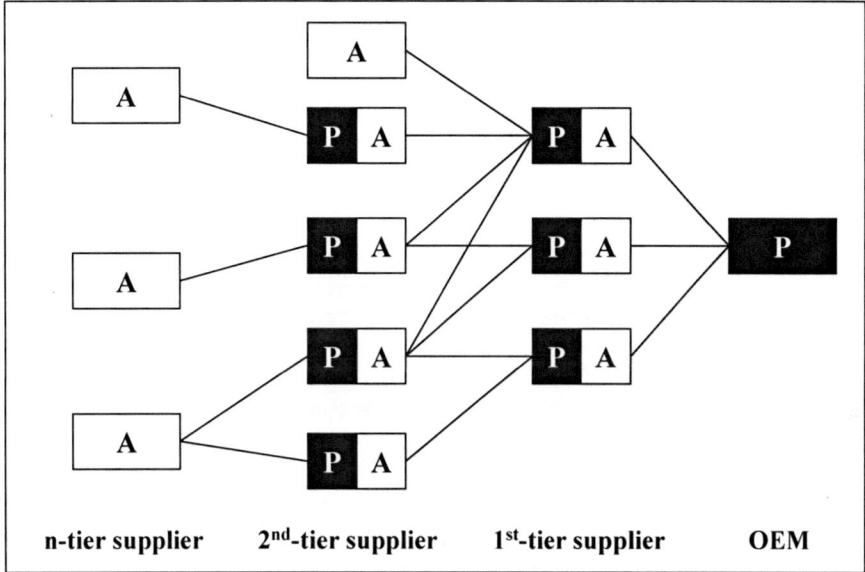

Fig. 15.1 Principals and agents in a supply network

Since the Principal-Agent Theory explains the functional and dysfunctional behaviour of partners in Principal-Agent-Relationships, it is able to explain the opportunistic behaviour of partners (Jensen and Meckling 1976). Therefore, it provides a valuable theoretical basis for the model of this chapter. Regarding the second objective of this chapter the Principal-Agent Theory contributes to determining the central hypotheses.

H1: *An increasing level of asymmetric information leads ceteris paribus to an increasing occurrence of opportunistic behaviour in the supply network.*

The term asymmetric information basically refers to the situation that "it is difficult or expensive for the principal to verify what the agent is actually doing" (Eisenhardt 1989, p. 58).

Addressing the first objective of the chapter, an empirical pre-study was conducted to assess the validity of the model. This pre-study identified instruments which were being explicitly employed in practice to reduce opportunism. Initially, a systematic review and analysis of the literature was undertaken to identify the range of possible instruments that may be utilized (see e.g., Das and Teng 1999; Das and Rahman 2001; Rokkan and Buvik 2003).

Basically, two types of instruments are found. The first type of instrument aims at the gathering of information on the partners, e.g., communication and partner selection. The second type of instruments aims at increasing the costs of opportunistic behaviour, e.g., mutual hostages and sanctions.

This list of instruments served as the basis for the design of the semi-structured interviews involving executives working in supply networks. These interviews were utilized to reduce the list of those instruments that are used most frequently in practice.

Ten executives were interviewed for ~2 h each in face-to-face interviews, employing a semi-structured interview schedule. The interview participants were chosen to minimize industry-bias. Hence, executives from the automotive, IT, media, pharmaceutical, and production equipment sectors were surveyed. The results from these interviews identify the following instruments as those most frequently used in practice:

- *Communication*: Communication is the formal and informal exchange of information between the partners within the supply network. There are many different forms of communication, e.g., oral communication or written communication. All forms of communication have in common that they reduce the degree of asymmetric information. But, the size of this effect depends on the frequency and quality of the communication.
- *Partner selection*: Partner selection is the examination of the match between the potential partners in a supply network (Das and Teng 2003). From a sequence perspective the selection of appropriate partners is the first instrument to prevent the opportunistic behaviour of suppliers. The literature suggests many recommendations concerning the selection criteria to use (for an overview see Seiter and Isensee 2007). By using a set of criteria the partner selection process seeks to ensure partner "fit". There are different types of fit, e.g., strategic fit, and resources fit (Das and Teng 1999). However, the most important fit is the fit of the partners' objectives. Only if all partners in the supply network can reach their goals simultaneously, will dysfunctional behaviour become unlikely (Das and Teng 1999).
- *Inter-organizational cost accounting*: A typical form of opportunistic behaviour a supplier can practice is the abuse of the information asymmetry. For example, to pretend that costs are higher and accordingly seek a higher price. In practice different forms of inter-organizational cost accounting may be implemented to avoid such behaviour. The forms vary from partial approaches to full approaches like open-book accounting (Kajüter and Kulmala 2005).
- *Inter-organizational planning*: Another way to prevent opportunistic behaviour is to reduce the extent of asymmetric information by sharing planning data. Several concepts have emerged in practice like "collaborative planning, forecasting and replenishment" or "supply chain planning". The diffusion of inter-organizational planning is also supported by a great variety of software solutions that support the structured planning along the supply network.
- *Sanctions*: Sanctions, in the event of the occurrence of opportunistic behaviour, are one of the first instruments mentioned by each interviewee.

Normally, sanctions are defined in the formal bilateral contracts between buyer and supplier (Wuyts and Geyskens 2005). Especially, in cases where product-specific information is shared in the partnership between buyer and supplier, the contracts include paragraphs preventing the abuse of this information. This is more often the case when it comes to international supply networks.

These five instruments serve as independent variables in the model developed for this study. Being the main enabler for opportunistic behaviour, asymmetric information is used as a mediating factor. In this way it can be tested whether or not these instruments only have a direct effect on opportunistic behaviour or whether they may also have an indirect effect by influencing the level of asymmetric information. The result is a three-stage model which represents a more complete approach than a more simplified two-stage model without the mediating factor (see Fig. 15.2).

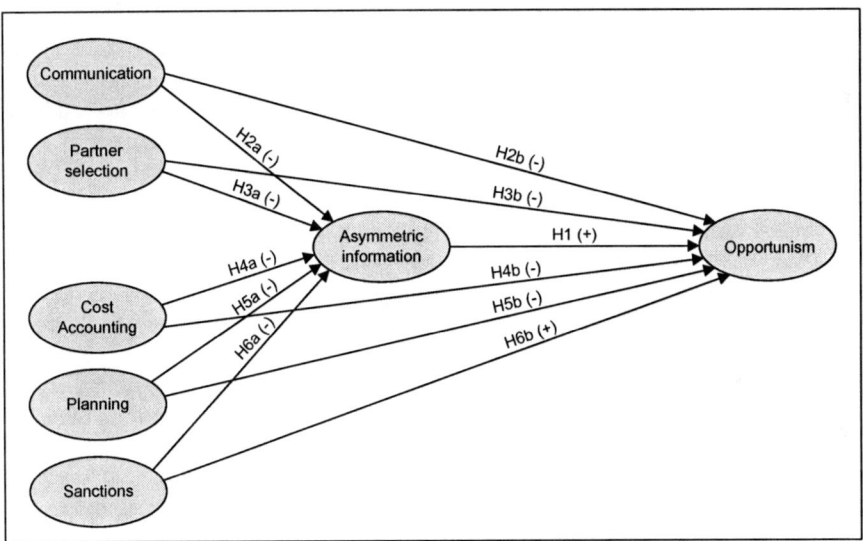

Fig. 15.2 Conceptual model

The first stage of this model is represented by our central hypotheses (H1). The structure and substance of hypotheses H2–H6 is basically the same only changing the independent variable. Direct and indirect effects are differentiated by introducing the letters "a" and "b", with "a" marking the indirect effect and "b" the direct effect of the examined instruments. To determine the direction of the predicted effects "+" and "–" are introduced. Hence, "H2b (–)" represents the hypotheses: *the higher the quality of communication the lower the level of opportunistic behaviour within the supply chain.*

The variables of this model are operationalized in the following sections, utilizing the data from a multi-industry survey. This will also permit the testing and evaluation of the model.

15.3 Research Methodology

15.3.1 Instrument

Structural equation modelling was used to test the model presented in Fig. 15.2. This allows the capture of measurement error associated with the variables and the segregation of the direct and indirect effects, addressing specifically the objectives of this study (Anderson and Gerbing 1988). The software packages SPSS 12 and AMOS 5 were used (Byrne 2001) for the data analysis. AMOS is a similar but more user-friendly package than the more frequently used LISREL.

15.3.2 Sample and Data Collection

The sample comprised all German organizations that are classified as large-scale enterprises (LSE) according to the definition of the European Union (Commission of the European Union 2003).

Questionnaires were sent to 5,717 executives in these companies, who had been identified as appropriate key respondents. The questionnaire had been pre-tested with ten executives from relevant companies and six academic experts. This step allowed for the assessment and evaluation of the presentation and content validity of items, ensuring that executives understand the instructions, questions, and response scales of the questionnaire as they were intended. To reduce autocorrelation effects, questions forming one construct were placed in separate sections of the questionnaires. The survey followed the guidelines prescribed in Dillman (1978).

To check for non-response bias, the respondents were analyzed based on company characteristics in terms of industries, sales turnover and number of employees. This analysis did not reveal any significant bias indicating that non-response bias is not a problem. After excluding some questionnaires due to a high proportion of missing data the final sample size was 104 (a 2.1% response rate). Given the high amount of survey-based research in Germany and the chosen sample this response rate is understandable and acceptable.

15.3.3 Scale Development

The measurement of the variables employed Likert-type scales where possible employing items from existing scales drawn from other studies. However, this was not possible for all measures, and consequently several new measurement items were developed. Appropriate measurement development techniques were used (Rossiter 2002). The development of these items was derived from extensive field studies including semi-structured interviews with the ten executives during the pre-study. These interviews provided the basis for the refinement and the definition of the variables and the identification of the appropriate wording for the target group. Table 15.1 shows the items or indicators relating to each of the constructs. Reverse coded indicators are marked with [R].

Table 15.1 Construct measurement

Construct	Indicators
Opportunism	The reliabilty of the partners (e.g. regarding agreements) is very high. [R]
	The partners do their duty even if they are not supervised. [R]
	All partners do their best to guarantee the success of the supply network. [R]
Asymmetric information	The partners know each other very well. [R]
	Each partner knows the characteristics and qualities of the other partners very well.
Communication	The quality of communication between the partners is very high.
	The frequency of communication between the partners is just at the right level.
	The partners inform each other about relevant events and changes in the network.
Partner selection	Level of partner selection effort.
Cost accounting	Implementation level of interorganizational cost accounting.
Planning	Implementation level of interorganizational planning.
Sanctions	Implementation level of interorganizational sanction systems.

15.4 Results

15.4.1 Analysis of the Measures and Model Fit

In this study, reflective measurement models were used as the indicators are determined by the construct (Diamantopoulos and Siguaw 2006). To evaluate the quality of the measurements two criteria were examined: reliability and validity.

Reliability of the measurements was evaluated using Cronbach's Alpha (Churchill 1979) as well as explorative and confirmative factor analysis. Besides reliability, a measurement model must show sufficient validity. Construct validity was evaluated using factor reliability and the average measured variance (Homburg and Giering 1998). Additionally, a test for significance of the factor loadings was performed (Homburg and Giering 1998). Finally and importantly, the discriminant validity of the constructs was tested using the Fornell-Larcker-Test (Fornell and Larcker 1981). All measurements show good or very good values, so that the measurements can be seen as reliable and valid.

There is no consensus which indices should be applied to test the quality of the structural models. Therefore, a variety of fit indices has to be used to evaluate the model fit. Here, six indices are applied each representing a different approach of evaluating the goodness of fit. All indices show exceptionally good model fit by showing better values than required for good fit (see Table 15.2).

Table 15.2 Model fit indices

Chi-square/df	1.04	≤3
Root mean square error of approximation	0.02	≤0.08
Goodness of fit Index	0.95	≥0.90
Adjusted goodness of fit index	0.90	≥0.90
Comparative fit index	1.00	≥0.90
Tucker-Lewis index	1.00	≥0.90

15.4.2 Structural Model Results

The results of the statistical test are presented in Fig. 15.3. Generally, the hypotheses are supported. First of all, empirical evidence provided support for the central hypothesis of this study (H1). The level of asymmetric information has a positive effect ($\beta = 0.24$, $p \leq 0.05$) on opportunistic behaviour. Therefore, asymmetric information could serve as a mediating variable for indirect effects of the independent variables.

Communication has a strong negative effect both directly ($\beta = -0.52$, $p \leq 0.001$) and indirectly on opportunism (H2a/b). High quality communications reduces the level of asymmetric information and the degree of opportunism. The same is true for inter-organizational cost accounting. However, both the direct effect ($\beta = -0.25$, $p \leq 0.01$) and the indirect are not as strong as for communication (H4a/b). No significant effects were found for inter-organizational planning systems (H5a/b) and sanctions (H6a/b). Both variables were therefore eliminated from the model.

A surprising result was found for partner selection. Significant direct and indirect effects were found (H3a/b). However, they were not negative effects as

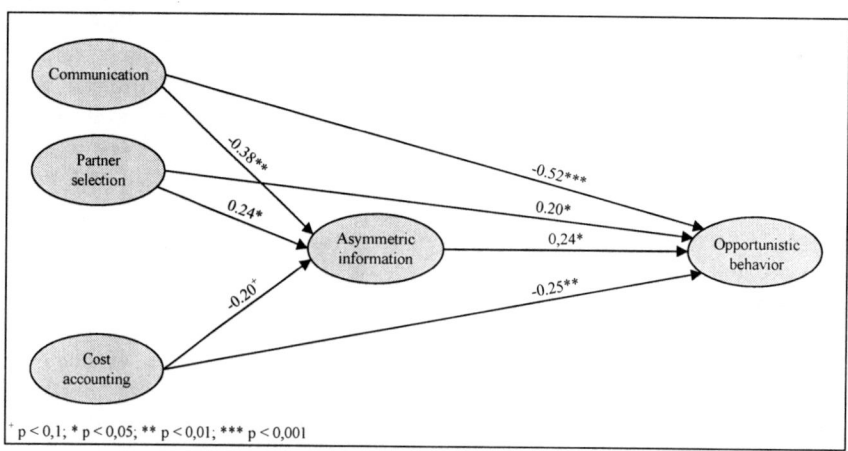

Fig. 15.3 Conceptual model with the results of the empirical test

postulated but positive. High efforts in partner selection show a positive correlation ($\beta = 0.24$, $p \leq 0.05$) to asymmetric information and also a positive correlation ($\beta = 0.20$, $p \leq 0.05$) to opportunism (for a possible explanation see Sect. 15.5 that follows).

The model explains 57% of the variance of opportunistic behaviour thus showing the exceptionally good quality of the model.

15.5 Discussion

Summarizing, the findings suggests that good quality communications and inter-organizational cost accounting both help to reduce opportunistic behaviour within the supply network. In contrast, no significant effects regarding possible sanctions and inter-organizational planning systems were found.

These non-significant results and the unexpected results for partner selection were discussed with a focus group consisting of a sub-group of the initial group of ten executives. The group came to the conclusion that sanctions cannot provide perfect protection as all future circumstances would have to be anticipated and planned for accordingly. Since this is more of a general problem, this may be one reason why the implementation of sanctions does not demonstrate the postulated effect. The absence of the anticipated effect of inter-organizational planning could be grounded in the construct itself. In practice several different types of planning systems do exist, some very simple and others very complex types. The effective operationalization of this construct may require the indicators to be more specific about the specific types of planning systems used.

The positive correlation between partner selection and asymmetric information as well as opportunism can be explained by a specific procedure which can often be seen in practice. In many supply networks the initial partner selection is undertaken with considerable care, attention and effort. However, once a potential partner passes this stage, further testing at later stages is rarely undertaken. Some partners could take advantage of this situation by behaving opportunistically having successfully passed the initial phase of partner selection. The survey captures only the initial partner selection effort but not whether partners are tested continuously. Therefore, this unexpected result may possibly be due to the need to define the indicators more closely to include this possibility.

Another explanation is the existence of a non-observed factor which influences the partner selection effort and the level of opportunistic behaviour both in the same way. The discussion in the focus team led to several possible factors, e.g., the length of time the partners know each other. This non-observed variable may have a negative effect on partner selection effort and on opportunistic behaviour, thus resulting in a positive correlation of those two constructs. The discussants agreed that the longer they know a partner, (e.g., from previous partnerships), the lower is the partner selection effort. Combined with the fact that the longer they know a partner the less likely is the occurrence of opportunistic behaviour. This may cause the occurrence of quasi-correlation.

15.6 Managerial Implications

The managerial implications were deduced from the discussion of the empirical results within the focus group.

The model (Fig. 15.3) shows that communication is a central factor for preventing opportunistic behaviour in supply networks. For this reason the improvement of communication should be a primary objective. Therefore, the content of the communication has to be differentiated from the communication media employed. In practice the focus has been more on the latter. Information and communication technologies are implemented to improve communication. But this only represents one aspect, the effective transmission of the message. The communication content itself must not be neglected. Therefore, it is necessary to match information demand and information supply. Too much communication can be counterproductive as well as too little. Finally, organizational action can contribute to improving communication. One possibility is to implement regular meetings of the employees of the various companies within the supply network. These employees can belong to the same functions (e.g., quality assurance) or to different functions. Finally, it is important to choose the right communication channel. According to the Media Richness Theory the channels have to be broader the more implicit, complex and important the communication content is (Daft and Lengel 1986).

The second implication refers to the implementation of inter-organizational cost accounting systems. The implementation can be carried out in several steps. There is no necessity for a complete implementation in one step. Even initial steps may show positive effects. A first initial step could be a common understanding of cost accounting. At that stage partners agree on the meaning of basic concepts, e.g., standard costs. A subsequent step could be the implementation of inter-organizational target costing. In this way, suppliers can be integrated into the cost accounting of the buyer. Finally, a common cost management approach can be implemented, e.g., consisting of joint cost optimization programs or open-book accounting.

Referring to partner selection, no simple coherence between partner selection and opportunistic behaviour has been observed. The discussion with the industry representatives has suggested that partner selection is most often a non-recurring process taking place at the beginning of the relationship. In most cases, the partners are tested in a very detailed way in the beginning but subsequently there are no further reviews or only reviews that are not as detailed as the initial ones. A possible solution can be the adoption of a continuous partner assessment process, requiring the partners to pass a review for example once a year; with the possibility of deselection should they fail such a review.

15.7 Conclusions

This research is a first step to investigate the effects of managerial instruments on opportunistic behaviour in supply networks. Based on the results of this study further research is necessary. A research design to analyze the cause-and-effect relations in detail seems to be the most promising route initially, e.g., single supply network case studies.

The role of partner selection in particular has to be investigated further. It appears that the degree of effort expended on this activity is less crucial than the actual procedure employed and the frequency of reviews. The investigation of the structure and processes of partner selection and the effects of the specific steps of that process may prove interesting and valuable. Only with such research will it be possible to give further recommendations for the design of partner selection.

These preliminary findings have to be validated and investigated in more detail in further studies. For example studies which focus on specific industries or specific sizes of companies seem to be appropriate. The context of small and medium sized enterprises is an especially important area for further research as small and medium-sized enterprises are more highly dependent in terms of their existence on efficient integration in supply networks, since increasingly they need to respond to pressures of internationalization and specialization.

This study has several limitations. Firstly, the sample size is sufficient but a sample with more cases could have made the results more valid. Therefore, the results have to be evaluated in further research. There are also limitations related to the measures that were used. Due to restrictions on the length of the questionnaire the number of items was limited. This aspect probably reduced construct validity to some degree compared to more complete scales. The non-existence of appropriate measures also made it necessary to develop new items. Although the scales were developed on the theoretical and empirical basis the scales and indices need modification and refinement in future research.

Acknowledgements Financial support for this study was provided by the Foundation of German Businesses and the Department of Management Control of the University of Stuttgart. I am grateful to the executives who generously contributed their time.

References

Anderson, J. C. and Gerbing, D. W. (1988), Structural Equation Modelling in Practice – A Review and Recommended Two-Step Approach, in: Psychological Bulletin, 103 (1988) 3, pp. 411–423.
Brindley, C. (ed.) (2004), Supply Chain Risk, Ashgate, Hampshire and Burlington 2004.
Byrne, B. M. (2001), Structural Equation Modelling with AMOS, Mahwah, NJ.
Churchill, G. A. (1979), A Paradigm for Developing Better Measures of Marketing Constructs, in: Journal of Marketing Research, 16 (1979) 2, pp. 64–73.
Commission of the European Union (2003), Recommendation 2003/361/EC of 6 May 2003 concerning the definition of micro, small and medium-sized enterprises.
Daft, R. L. and Lengel, R. H. (1986), Organizational Information Requirements, Media Richness and Structural Design, in: Management Science, 32 (1986) 5, pp. 554–571.

Das, T. K. and Rahman, N. (2001), Partner Misbehaviour in Strategic Alliances – Guidelines and Effective Deterrence, in: Journal of General Management, 27 (2001) 1, pp. 43–70.

Das, T. K. and Teng, B.-S. (1999), Managing Risks in Strategic Alliances, in: Academy of Management Executive, 13 (1999) 4, pp. 50–62.

Das, T. K. and Teng, B.-S. (2000), Instabilities of Strategic Alliances – An Internal Tension Perspective, in: Organizational Science, 11 (2000) 1, pp. 77–101.

Das, T. K. and Teng, B.-S. (2003), Partner Analysis and Alliance Performance, in: Scandinavian Journal of Management, 19 (2003) 3, pp. 279–308.

Diamantopoulos, A. and Siguaw, J. A. (2006), Formative Versus Reflective Indicators in Organizational Measure Development – A Comparison and Empirical Investigation, in: British Journal of Management, 17 (2006) 3, pp. 263–282.

Dillman, D. A. (1978), Mail and Internet Surveys – The Tailored Design Method, 2nd edition, New York 1978.

Eisenhardt, K. M. (1989), Agency Theory – An Assessment and Review, in: Academy of Management Review, 14 (1989) 1, pp. 57–74.

Fornell, C. and Larcker, D. F. (1981), Structural Equation Models with Unobservable Variables and Measurement Error – Algebra and Statistics, in: Journal of Marketing Research, 18 (1981) 3, pp. 39–50.

Gaudenzi, B. and Borghesi, A. (2006), Managing Risks in the Supply Chain Using the AHP Method, in: The International Journal of Logistics Management, 17 (2006) 1, pp. 114–136.

Hallikas, J. and Virolainen, V.-M. (2004), Risk Management in Supplier Relationships and Networks, in: Brindley, C. (ed.) (2004), Supply Chain Risk, Hampshire and Burlington 2004, pp. 43–65.

Hendricks, K. B. and Singhal, V. R. (2005), An Empirical Analysis of the Effect of Supply Chain Disruptions on Long-Run Stock Price Performance and Equity Risk of the Firm, in: Production and Operations Managemement 14 (2005) 1, pp. 35–52.

Homburg, C. and Giering, A. (1998), Konzeptualisierung und Operationalisierung komplexer Konstrukte - Ein Leitfaden für die Marketingforschung, in: Hildebrandt, L. and Homburg, C. (eds.) (1998), Die Kausalanalyse - Ein Instrument der empirischen betriebswirtschaftlichen Forschung, Stuttgart 1998.

Jap, S. D. and Anderson, E. (2003), Safeguarding Interorganizational Performance and Continuity Under Ex Post Opportunism, in: Management Science, 49 (2003) 12, pp. 227–245.

Jensen, M. C. and Meckling, W. H. (1976), Theory of the Firm – Managerial Behaviour, Agency Costs and Ownership Structure, in: Journal of Financial Economics, 3 (1976) 4, pp. 305–360.

Joshi, A. W. and Stump, R. L. (1999), Determinant of Commitment and Opportunism – Integrating and Extending Insights for Transaction Cost Analysis and Relational Exchange Theory, in: Canadian Journal of Administrative Science, 16 (1999) 4, pp. 334–352.

Kajüter, P. and Kulmala, H. I. (2005), Open-Book Accounting in Networks – Potential Achievements and Reasons for Failure, in: Management Accounting Research, 16 (2005) 2, pp. 179–204.

Pratt, J. W. and Zeckhauser, R. J. (1985), Principals and Agents: An Overview, in: Pratt, J. W. and Zeckhauser, R. J. (eds.) (1985), Principals and Agents – The Structure of Business, Boston, pp. 1–35.

Rokkan, A. I. and Buvik, A. (2003), Inter-firm Cooperation and the Problem of Free Riding Behaviour – An Empirical Study of Voluntary Retail Chains, in: Journal of Purchasing and Supply Management, 9 (2003) 5/6, pp. 247–256.

Rossiter, J. R. (2002), The C-OAR-SE Procedure for Scale Development in Marketing, in: International Journal of Research in Marketing, 19 (2002) 12, pp. 305–335.

Seiter, M. and Isensee, J. (2007), Partner Selection in R&D Cooperations – An Empirical Study in the Software Industry, Research Paper, International Performance Research Institute, Stuttgart 2007.

Tirole, J. (1999), Incomplete Contracts – Where Do We Stand? in: Econometrica, 67 (1999) 4, pp. 741–781.

Williamson, O. E. (1975), Markets and Hierarchies – Analysis and Antitrust Implications, New York.

Williamson, O. E. (1991), Comparative Economic Organization – The Analysis of Discrete Structural Alternatives, in: Administrative Science Quarterly, 36 (1991) 2, pp. 269–296.

Wuyts, S. and Geyskens, I. (2005), The Formation of Buyer-Supplier Relationships – Detailed Contract Drafting and Close Partner Selection, in: Journal of Marketing, 69 (2005) 10, pp. 103–117.

Chapter 16: SCRM and Performance – Issues and Challenges

Bob Ritchie* and Clare Brindley

*Corresponding author: Lancashire Business School, University of Central Lancashire, UK

16.1 Introduction

There is little doubt that risk, or at least our perception of risk, is becoming more prevalent in almost every dimension of our lives. Not only do we perceive greater exposure, increased likelihood and more severe consequences, we also become aware of risks previously unknown. As with individuals, organizations are continuously receiving information inputs suggesting new risks, enhanced exposure to existing risks and escalating costs associated with compensation should such risks materialize. A study of the views of 500 financial executives in Europe and America (FM Global 2007) concluded that they perceived an increase in overall business risks in the foreseeable future, with supply chain related risks featuring as one of the top three risks alongside competition and property-related risks. Several authors (e.g., Smallman 1996; Giannakis et al. 2004) have evidenced the emergence of risk management as an important contributor to most fields of management decision and control, including Supply Chain Risk Management (SCRM).

Underlying the increasing responsiveness of many organizations to engage in 'risk management', howsoever defined, is the need to evidence the justification for such an investment in terms of benefits to corporate performance. In other words, is Risk Management cost-effective? The emerging field of Supply Chain Management is an appropriate field to evaluate such issues, since it has the capacity to demonstrate a diversity of risks and risk management responses as well as producing an impact across most dimensions of an enterprise's performance. Christopher and Lee (2004) recognize the increasing risk in organizational supply chains and identify the need for new responses to manage these. Brindley (2004) suggests that global competition, technological change and the continuous search for competitive advantage are the primary motives behind organizations turning towards risk management approaches concerning their supply chains.

The primary purpose of this chapter is to explore the interaction between risk and performance, more specifically to start to address the question of how engagement in risk management activities might impact on corporate performance. Initially, the constructs of performance and risk are defined and mapped against each other, seeking to provide new perspectives for researchers and practitioners. The approach involves the review of the conceptual and empirical work in the supply chain management field, the SCRM field and performance. Risk in the supply chain is also explored in terms of risk/performance sources, drivers, consequences and management responses. The duality of risk management and performance management are examined in the context of a framework and illustrated in terms of the supply chain context. Two key challenges facing the research and practitioner communities are addressed: (1) the ability to prescribe strategies to address particular risk drivers and (2) the interaction of risk management with performance. An empirical case is used to illustrate the application of the Framework and the associated issues.

An Agency Theory approach has been employed in describing the roles, interactions and relationships within the supply chain, employing the terms *principal* and *agent* (Eisenhardt 1989). The value of this approach is demonstrated in situations where contractual arrangements, risk sharing and other mutually supportive relationships evolve between the parties, the situation found in many supply chains. Melnyk et al. (2004) supported the Agency Theory approach in analyzing supply chains since it emphasizes the primary importance of performance measurement or 'metrics' rather than the notion of the 'contract' itself. The *principal* may be located at different stages or nodes in the vertical supply chain (e.g., manufacturer or logistics organization). At each node we might expect to find a *Principal-Agent* relationship and most probably a multiple of these between the *principal* and several *agents*. It is recognized that certain *principals* within the chain may be influential beyond their immediate contact with agents or more distant supply chain members. In practice the principal organization may seek to specify the performance criteria and identify the associated risks in relation to its portfolio of individual suppliers or distributors. Consequently, each agent will seek to negotiate an agreement (e.g., contract) in terms of performance, risk sharing and reward outcomes. It is predicted that the principal's primary focus is likely to be towards the development and implementation of outcome-related performance measures (Melnyk et al. 2004) whilst seeking to manage or minimize the risk exposure. Such measures may include behaviour-related performance measures, shared information systems and other control systems to monitor and manage the interface with its agents (Eisenhardt 1989). Following negotiation and agreement on the performance levels and risk sharing, the post-contractual process then becomes one of monitoring, controlling and managing performance and risks. Most businesses will be engaged in a multiple of such agreements. However, the extent to which these are formally constructed and their degree of specificity is likely to vary, depending on the importance of the relationship and possibly the cultural norms in the industry sector or country.

16.2 Risk and Performance

Conventionally in the financial decision context, it has been accepted (e.g., Knight 1921) that risk and performance are directly and typically positively related. The higher the risk taken then the higher would be the expectation of higher rewards or financial returns. Figure 16.1 summarizes the possible range of outcomes between perceived risk and expected performance. Cell A demonstrates the relationship between low risk perception and the expectation of low returns in terms of performance. Cell D, on the other hand correlates potentially high rewards/performance with higher levels of perceived risk. The remaining cells suggest that either low risk situations may potentially generate high performance outcomes (Cell B) or alternatively, high risks may often only produce modest or low levels of performance (Cell C). Only four out of the infinite set of potential risk-performance outcomes are illustrated as exemplars.

The relationship in practice is perhaps less predictable than this might suggest. For example, the position prior to the investment decision (e.g., the *ex-ante*) will usually be different from the position at the conclusion of the investment (e.g., the *ex-post*). The realization of an exceptional outcome, from a range of probable anticipated outcomes, may prove financially advantageous but may have exposed the organization to significantly higher levels of risk in the process of achieving this. Hence, the time perspective is a critical dimension of any decision involving risks – 'decisions taken in the *ex-ante* timeframe reap their rewards in the *ex-post* timeframe'. There is another feature worthy of note relating to the timeframe. This concerns the anticipated length of time from the decision being taken to the realization of the outcomes. It may be anticipated that the shorter the anticipated elapsed period from decision to outcome realization, then the higher the predictability and thus lower the risk (e.g., short term decisions may on balance be less risky than long term decisions).

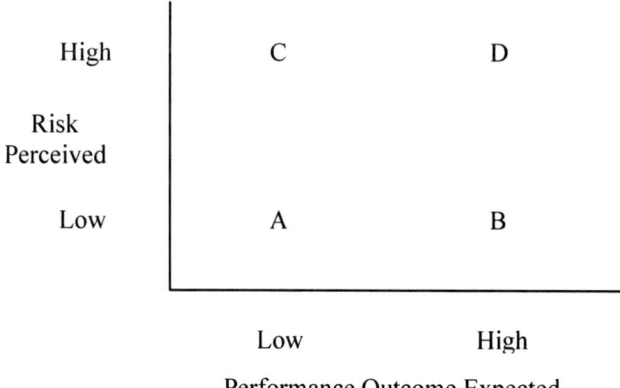

Fig. 16.1 Risk/performance relationship

Another important feature associated with risky decisions is the recognition that individual investment decisions are typically part of a portfolio of activities. The risk/performance profile ascribed to a particular investment decision needs to be viewed in the context of the portfolio of activities and not simply in isolation. Developments in Portfolio Theory in the Financial Economics (e.g., Capital Asset Pricing Method – Ball and Brown 1968) field provides some interesting and useful guidance for practical decisions but fails to reflect the 'messier, partial and fragmented' situation found in most organizations. In essence, the organization may have a combination of the four categories of investment identified in Fig. 16.1, although there may well be predominance of a particular category, reflecting the overall preparedness to undertake risk in response to enhancing performance.

These four elements, the *risk-performance relationship, portfolio of investments*, the *timeframe* and the *risk-preparedness* of the organization, although initiated within the context of financial markets and relating primarily to security transactions and shareholder wealth performance, are generally held to apply more widely within business decision making. The remainder of the discussion in the present chapter will focus on supply chain risks and their management. Several authors (e.g., Child and Faulkner 1998) have argued that decisions relating to changes in the supply chain structure and relationships ought to involve the analysis and evaluation of the associated potential outcomes in terms of benefits, costs and risks. Lonsdale and Cox (1998) likewise contend that performance and risk are inextricably interconnected and thus require the development and application of supplier management tools and controls to maximize performance whilst controlling the consequential risks.

16.3 Risk – Uncertainty and Risk Management

Definitions of the term risk are almost infinite (Ritchie and Marshall 1993), usually varying according to the specific situations, decision contexts and problems. Sitkin and Pablo (1992, p. 9) provide a very generalized definition of risk as *'the extent to which there is uncertainty about whether potentially significant and/or disappointing outcomes of decisions will be realized'*. Zsidisin (2003, p. 15) provides a definition more focused on the supply chain context which is the subject of the present chapter, defining risk as *'the potential occurrence of an incident or failure to seize opportunities with inbound supply in which its outcomes result in a financial loss for the [purchasing] firm'*. Most generalized definitions (e.g., MacCrimmon and Wehrung 1986) of risk have in common three dimensions:

> *Likelihood of occurrence* of a particular event or outcome
> *Consequences* of the particular event or outcome occurring
> *Exposure or Causal pathway* leading to the event

In the context of Decision Theory, these components are readily identifiable and measurable, lending themselves to formulaic and precise resolution. The practical application within risk management yields a totally different context. The *likelihood of occurrence*, more usually termed the probability, can be expressed in objective terms or in subjective terms, each being capable of measurement, although utilizing differing scales, often derived from experience and intuition. *Consequences* are typically expressed as a multiple of simultaneous outcomes, many of which interact with one another (e.g., failure of a new product launch may generate consequences for the organization's reputation, financial performance and the standing of the individual product champion). Consequences should not simply be regarded as only or primarily negative, since the essence of risk taking is the potential opportunity to produce positive outcomes (Blume 1971). The third dimension of the risk construct, the *causal pathway*, has particularly important implications for risk management. Understanding the nature, sources and causes of factors that generate the events or circumstances which might influence type and scale of consequences (e.g., both positive and negative), and the likelihood of their occurrence are fundamental requirements for effective risk management. The term *risk driver* is used in the present discussion to represent these sources and causal pathways and will be explored in detail later in the chapter. Risk management seeks to address all three of these dimensions concurrently, analyzing the potential risk sources (existing, novel and unexpected), seeking to understand the forces that may drive a particular sequence of events and how these might be managed to improve the chances of positive outcomes in terms of performance and by corollary avoid negative consequences. Such analyses may enable a more pro-active management approach (e.g., consumer liability insurance, securing alternative suppliers) to modify if not necessarily to eliminate the potential negative consequences should these occur. This diversity of potential outcomes and measurement systems reflects the real-life complexity and flux of what is often perceived as a simple theoretical construct.

The associated construct of uncertainty is viewed by many authors as a special case of the risk construct (Paulsson 2004), in which there is insufficient information (Rowe 1977), knowledge or understanding to enable the decision taker to identify all of the potential outcomes (Ritchie and Marshall 1993), their consequences or likelihood of occurrence (MacCrimmon and Wehrung 1986). In its extreme form, uncertainty relates to the situations where there may be either a total absence of or possibly lack of awareness of a potential occurrence, whether generating positive or negative outcomes. The term uncertainty is often associated with the timescale of the parameters of the decision situation, usually relating to some distance into the future. The terms risk and uncertainty are frequently used interchangeably, as typically risk contexts involving decisions are often somewhere in the middle of the risk-uncertainty spectrum (e.g., neither pure 'objective' risk taking nor complete uncertainty).

16.4 Performance

As with the term *risk*, the term *performance* lends itself to an almost infinite variety of definitions, many of which relate to specific contexts or functional perspectives. Anthony (1965) provided a generic and now well-established definition of performance, dividing this construct into two primary components, efficiency and effectiveness. Efficiency addresses performance from a resource input-output perspective such that the greater the volume of outputs for a given volume of inputs then the greater the efficiency. Effectiveness addresses performance related to the degree to which the planned outcomes are achieved (e.g., achieving the objective of avoiding supply disruptions during a given period may be viewed as an effective outcome). However, these two dimensions of performance need not necessarily operate in unison. For example, the avoidance of supply disruptions during the period may have required maintaining high buffer stocks or special incentive payments to the supplier, both of which may prove inefficient in terms of profitability but highly effective in terms of customer service levels. Many of the earlier definitions of performance tended to focus on the efficiency dimension, featuring financial performance as the primary outcome measure. Subsequently, more encompassing definitions of performance have evolved, most notably the Balanced Scorecard (Kaplan and Norton 1992, 1996) which incorporates not only the Financial Perspective but also the Internal Perspective, the Customer Perspective and the Innovative and Learning Perspective. There is a continuing pursuit of performance measures and metrics in all fields of business activity, seeking to capture in a readily understood form, the performance of the business unit in terms of effectiveness and efficiency. For example, Melnyk et al. (2004, p. 210) reflecting on metrics and performance measurement concluded that 'performance measurement continues to present a challenge to operations managers as well as researchers of operations management'.

16.4.1 Timeframe – Risk and Performance

Another feature of managing risk and performance is what we have defined as the timeframe. Conventionally, the timeframe has been divided into the *ex-ante* and the *ex-post* perspectives. Risk management is primarily located in the ex-ante stage of the timeframe since its activities are designed to identify risks; assess their potential likelihood and impact on performance; undertake steps to reduce or minimize the likelihood; and institute actions to ameliorate the impact should they occur. Most of these activities are undertaken prior to the time period in which the event is likely to take place. Risk management may still perform a role in the ex-post timeframe, but this is more concerned with evaluating the reasons for the risk and performance outcomes, instituting corrective actions to resolve undesirable consequences and accumulating knowledge and experience to aid future decisions and risk management strategies. The success of risk management is ultimately viewed in the ex-post timeframe, reflecting factors such as the total avoidance of potential risks, the avoidance of any detrimental impact, minimizing the detrimental outcomes and ensuring the benefits of the risk management actions are not outweighed by the costs involved.

A further dimension of risk management relating to the timeframe is the extent to which the risks being managed can be identified as singular 'one-off' risks or a continuous flow of risks. Certain industry sectors may conventionally be focused on a single activity or project; (e.g., transportation of a particularly hazardous cargo) others are more concerned with a continuous flow of products and services throughout each of the nodes in the supply chain. The former category may distinguish between the ex-ante and ex-post positions in the timeframe but the latter are less likely to recognize this difference.

Risk, or more precisely the perception of risk, is conditioned by the passage of time. For example, early stage views may be tainted with significant levels of uncertainty, largely reflecting a lack of, or imprecise information and knowledge (e.g., what are the possible risks – who are the players and what are the drivers?). The process of Risk Management involving the investment of time, information gathering and individual mental dexterity may help to resolve the risks perceived and re-affirm preventative actions and amelioration. Gaining information, knowledge and understanding may lead to enhanced confidence and reduced perceptions of risk. There is a corollary to this anticipated outcome in that gaining more information and knowledge about possible risks and outcomes may equally diminish confidence and heighten risk perceptions concerning risks and consequences previously unknown. In the 'one-off' decision case we can envisage that the progression of time combined with the investment in risk management may ideally lead to a reduction in risk perception and increased confidence, although recognizing that this is unlikely to eliminate the risks but may simply provide assurance that they are adequately covered. In the continuous flow context associated with many supply chain activities and processes, the passage of time either without any incidents or incidents that have been well handled and foreseen may again build the confidence level of the risk management team. The difference in this latter context is that responses have to be made to risk outcomes as they occur including making changes immediately to avoid their recurrence in the continuous flow in the next cycle. Time has been demonstrated as an evident, if somewhat implicit, parameter in the risk management process.

This timeframe has also been employed as the basis for subdividing decisions into different categories, for example, operational, tactical and strategic. Such categorization reflects Paulsson's (2004) approach to differentiating supply chain risks as Operational Disturbance, Tactical Disruption and Strategic Uncertainty. Kleindorfer and Wassenhove (2003), however, subdivided supply chain risks into only two categories Supply Co-ordination Risks and Supply Disruption Risks, dividing Paulsson's tactical level decisions between the two categories. Whilst such categorization may prove helpful conceptually, Mintzberg and Waters (1985) highlighted the difficulty in differentiating these in practice, suggesting that strategic decisions are in essence the aggregation of a sequence of operational and tactical decisions, leading to some common planned or emergent pattern. Similarly, the differentiation between tactical and operational decisions may prove somewhat arbitrary in practice. Nevertheless, in the present discussion of supply chain risk management it is helpful to continue this differentiation in exploring the different types of risk exposure and the potential risk management responses.

16.5 Performance and Risk Metrics

Questions arise as to the optimal level of investment in SCRM and how the actions taken might be measured in terms of changes in performance (e.g., the resulting risk and performance profiles for the organization as a whole). The measurement of corporate performance and risk may be addressed from a number of different although not mutually exclusive perspectives. Each stakeholder may seek a different balance in terms of the timeframe (e.g., short-term versus long-term performance), prioritize different criteria (e.g., profitability, cash flow or customer service) and demonstrate different attitudes to the exposure to risk (e.g., avoidance of high risk exposure). Stainer and Stainer (1998) identified eight different categories of stakeholder (e.g., shareholders, suppliers, creditors, employees, customers, competitors, Government and society) together with their differing performance expectations from the business. A basic tenet of performance for most, if not all, of these stakeholders concerns both profit performance and the risks associated with achieving this performance. Mathur and Kenyon (1997) highlighted the reason for the primacy of financial performance metrics on the basis that failure to achieve the minimum level would result in the demise of the business and the consequent loss of the shareholder's investment with the further impact on the potential benefits to all other stakeholders (e.g., Government, employees). Whilst the primacy of financial performance measures, especially this narrower profitability measure may be challenged on the basis that 'it should relate to the ultimate outcome of a better society for all' (Stainer and Stainer 1998, p. 7), there is little doubt that financial performance outcomes remain the primary metrics for commercial situations such as SCRM. Accepting the importance of financial metrics, Marsden, (1997) argues for a more long-term perspective rather than the short-termism, often implied by many financial performance measures.

A number of researchers have developed models of corporate performance which incorporate the risk measure. Bettis (1982), for example, developed a model linking performance to the three independent variables Industry Characteristics (IC), Strategic Decisions (S) and Risk (R):

Performance $= f(\text{IC}, \text{S}, \text{R})$

Risk in turn was viewed as being determined by the two variables Industry Characteristics and the Strategy developed. Bettis and Hall (1982) and Bowman (1980) supported this view that risk is essentially an endogenous variable, with risk exposure being largely determined by the strategic choices of the business. Similarly, Ritchie and Marshall (1993, p. 165) concluded that 'a well devised strategy could simultaneously reduce risks and increase returns'. The performance metrics focus of these studies essentially employed a financial definition of performance (e.g., return on assets (ROA); return on investment (ROI)) and generally sought to maximize this over the longer term period as opposed to the short term such as the current year.

The parallel development of Portfolio Theory and more specifically the Capital Asset Pricing Model (CAPM) (Ball and Brown 1968) provides a further contribution to the conceptual development and practical application of the risk-performance interaction. CAPM generally seeks to explain the relationship between the financial performance of the organization (e.g., in terms of reported earnings) and the share price in the securities markets. Risk, measured as the degree of variability of the returns over a period of time (e.g., greater the frequency of variations and their amplitude then the higher the risk), represents a critical variable. The risk is subdivided into two components, *systematic risk* and *unsystematic risk*. The former relates to the risks experienced by all organizations as a function of the environment within which they operate (e.g., macro-economic conditions, political situation and competitive structure within the industry). Systematic risk represents the risk exposure that cannot be avoided by the organization irrespective of the actions or strategies that might be instituted. This does not mean that risk management actions will not enable the company to modify the consequences or ameliorate the impact (see Kleiderhorfer and Saad 2005). Unsystematic risk relates to those risks that are company-specific and are generally within the control of the business itself in terms of the strategies it formulates and their effective implementation. The potential contribution of CAPM to the field of risk management in general (e.g., Ritchie and Marshall 1993) and more especially SCRM (e.g., Kleiderhorfer and Saad 2005) is increasingly being recognized.

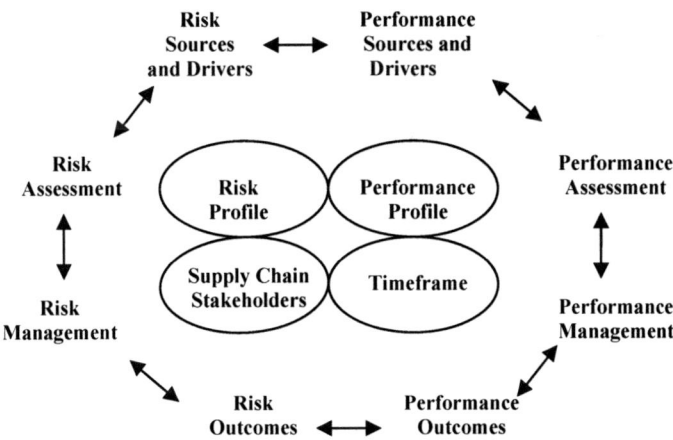

Fig. 16.2 Supply chain risk management framework – process overview

The SCRM Framework (Fig. 16.2) may be subdivided into two components, the enduring, aggregate and strategic performance components at the core of the Framework and the ongoing components of the risk and performance management processes in the outer ring. Dealing with the strategic performance:

- The *Performance Profile* represents a multi-dimensional view of the organisation (e.g., equivalent say to the Balanced Scorecard (Kaplan and Norton 1992, 1996) incorporating Financial, Internal, Customer, Innovative and Learning perspectives. This profile reflects the strategic aims of the organization and also recognises the need for flexibility in balancing the various, often incompatible, performance criteria.
- Similarly, the *Risk Profile* represents the nature of the risk exposure that the organisation is prepared to endure. It is important to recognise that this profile comprises a portfolio of different risks which are linked to the portfolio of projects/activities/investments represented by the performance profile.
- The *Timeframe*, as discussed earlier, indicates the importance of locating the point in time at which risk and performance are viewed, assessed and managed.
- Introducing the *Supply Chain Stakeholders* as a component of the organisation's performance is designed to emphasise the inter-dependence of the members of the supply chain. Expectations of performance and risk profiles for the organisation must be cognisant of the other partners and stakeholders in the supply chain.

The outer ring (Fig. 2) represents the ongoing processes of risk and performance management. On one side the Risk Management incorporates the sequence of identifying risk sources and drivers – assessing their potential impact, instituting appropriate risk management responses and evaluating the impact on performance criteria. The performance management sequence parallels that for risk management. Three important points relate to the interpretation of the risk and performance management process elements in the Framework:

- The sequence of actions within each of the processes is not a linear progression; it often requires the re-iteration of previous stages to achieve greater understanding or verification.
- The risk and performance processes are inextricably joined in practice with decision makers continuously involved in balancing risk and performance information.
- The more tactical or operational activities involved are informed by and also inform the core components, namely the performance profile, risk profile and stakeholder profile, albeit often from a different timeframe.

16.6 Risk-Performance: Sources, Profiles and Drivers

Risk in a supply chain is evidently influenced by a large number of different sources and factors. Categorizing these sources and factors may prove helpful in understanding the nature of SCRM. One possible categorization based on the model initially developed by Ritchie and Marshall (1993) suggests a categorization into seven groups of sources and risk drivers:

$$\text{Risk} = f(E_r, I_r, SC_r, SS_r, O_r, P_r, DM_r)$$

where:
- E_r = environmental variables
- I_r = industry variables
- SC_r = supply chain configuration
- SS_r = supply chain stakeholders
- O_r = organizational strategy variables
- p_r = problem specific variables
- DM_r = decision-maker related variables

This formulation provides the basis for the variables grouped in Fig. 16.3, indicating that these variables not only influence risk (systematic and unsystematic) but also potential performance. Factors from the seven sources (Fig. 16.3), singly or in combination, determine the risk and performance profile for the organization at that point in time and for that particular decision or set of decisions. It is

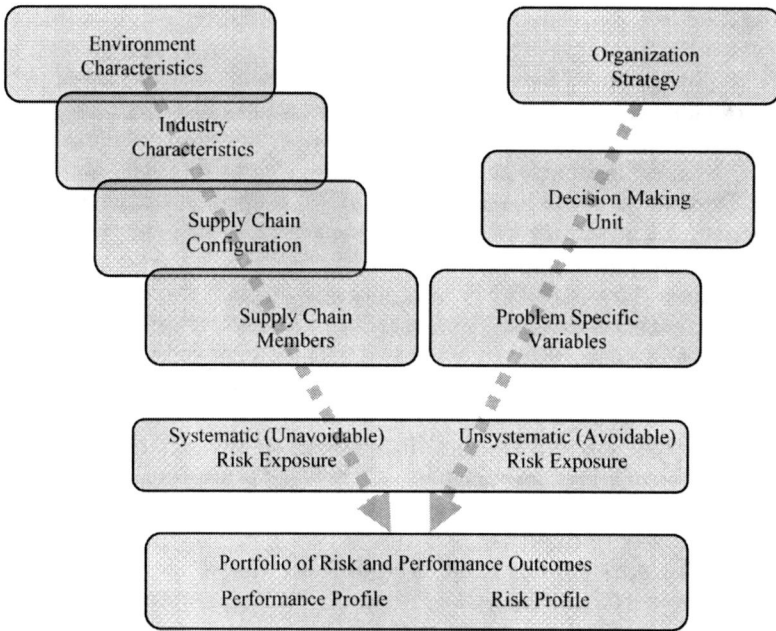

Fig. 16.3 Risk and performance: sources and drivers

recognized that any one of the seven sources may generate new risks at any time on a continuous basis, affecting both the risk and performance profiles of the organization. Such profiles represent a dimension of the portfolio of investments and activities for the organization and it is the aggregate performance that is ultimately of concern to the management. *Systematic risk* and *unsystematic risk* are included in the profile to illustrate the different nature of the risk sources (e.g., environment sources tending to be systematic and problem specific more unsystematic), although recognizing that it is not always possible to differentiate between these, as the passage of time itself may alter the position (e.g., simply delaying action may well enhance the systematic risk).

There are potentially an almost infinite number of factors exposing the business to undesirable consequences in terms of performance and risk. The organization needs to establish which are critical and which are less so, accepting that only a very small proportion of the total will fall within this high risk category. The term *driver* has been introduced to differentiate those factors most likely to have a significant impact on the exposure (e.g., likelihood and consequences) to undesirable performance and risk outcomes. Performance drivers may well offer opportunities to enhance performance outcomes, albeit by incurring increased risk. For example the decision to develop a new direct channel to the consumer, bypassing existing distribution channel members would expose the business to new risks both from the reaction of the consumer and the retaliatory actions of the other channel members, although possibly improving potential performance outcomes. There are likely to be key risk and performance drivers associated with each of the seven sources listed in Fig. 16.3. Not all of these are likely to be equally important and the composition of drivers may well vary over time and changes in the supply chain situation.

It should be emphasized that the Framework relates not solely to those risks that pose a major threat to the survival and future development such as crises. More importantly, the Framework addresses those risks that have an influence on the ongoing 'normal' performance of the business in terms of effectiveness and efficiency. However, it would make sense to direct management attention only to the appropriate drivers which are likely to have a significant impact on the performance or risk profile.

Another feature associated with SCRM concerns the interrelationships within the typical supply chain network. This means that risk and performance changes may affect many, if not all members associated with the supply chain. Potentially, all members within a network will be exposed to the risks, although the direct impact may be ameliorated or modified by the actions taken by others in the chain. Members throughout the supply chain may benefit if all partners engage in systematic SCRM activities, although one can see benefits also from encouraging others to undertake SCRM and its associated costs, whilst avoiding these from your own organization's perspective.

The underlying presumption of the SCRM Framework is that the risk and performance sources and drivers can be foreseen, identified, evaluated, prioritized and managed. The practical reality suggests considerably more 'fuzziness' relating to these drivers, their impact on performance and the effectiveness of possible management solutions. This begs the question as to why organizations bother to prepare themselves for risk encounters and invoke SCRM. Quite simply, organizations believe that the best approach is to accept that they will be exposed to risks and the best strategy is to become more aware and pro-active towards the risks and better prepared to respond more quickly should such risks materialize (Kovoor-Misra et al. 2000). The simultaneous presence of different drivers may have a compounding effect both on the exposure and potential consequences. For example, the dislocation of supplies due to transport failures may be compounded by inadequate communications between supply chain members and further exacerbated by poor management controls in the principal organization. Alternatively, the risk drivers may be counterbalanced by high levels of performance, enabling the management to effectively control the worst consequences of supply disruption.

The process of identifying, assessing, prioritizing and evaluating sources and drivers will yield an assessment of the *ex-ante* profile in relation to risk and performance consequences for the organization at a particular point in time. The organization may assess dimensions of this profile which it considers unacceptable or undesirable and consequently seek to consider actions to address these.

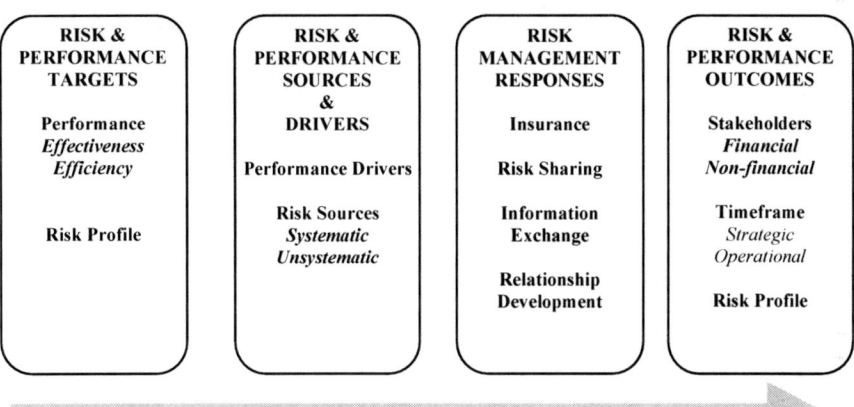

Fig. 16.4 SCRM framework

16.7 Risk Management Responses

The range of risk management responses is typically extensive, although some exemplars of risk management responses (Fig. 16.4) include risk insurance, information sharing, relationship development, agreed performance standards, regular joint reviews, joint training and development programmes, joint pro-active assessment and planning exercises, developing risk management awareness and skills, joint strategies, partnership structures and relationship marketing initiatives. A number of supply chain groups have displayed some degree of progression in the risk management responses, leading from the more individualistic and independent responses (e.g., insurance, establishing supplier service levels) to the more co-operative responses (e.g., sharing strategic information, relationship development and partnering). Similar developmental trends were discovered by Kleindorfer and Saad (2005) when examining disruption risks in supply chains in the US chemical Industry. Progressing to more integrated and comprehensive approaches may be the result of the nature and severity of the potential risk consequences, the number of members involved in the supply chain and enhancements in terms of confidence and trust between the parties engaged in the supply chain. The initiation of such developments is often undertaken by the larger organizations in the supply chain, since they may have the resource base and expertise, as well as the comparative power to encourage participation by smaller enterprises.

Risk management needs to target both the systematic and unsystematic risk sources. Whilst the organization may not be able to control sources of systematic risk (e.g., changes in interest rates or political instability), the adoption of particular management approaches may enable the organization to better understand the risks and to respond more quickly and effectively to modify or ameliorate the impact on performance (e.g., reducing financial risk exposure in advance by limiting borrowing). Similarly, improving the awareness and understanding of unsystematic risks can enable the development of strategies which either avoid or minimize exposure to identified risks. Equally, developing relationships and partnerships with key members of the supply chain may provide a more generic shield against exposure to other unsystematic risks emanating from elsewhere in the supply chain.

16.8 Risk and Performance Outcomes

The risk and performance outcomes may be viewed from three important perspectives, *stakeholders, timeframe* and *risk profile* or portfolio.

1. *Stakeholders*: The risk and performance outcomes need to reflect the different expectations of the variety of stakeholders involved in most organizations. This often diverse group may have differing ex-ante expectations concerning both performance criteria and the nature of the risk profile. The risk management task is to identify and prioritize these expectations (e.g., which are the most critical groups to satisfy) and to develop appropriate strategies to achieve these. Certain stakeholder groups may respond negatively if the ex-post performance outcomes fail to meet their ex-ante expectations (e.g., shareholders selling shares if returns are below expectation). Examples of stakeholders and possible expectations include:

    Suppliers: *Assurances on payment and future orders*
    Shareholders: *Return on Share Capital*
    Lenders: *Security Assurance on Lending*
    Employees: *Assurances on future employment*

Although financial and quantitative performance measures tend to be predominant there are developments discussed earlier (e.g., Balanced Scorecard) that are resulting in more qualitative and subjective measures of performance to provide a more holistic perspective.

2. *Timeframe*: The timeframe in which the performance/risk outcomes are assessed varies not just between the ex-ante and ex-post perspectives as discussed earlier in the chapter, but also in relation to short term as opposed to long term performance. At the operational or transaction level it may be appropriate to measure and monitor risk and performance (e.g., changes in the level of unpaid accounts), on the basis that over time these may prove a significant risk source. Measuring risk and performance at the organizational level typically involves more aggregated and longer-term measures, although investors are often criticized for adopting an unnecessarily short term perspective of risk and performance.

3. *Risk Profile*: Changes in performance outcomes are usually paralleled by changes in the risk profile. Deterioration in financial performance will usually be reflected in a higher risk profile although the two may not always be strongly correlated. Viewing the wider portfolio of investments and activities of the business including the 'upside' and 'downside' risks may well produce levels of confidence not evident from simply assessing the historical financial performance.

16.9 Empirical Application – Illustrative Case

In parallel with the conceptual development of the Framework, empirical investigations were conducted seeking to collect, structure and analyze the practice based evidence. This was necessary to ensuring greater correlation or fit between the Framework and practice, enhancing its robustness and improving the quality of guidance that may be proffered. The in-depth case analysis approach addressed four questions:

1. To what extent are the supply chain members aware of SCRM?
2. How do the supply chain members identify, evaluate and prioritise the risk drivers?
3. How do they perceive risk and the risk/performance interaction?
4. What are the primary risk management responses employed?

The ongoing programme of investigative empirical research into supply chain risk management has been active over the last six years, involving a number of organizations. This longitudinal research has enabled the assessment of the evolutionary development of SCRM and the risk management responses employed (e.g., transfer of information, knowledge and learning between members). The empirical approach is essentially qualitative, utilizing external contextual information, internal documentation and interviews with key staff within the supply chain organizations (e.g., principals and agents). The case study methodology was particularly appropriate as 'the focus [was] on a contemporary phenomenon within some real-life context' (Yin 1994, p. 1). The case study approach developed was primarily an 'illustrative' case study, designed to explore new and innovative practices (Scappens 1990). The analytical approach employed mirrored that of well-documented practice (e.g., Baker 1976; Mitchell 1983; Smith and Dainty 1991) and involved assembling the raw material (e.g., contextual data, working documents, interviews and observation notes), structuring it on the basis of the conceptual Framework, reviewing the evidence, checking assumptions made, then constructing the detailed case record. The usual difficulties encountered by longitudinal studies (e.g., changes in the external and internal contexts, changing personnel and changes in the strategies adopted) required the exercise of care to ensure the appropriate interpretation and use of the evidence. Only one of the cases is reported in this chapter, as the intention is to illustrate the application of the Framework in a given case setting and not in any sense to draw generalized conclusions.

16.9.1 Case Study

The case company is a multinational organization operating in 140 different countries and manufacturing and selling a range of high value agricultural equipment, such as tractors under a variety of brand names. The focus of the investigation was the principal's supply chain interactions with its distribution channels and dealerships. These independent agents, although small in number, around 40 covering all of England, represent a critical interface between the principal and its ultimate consumers, the farming community, since the principal itself had no direct presence in the marketplace.

16.9.2 Supply Chain Members' Awareness of SCRM

The principal demonstrated an awareness of the risks associated with the downstream supply chain at the start of the case investigation, some six years previously. The approach at this time reflected more the conventional 'arms-length' management of its distribution agencies, achieved through contractual agreements, regular monitoring of sales performance and 'fire-fighting' problems as they arose. In essence, SCRM involved risk identification, limited analysis and no evidence of pro-active risk management, with the possible exception of the selection of new agents. More recently, the principal's management has recognized that a more pro-active approach to managing the interaction with its distributors might reduce exposure to risks and enhance performance. Given the international scale of the company, this change in attitude towards SCRM paralleled similar developments elsewhere in the organization's global operations. The distributors themselves proved broadly responsive to the initial proposals to establish closer links between themselves and the principal. Underlying concerns about the motives and the possible implications for their independence were outweighed by recognition of the potential risk resolution (e.g., more advanced information concerning the principal's strategy and sales forecasts) and re-assurances about performance levels.

16.9.3 Risk Drivers – Identification, Evaluation and Prioritization

The primary risk sources and drivers associated with the strategic developments in the supply chain were a composition of environmental and industry characteristics. Such factors include the drive for greater efficiency, leading to larger farming units and increased intensity of land use. The consequence has been a reduction in the number of potential customers; technological developments yielding new products, higher added value and significant reductions in costs, more frequent replacement (new models on average every 3 years), intensified competition and product differentiation, and the aftermath of the Foot and Mouth epidemic. These risk sources had implications for customer's investment potential, influencing the potential future demand and making the market more price-sensitive. Many of these risk sources and drivers may clearly be categorized as systematic and were viewed by the management as largely outside of their control. The competitive pressure to realize scale economies and cost reductions throughout the manufacturing and supply chain resulted in increased centralization of production facilities. This resulted in the Company ceasing to manufacture in the UK. Strategic decisions such as this represent a form of unsystematic risk, since the Company has some choice in whether or not to take the decision and hence incur the risks.

In the more immediate term the risk sources and drivers were more directly related to the Supply Chain interactions between principal and agents (e.g., stockholding of finished products at agent's site; delivery times for made-to-order products; stockholding levels of parts; service quality and support; sales performance; retention of skilled labour within the dealership). More operational risk drivers such as those associated with effective stock management, credit control, and sales order processing had the capacity to generate significant levels of risk and attention, albeit at a much more localized level (e.g., involving a single customer and distribution agent). Priority and the focus of management attention was primarily located within the tactical or more medium term level risks, identifying what was driving these, assessing the potential impact on performance for both parties and considering responses to mutual resolution. Such risks were predominantly unsystematic and hence management perceived these to be within their sphere of influence.

16.9.4 Perception of the Risk/Performance Interaction

The downstream relationship with their customers, via their dealership network, represented the Principal's main group of supply chain risk sources and drivers. They were critically aware of the importance of the dealerships in terms of service and support quality and that this was a significant element in customer buying behaviour, overall customer satisfaction and in generating repeat business. The situation in effect was a multi-dimensional set of supply chain interactions, involving the Principal, the agents and the customers. The effectiveness of the SCRM approach should help to reduce the risk and consequently enhance performance outcomes for the agents and the Principal. For example, reducing the likelihood of the customer experiencing poor service should ensure sufficient returns on their investments for the dealerships and for the principal itself. The caveat in terms of enhancing performance requires the development of an efficient SCRM approach, ensuring that the costs of managing the risks does not outweigh the increased benefits (e.g., profitability) for both the distributors and the agents.

16.9.5 Risk Management Responses

The management at the Principal organization developed three strands in their SCRM response to the key risk sources and drivers perceived and evaluated. All three strands involved developing a much closer relationship with all 40 distributors, involving changes to the structures, processes, systems and interactions that governed their principal-agent relationship at that time. These three initiatives comprised:

1. The development of Business Plans for each dealership in conjunction with the principal's representatives. A system of bonus payments to the dealership was linked to the achievement of agreed and planned performance outcomes. There was a combination of financially-oriented targets (e.g., achievement of target liquidity ratios to assess liquidity risk exposure) and other service quality related targets (e.g., percentage of parts that are supplied direct from the agent's own stock without the need to order; staff turnover levels). Although not always related specifically to possible risk and performance drivers, the approach positioned the management at both the principal and agent to respond more effectively to unexpected risks that arose.
2. The development of shared Management Information Systems, based on the principal's own extensive database plus regular market research reports (e.g., official UK vehicle registrations for agricultural equipment; competitor sales analysis by brand and model within individual post codes, areas and regions). This type of information on market developments, market shares and competitor activities would normally not be available to the distributors due to the cost of acquisition.
3. The most far-reaching initiative was the establishment of an agreed set of Performance Standards covering all the key dimensions of the dealership's operations (e.g., staff training and development programs; succession planning for senior management). These Performance Standards and the associated target levels of achievement were agreed following a process of negotiation between the two parties. The performance measures are largely non-financial and reflect long-term performance targets. Attainment above the agreed standards of performance is rewarded with bonus payments whilst below

target involves the commitment by the agent to initiate a program of remedial actions agreed with the Principal's representative.

The Principal organization sought to manage the risks in the downstream supply chain through a process of close collaboration, effective communications and the building of trust within the relationships. The development of Performance Standards was not seen by the distributors as the imposition of a risk management strategy to protect the principal but rather a collaborative approach that resolved the risks within the chain to the mutual benefit of all partners.

16.10 Conclusions and Future Developments

Risk management approaches have arguably been implicit in the management of supply chains over a long period of time. It is not until more recently (see Mentzer 2001; Brindley 2004) that a concerted effort has been made to study these in a more logical and coherent manner. This was evidently a result of significant pressures experienced both contextually and from within the supply chain to modify and in some situations radically change the structures, modes of operations and relationships. These processes of change appear to be more radical and revolutionary than evolutionary. This in itself may engender an increased sense of uncertainty and risk throughout the various stages of the supply chain.

The case illustrated the desire to incorporate metrics involving the risk-performance interface, a willingness shared by the agents as well as the Principal organization. There remain questions about the robustness and reliability of some of the more qualitative measures employed, although the Principal in the case expressed reasonable confidence in these as tools to aid risk management. Shepherd and Gunter, (2006, p. 253) reviewing the development of supply chain metrics more generally concluded that researchers 'should consider developing measures of supply chain performance as a whole, rather than measures of inter-organizational performance' and that there was a need to address 'the paucity of qualitative metrics and non-financial measures'.

The chapter aimed to develop a deeper understanding of the main constructs underpinning risk and performance within the supply chain, especially the inherent linkage between these two. The review of the literature and the development of a Framework provided the platform for the empirical assessment within the context of supply chains. Responding to the two key research challenges posed earlier:

The possibility of prescribing effective risk management strategies or responses and gaining an understanding of the risk – performance interaction.

The development of the conceptual Framework and the evidence from the empirical work suggests scope for the development of guiding principles. However, it is accepted that each supply chain situation will be unique and a contingency based approach may prove valuable.

How do risk and performance drivers interact? Can risk reduction only effectively be achieved by the deterioration in some aspects of performance?

The limited empirical evidence presented in the chapter suggests that organizations are increasingly recognizing the interaction between performance and risk. There is a trend towards risk and performance metrics which are not primarily quantitative, financially oriented and short-term. Other metrics are evolving, although these are designed to tackle some of the broader risk-performance issues and are strategically oriented. The development of improved relationships leading to more formalized associations is an evident trend in many supply chain situations. Improved relationships leading to trust between principal and agent should help to improve the effectiveness of SCRM. The potential exists to improve the measurement of the risks and performance consequences in most settings. The challenge is to develop multi-dimensional risk-performance measures which are both measurable and meaningful in the practical context.

Future research in the SCRM field needs to engage more effectively the practitioner community in the development of effective and efficient approaches in a number of aspects:

- *Risk identification, categorization and evaluation.* Improved frameworks which assist in identifying and categorizing different risk sources and causal pathways will assist in greater awareness and sensitivity to supply chain risks. Improving the understanding of the nature and scale of risks and especially their impact on aggregate performance should aid in focusing management attention to the most critical areas. Initial work by Brindley and Ritchie (2001), Paulsson (2004) and Kleindorfer and Wassenhove (2003) have provided a useful starting point for such developments.
- *Risk Management Responses and Practices.* The evidence to date (e.g., Ritchie and Brindley 2004) suggests a degree of commonality in the supply chain risk situations encountered by different organizations. Considerable benefits may be gained from practice-related research into the risk management responses and practices being employed in different sectors and in response to differing situational risk scenarios. The development of effective dissemination pathways providing the opportunity to provide guidance to practitioners to enable them to share their own experiences is a further important area for development.
- *Metrics.* Increased attention is being devoted to the development of appropriate metrics to provide effective and appropriate measures of performance (e.g., Ambler 2000; Shepherd and Gutner 2006). This chapter has argued for the recognition of the duality of performance and risk. Hence, metrics need to be developed which enable the decision maker to assess the impact on the risk profile as well as performance outcomes. The illustrative case study demonstrated the development of metrics in a particular setting designed to achieve this duality of approach. There remains further scope within the supply chain management and the associated risk management activities to develop meaningful, robust and practically relevant metrics.

This chapter, in pursuing the aim of enhancing understanding of the risk performance interface, has demonstrated the contribution that risk management approaches may make towards more effective, efficient and resilient supply chains.

References

Ambler, T. (2000), Marketing and the Bottom Line, Pearson Education, UK.
Anthony, R.N. (1965), Planning and Control Systems: A Framework for Analysis, Harvard University School of Business Administration, Division of Research, Boston.
Baker, M.J. (1976), "The written analysis of cases", Quarterly Review of Marketing, Vol. 1, pp. 1–6.
Ball, R. and Brown, P. (1968), "An empirical evaluation of accounting income numbers", Journal of Accounting Research, Vol. 6, No. 2, pp. 159–178.
Bettis, R.A. (1982), "Risk considerations in modelling corporate strategy", Academy of Management Proceedings, pp. 22–25.
Bettis, R.A. and Hall, W.K. (1982), "Diversification strategy, accounting determined risk and accounting determined return" Academy of Management Journal, Vol. 25, No. 2, pp. 254–264.
Bowman, E.H. (1980), "A risk/return paradox for strategic management", Sloan Management Review, Vol. 21, No. 3, pp. 17–31.
Blume, M.E. (1971), "On the assessment of risk", Journal of Finance, Vol. 26, No. 1, pp. 1–10.
Brindley, C.S. (ed.) (2004), Supply Chain Risk, Ashgate Publishing Ltd., UK.
Brindley, C.S. and Ritchie, R.L. (2001), "The information-risk conundrum", Marketing Intelligence and Planning, Vol. 19, No. 1, pp. 29–37.
Christopher, M. and Lee, H. (2004), "Mitigating supply chain risk through improved confidence", International Journal of Physical Distribution and Logistics Management, Vol. 35, No. 4, pp. 388–396.
Eisenhardt, K.M. (1989), "Agency Theory: an assessment and review", Academy of Management Review, Vol. 14, No. 1, 57–74.
FM Global (2007), "Managing Business Risk – Through 2009 and Beyond", FM Insurance Company Limited, Windsor, Berks, UK.
Giannakis, M., Croom, S. and Slack, N. (2004), "Supply Chain Paradigms", in New, S. and Westbrook, R. (eds.), Understanding Supply Chains, Oxford University Press, UK, pp. 1–22.
Kaplan, R.S. and Norton, D. (1992), "The balanced scorecard measures that drive performance", Harvard Business Review, January/February, 71–79.
Kaplan, R.S. and Norton, D. (1996), The Balanced Scorecard, Harvard Business School Press, Boston, MA.
Kleindorfer, P.R. and Saad, G.H. (2005), "Managing disruption risks in supply chains", Production and Operations Management, Vol. 14, No. 1, pp. 53–68.
Kleindorfer, P.R. and Wassenhove, L.K. (2003), "Managing Risk in Global Supply Chains", Paper presented to Wharton Insurance and Risk Management Department Seminar, 27th February 2003, Wharton University.
Knight, F.H. (1921), Risk, Uncertainty and Profit, Houghton Mifflin Company, Boston, MA, and New York, NY.

Kovoor-Misra, S., Zammato, R. and Mitroff, I.I. (2000), "Crisis preparation in organisations: prescription versus reality", Technological Forecasting and Social Change, Vol. 63, pp. 43–62.

Lonsdale, C. and Cox, A. (1998), Outsourcing: A Business Guide to Risk Management Tools and Techniques, Eastgate Press, Boston, MA.

MacCrimmon, K.R. and Wehrung, D.A. (1986), Taking Risks: The Management of Uncertainty, Free Press, New York, NY.

Marsden, C. (1997), "Corporate citizenship", Faith in Business, Vol. 1, No. 4, pp. 3–15.

Mathur, S.S. and Kenyon, A. (1997), Creating Value – Shaping Tomorrow's Business, Oxford, Butterworth-Heinemann Ltd., pp. 25–43.

Melnyk, S.A., Stewart, D.M. and Swink, M. (2004), "Metrics and performance measurement in operations management: dealing with the metrics maze", Journal of Operations Management, Vol. 22, 209–217.

Mentzer, J.T. (eds.) (2001), Supply Chain Management, Sage Publications Ltd., USA.

Mintzberg, H. and Waters, J.A. (1985), "Of strategies, deliberate and emergent," Strategic Management Journal, Vol. 6, pp. 257–272.

Mitchell, J.C. (1983), "Case and situation analysis", The Sociological Review, Vol. 31, cited in Smith and Dainty (eds.) (1991), The Management Research Handbook, Routledge, London.

Paulsson, U. (2004), "Supply Chain Risk Management", in Brindley, C. (ed), Supply Chain Risk Management, Ashgate, UK, pp. 79–96.

Ritchie, R.L. and Brindley, C.S. (2000), "Disintermediation, disintegration and risks in the SME global supply chain", Management Decision, Vol. 38, No. 8, pp. 575–583.

Ritchie, R.L. and Marshall, D.V. (1993), Business Risk Management, Chapman Hall, London.

Rowe, W.D. (1977), Anatomy of Risk, Wiley, New York.

Scappens, R.W. (1990), "Researching management accounting practices: the role of case study methods", British Accounting Review, Vol. 22, pp. 259–281.

Shepherd, C. and Gunter, H. (2006), "Measuring supply chain performance: current research and future directions", International Journal of Productivity and Performance Management, Vol. 55, No. 3/4, pp. 242–258.

Sitkin, S.B. and Pablo, A.L. (1992), "Reconceptualising the determinants of risk behaviour", Academy of Management Review, Vol. 17, No. 1, pp. 9–38.

Smallman, C. (1996), "Risk and organizational behaviour: a research model", Disaster Prevention and Management, Vol. 5, No. 2, pp. 12–26.

Smith, N.C. and Dainty, P. (eds.) (1991), The Management Research Handbook, Routldege, London.

Stainer, A. and Stainer, L. (1998), "Business performance – a stakeholder approach", International Journal of Business Performance Management, Vol. 1, No. 1, pp. 2–12.

Yin, R.K. (1994), Case Study Research: Design and Methods, Sage, Newbury Park, CA.

Zsidisin, G.A. (2003), "Managerial perceptions of supply risk", Journal of Supply Chain Management. A Global Review of Purchasing and Supply, Vol. 39, No. 1, pp. 14–25.

Chapter 17: Dominant Risks and Risk Management Practices in Supply Chains

Stephan M. Wagner* and Christoph Bode

*Corresponding author: Swiss Federal Institute of Technology Zurich, Zurich, Switzerland

17.1 Introduction

Supply chains are inherently susceptible to risky events. Earlier articles in supply chain management by Kraljic (1983) and Treleven, and Schweickhart (1988) stressed the importance to consider the risks arising from interconnected flows of material, information and funds in inter-organizational networks. However, during the last several years, the interest in this topic has significantly gained momentum. A large body of recent literature reports on events that disrupted supply chains and on their negative impact on businesses. These reports are paralleled by numerous articles from researchers and practitioners proposing best practices, guidelines, and concepts for risk management strategies that aim to ultimately create resilient supply chains. But what actually fuelled this recent attention to supply chain risks and their management? There are arguably at least two significant factors.

First, there is substantiated evidence that *the frequency of catastrophic events* such as natural hazards is increasing (Coleman 2006). Elkins et al. (2005) state that there has been an increase both, in the potential for disruptions and in their magnitude. And according to Munich Re's (2007) annual report on natural hazards, the comparison of the last 10 years with the 1960s reveals a significant increase in the number of natural hazards. The series of memorable crises and catastrophes that occurred in the past years underscores this development. Natural disasters such as hurricane Katrina devastating New Orleans in 2005, terrorist acts such as the World Trade Center attack from September 11, 2001, and epidemics like SARS in South-East Asia in 2003 are violent reminders that we live in an unpredictable and increasingly unstable world.

Second, *the vulnerability of modern supply chains* seems to have increased. Almost all industries have witnessed a remarkable change in their business environment, in particular due to increased competitive pressure and the globalization of markets. This resulted in a massive pressure to make intra-firm business processes and inter-firm supply chains either more efficient or responsive. Many firms reacted to this development by outsourcing or off shoring large portions of their manufacturing activities, sourcing in low-cost countries, lowering inventories, or collaborating more intensively with other supply chain actors (Christopher and Peck 2004; Fisher 1997; Hult et al. 2004). However, these developments ultimately led to an increased inter-firm dependence and, generally speaking, to an amplification of the vulnerability of supply chains to the impact of business disruptions (e.g., Gilbert and Gibs 2000; Kleindorfer and van Wassenhove 2004; Sarathy 2006). Certainly, modern supply chain management initiatives are powerful concepts for making operations leaner and more efficient in a stable environment but make supply chains more fragile (Zsidisin et al. 2005a). This argumentation is supported by findings from organizational scientists. There is evidence that increasingly complex and technology-oriented processes in organizations make errors almost inevitable (Lin et al. 2006).

Several researchers emphasize that as a consequence of this development which is characterized by a relatively unstable state of the world and an increased susceptibility of supply chains to disruptions, companies are compelled to tackle supply chain risks just as seriously as they tackle other business risks (Elkins et al. 2005).

Although risks are inherent in supply chains, their impact as well as their appropriate management have been receiving increasing attention, current knowledge is still limited and most articles on supply chain risks are anecdotal or case study-based (Hendricks and Singhal 2005a). Results from large-scale empirical research are scarce and mostly of descriptive nature (Jüttner 2005; Peck and Jüttner 2002; Zsidisin and Ellram 2003). This contribution tackles this lack of evidence and presents the results of a large-scale empirical study ($n = 760$) conducted in Germany among executives in supply chain management, logistics, and purchasing. The goal of this survey was to reveal (1) the *dominant and most prevalent types of supply chain disruptions* that have affected firms operating in Germany during the last years, (2) *how* these disruptions affected the surveyed firms, and (3) *what supply chain risk management practices* (measures and activities) the firms are currently pursuing to deal with supply chain risks.

The rest of this chapter is organized as follows: In Sect. 2, we firstly depict our understanding of the terms risk, risk source, as well as disruption, and secondly present the applied classification of supply chain risks and supply chain risk management practices. Section 3 presents the findings of the empirical study. Finally, Sect. 4 discusses the results, and presents implications for managerial practice.

Chapter 17: Dominant Risks and Risk Management Practices in Supply Chains

17.2 Nomenclature and Conceptual Framework

Recently, several researchers have advanced the conceptual clarity of the nomenclature used in the domain of supply chain risk management – yet, there is still no commonly agreed nomenclature. Due to this reason, the purpose of this section is to outline a consistent nomenclature which represents the basis for the survey in the third section.

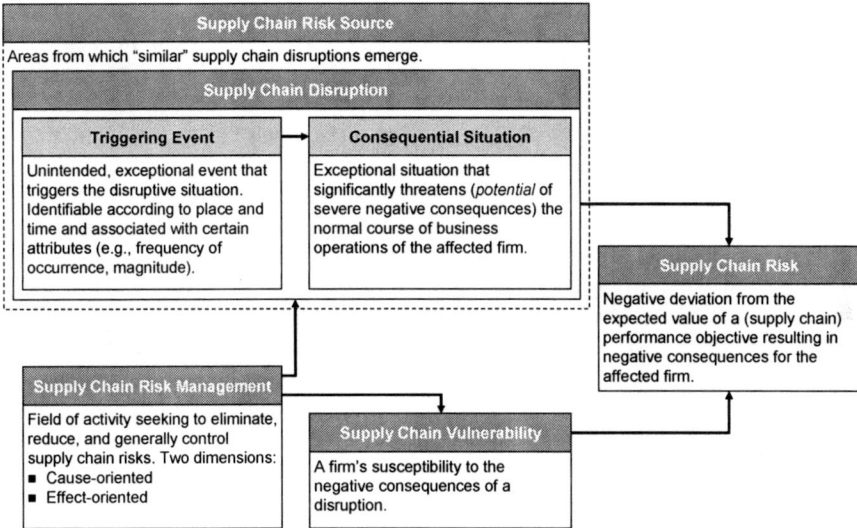

Fig. 17.1 Nomenclature and conceptual framework

In essence, we distinguish four interrelated terms: Supply chain risk, supply chain disruption, supply chain risk source, and supply chain vulnerability. In the following, these terms will be briefly derived from the pertinent literature and discussed. Figure 17.1 illustrates how these concepts are connected with each other.

17.2.1 Supply Chain Risk

In the field of supply chain management, several publications have addressed the question of how to define supply chain risk. Two different approaches can be distinguished: (1) risk as both danger and opportunity and (2) risk as purely danger.

The first approach is in line with common practice in many fields of business research such as finance. Here, the fluctuations around the expected value (mean) of a performance measure are used as proxy for risk, where risk is equated with variance and covers both a "downside" and an "upside" potential. Following these considerations and in analogy to the general definition of March and Shapira

(1987, p. 1404) – risk is the "variation in the distribution of possible outcomes, their likelihoods, and their subjective values" – Jüttner et al. (2003, p. 200) define supply chain risk as a "variation in the distribution of possible supply chain outcomes, their likelihood, and their subjective value."

In contrast, in most dictionaries as well as in the field of insurance, risk is viewed as the chance of injury, damage, or loss (Webster 1983). The notion that risk inherits primarily negative consequences corresponds to the common human perception. For instance, March and Shapira (1987) empirically examine how managers perceive risk and react to it. They find that the majority tend to overrate the "downside" potential of risk. Several scholars in the supply chain management and supply management field have adopted this view. Harland et al. (2003, p. 52), for instance, discuss several definitions and conclude that supply chain risk is associated with the "chance of danger, damage, loss, injury or any other undesired consequences."

Considering the impact of recent disruptions on supply chains, we find that the latter notion of risk as being purely negative corresponds best to supply chain business reality. In addition, businesses usually consider their goals (e.g., turnover or production volume) not so much as a target point but as lower limits of half-open ranges, e.g., to achieve *at least* a certain turnover. Hence, a goal deviation only occurs when falling below these thresholds. In insurance research, such a situation is called loss or damage (Knight 1921). Here, we follow this notion and understand risk as being the negative deviation from the expected value of a certain performance measure, resulting in negative consequences for the focal firm. Hence, risk is equated with the detrimental consequences of a supply chain disruption – the realized harm or loss.

17.2.2 Supply Chain Disruption and Supply Chain Risk Sources

A supply chain disruption is a quite vaguely defined concept. This holds also true for related concepts such as incident, accident, glitch, failure, hazard, crisis, or disturbance. Here, we define a supply chain disruption as the combination of (1) an unintended, anomalous event that materializes somewhere in the supply chain or the supply chain environment and (2) a consequential situation which significantly threatens the normal course of business operations of the affected firms in the supply chain. For the affected firms, it is an exceptional and anomalous situation in comparison to every-day business. There has been intensive research by organizational scientists on events that adversely affect organizations, how organizational crises emerge from those events, and how organizations react to them (Pearson and Clair 1998). Several helpful insights can be derived from this research and transferred to the supply chain risk context. For instance, similarly to the triggering event of an organizational crisis, the triggering event of a supply chain disruption is "identifiable according to place, time, and agents" (Shrivastava et al. 1988, p. 288). The disruption is associated with a certain probability of occurrence and characterized by its severity as well as direct and indirect effects (Kleindorfer and Saad 2005). Since the resulting detriment is usually a function of time, supply chain disruptions involve time pressure, implying that decisions for mitigation must be made swiftly (Hermann 1963).

Supply chain disruptions can materialize from various areas internal and external to a supply chain. Consequently, their nature can be highly divergent. For instance, the financial default of a supplier and a natural disaster destroying production capacity are situations with completely different attributes and therefore entail different effects on the supply chain. Addressing this issue and attempting to differentiate supply chain risks from other business risks, many scholars have proposed classifications in the form of typologies and/or taxonomies of risks (e.g., Chopra and Sodhi 2004; Christopher and Peck 2004; Hallikas et al. 2004; Jüttner 2005; Jüttner et al. 2003; Spekman and Davis 2004; Svensson 2000). The derived categories of supply chain disruptions are often labelled "supply chain risk sources." As such, Svensson (2000) named two categories (quantitative and qualitative), Jüttner (2005) delineated three categories (supply, demand, and environmental), and Chopra and Sodhi (2004) proposed nine categories (disruptions, delays, systems, forecast, intellectual property, procurement, receivables, inventory, and capacity). In the following, for brevity, we will call a negative deviation from the expected value of a performance measure for instance a "supply side risk" if the deviation results from a supply side (supply chain) disruption.

For our purpose, we will divide supply chain risk sources into five distinct classes: Demand side risk, supply side risk, regulatory, legal and bureaucratic risk, infrastructure risk, and catastrophic risk. While the first two risk source categories deal with supply-demand coordination risks that are internal to the supply chain, the latter three focus on risk sources that are not necessarily internal to the chain.

17.2.2.1 Demand Side Risks

Demand side risks result from disruptions emerging from downstream supply chain operations (Jüttner 2005). This includes disruptions in the physical distribution of products to the end-customer with particular issues being transportation operations (e.g., a truck driver strike) (McKinnon 2006) and the distribution network (e.g., a fire in a warehouse). Further, demand side risks can originate from the uncertainty surrounding the random demands of the customers (Nagurney et al. 2005). Disruptions occur here from a mismatch between a company's projections and actual demand as well as from poor supply chain coordination. The consequences from these disruptions include costly shortages, obsolescence, and inefficient capacity utilization.

An important issue in this context, affecting forecast quality and therefore demand side disruptions is the *bullwhip effect*, which is characterized by an amplification of demand volatility in the upstream direction of the supply chain. Lee et al. (1997) analyzed this phenomenon and identified delayed and distorted information, sales promotions, order batching, price fluctuations and rationing or shortage gaming as major causes. Other factors intensifying the bullwhip effect are over-reactions, unnecessary interventions, second guessing, and mistrust (Christopher and Lee 2004). Demand volatility still presents a major risk source for many firms. Spekman and Davis (2004) cite the example of Cisco Systems Inc. that wrote off US$ 2.5 billion in inventory in 2001 due to a lack of communication among its downstream supply chain partners.

17.2.2.2 Supply Side Risks

Purchasing organizations are exposed to numerous risks associated with their suppliers and their supply network. Supply side risks reside in purchasing, supplier activities and supplier relationships. These include supplier business risks, production capacity constraints on the supply market, quality problems, technological changes, and product design changes (Zsidisin et al. 2000).

Supplier business risks relate to the various events that affect the continuity of the supplier and result in the temporary or permanent perturbation or termination of the buyer-supplier relationship. This concerns particularly the threat of financial instability of suppliers and the consequences of supplier default, insolvency, or bankruptcy (Wagner and Johnson 2004). The financial default of a supplier, such as a supplier going out of business, is a particular but common supply chain disruption that can have severe consequences for the buying firm. Sheffi and Rice (2005) cite the example of the automobile manufacturer Land Rover that found itself in serious trouble after its only supplier of chassis frames, UPF-Thomson, suddenly and unexpectedly folded in 2001. Another type of disruption occurs when a supplier is vertically integrated by a direct competitor of the customer firm, forcing the termination of the relationship (Chopra and Sodhi 2004). In cooperative settings, opportunistic behaviour from suppliers has also been reported in literature as a source of supply risk (Wagner and Hoegl 2007). Particularly organizational lock-in is a threat where a purchasing organization is so dependent on a certain supplier that it has only a limited room for maneuvering.

Capacity constraints or shortages as well as poor logistics performance (delivery reliability) derive from unsolved problems in the supplier's production and operations management (Lee and Billington 1993). The bullwhip effect cited in the previous paragraph plays a role here as well and has to be tackled by the suppliers. Furthermore, poor quality in the purchased products or services is a significant risk and can have a cascading effect through the supply chain to the final customer (Zsidisin et al. 2000).

Finally, the inability of suppliers to adapt to technological or product design changes may have detrimental effects on the customer's costs and competitiveness (Zsidisin and Ellram 2003). With the increased importance and reliance on outsourcing, the cited risks are amplified (Giunipero and Eltantawy 2004).

17.2.2.3 Regulatory, Legal, and Bureaucratic Risk

In many counties, authorities (administration, legislation, regulatory agencies) are an important factor of uncertainty to the setup and operation of supply chains. Regulatory, legal, and bureaucratic risks refer to the legal enforceability and execution of supply chain-relevant laws and policies (e.g., trade and transportation laws) as well as the degree and frequency of changes in these laws and policies. This includes the ability to obtain approvals necessary for supply chain design activities and supply chain operation. Symptomatic for this risk source is that it is not internal to the individual supply chain or firm.

With the exception of government initiatives for security facilitation such as the Customs-Trade Partnership Against Terrorism (C-TPAT) or Authorized Economic Operators (AEO) certifications (Sarathy 2006; Zsidisin et al. 2005b), little attention has been paid to risks stemming from changing legal stipulations and conditions. Hendricks and Singhal (2003, 2005a, b) mentioned supply chain disruptions associated with actions or decisions of authorities. Administrative barriers (e.g., customs, trade regulations) may restrict the design and influence the operative performance of supply chains. Legal changes are often sudden and very difficult to anticipate. Examples are the newly introduced road pricing schemes for freight vehicles in various European countries which substantially affect transportation costs, as well as the environmental legislation with its requisites for product traceability and the setup of reverse logistics systems. In order to meet such environmental requisites, firms frequently get involved in more complex supply chains and incur higher supply chain costs.

17.2.2.4 Infrastructure Risks

The risk source "infrastructure" includes those disruptions that materialize from the infrastructure that a firm maintains for its operations. It includes socio-technical accidents such as equipment malfunctions, machine breakdowns, disruptions in the supply of electricity or water, IT failures or breakdown, as well as local human-centred issues (vandalism, sabotage, labour strikes, industrial accidents) (Chopra and Sodhi 2004; Spekman and Davis 2004).

IT related problems are highly relevant to supply chain management since a large portion of SCM functions builds on information processing and sharing. Organizations have become increasingly technology-dependent and, consequently, vulnerable to IT problems or breakdown (Chopra and Sodhi 2004). Causes of those events can be malicious intent by individuals or groups (cyber-attacks, virus attacks) as well as software bugs and hardware failures (Warren and Hutchinson 2000). Moreover, modern Enterprise Resource Planning (ERP) systems force firms to open their internal processes and databases both to their suppliers and customers which increases the exposure to IT related threats.

17.2.2.5 Catastrophic Risks

This class subsumes pervasive events that, when they materialize, have a severe impact in terms of magnitude in the area of their occurrence. This refers to natural hazards (force majeure), socio-political instability, civil unrest, economic disruptions and terrorist attacks (Kleindorfer and Saad 2005; Martha and Subbakrishna 2002).

In many regions of the world, natural hazards such as tsunamis, droughts, earthquakes, hurricanes, and floods are a constant threat to societies in general and to firms in particular (Helferich and Cook 2002). The negative consequences on supply chains are obvious since production facilities and transportation are highly vulnerable to natural disasters. Due to the globalization of markets and a surge in globe-spanning supply chain operations, local catastrophes have increasingly indirect global repercussions. Terrorism is a special topic which has been discussed extensively in a supply chain context. In the current state of the world, terrorism is a threat to global supply chains that has to be considered (Sheffi 2005). These attacks can impact supply chains either directly or indirectly. In addition, there are indirect consequences of terrorism that are not caused by an attack itself but by the reaction of governments and markets. For example Ford, Toyota and DaimlerChrysler experienced massive disruptions to the flow of materials into their North-American assembly plants within a few days after the terrorist attack of 9/11 (Sheffi 2001) due to border shut-downs.

17.2.3 Supply Chain Vulnerability and Its Drivers

While a supply chain disruption is the trigger that leads to the occurrence of risk, it is not the sole determinant of the final loss. It seems consequential that also the susceptibility of the supply chain to the harm of this situation is of significant relevance. This leads to the concept of supply chain vulnerability. The basic premise is that supply chain characteristics are antecedents of supply chain vulnerability and impact both the probability of occurrence as well as the severity of supply chain disruptions.

Although the literature offers numerous approaches to the construct "supply chain vulnerability," Peck (2005) still appraises its conceptual basis as immature. Christopher and Peck (2004, p. 3) define supply chain vulnerability as "an exposure to serious disturbance." Svensson (2000, 2002) published several contributions that shed light on the construct. He distinguishes between atomistic vulnerability (of a part of the supply chain) and holistic vulnerability (across the entire supply chain). In the literature on natural hazards and crisis management, vulnerability has been defined as a person's (or a group's) capacity to anticipate, cope with, resist, and recover from the impact of a natural hazard (Blaikie et al. 1994). In the context of maritime supply chains, Barnes and Oloruntoba (2005, p. 519) describe vulnerability as "a susceptibility or predisposition to ... loss because of existing organizational or functional practices or conditions."

In this contribution, this latter definition is applied and the atomistic perspective (supply chain vulnerability on the individual firm-level) is taken. We follow the notion that supply chain vulnerability is a function of certain supply chain characteristics and that the loss a firm incurs is a result of its supply chain vulnerability to a given supply chain disruption (Wagner and Bode 2006b).

Several publications argue that certain supply chain characteristics increase or decrease the vulnerability of the supply chain. Craighead et al. (2007) derive the propositions that supply chain density, supply chain complexity, and node criticality increase the severity of supply chain disruptions. Wagner and Bode (2006a) show empirically that customer dependence, supplier dependence, as well as single sourcing and global sourcing can amplify a firm's vulnerability to supply chain disruptions. Normal Accident Theory (NAT) can be a theoretic underpinning for research on supply chain vulnerability. This theory links the occurrence and impact of accidents (disruptions) to the structure of the organization and its technology (Perrow 1984, 1999). Two organizational attributes are argued to be relevant for both the probability of occurrence and the severity of adverse events: (1) (interactive) complexity of the system and (2) tight coupling of the elements in the system. In general, complex organizational systems – such as supply chains – are characterized by a large number of (varied) elements that interact in a non-simple way (Choi and Krause 2006). A system is tightly coupled if the components are interrelated in such a manner that there are few possible substitutions, time-dependent processes, and minimal slack or buffers (Perrow 1984). Given that this theory holds true for supply chains, more complex and tighter coupled supply chains are likely more prone to disruptions (Christopher and Lee 2001).

17.2.4 Supply Chain Risk Management

In general, enterprise risk management can be defined as the "field of activity seeking to eliminate, reduce and generally control pure risks" (Waring and Glendon 1998, p. 3). While the terminology can differ from author to author, a systematic risk management process usually comprises the stages of (1) risk identification, (2) risk analysis (including risk assessment and classification), (3) risk management in the narrow sense, e.g., risk treatment, and (4) risk monitoring. The overall objective of this process is to determine, implement, and monitor an optimal mix of measures to avoid, defer, reduce, or transfer all relevant risks. The determined mix is considered to be optimal if the remaining amount of risk is in line with the firm's risk preference and corporate strategy. This generic risk management process was transferred and adapted to the supply chain context by various authors (e.g., Hallikas et al. 2004; Ritchie and Brindley 2007).

However, for the purpose of this study, we are interested in specific practices of risk handling, e.g., the third stage of the outlined process. There is a large body of literature proposing measures and activities of supply chain risk management (e.g., Chopra and Sodhi 2004; Christopher and Peck 2004; Elkins et al. 2005; Johnson 2001; Lee and Wolfe 2003; Martha and Subbakrishna 2002; Rice and Caniato 2003; Zsidisin et al. 2005a). The proposed practices can be differentiated or classified according to various criteria. Tang (2006), for instance, identified four areas where supply chain risk management activities can take place: supply management, demand management, product management, and information management. Kleindorfer and van Wassenhove (2004) cite two groups of supply chain risk management activities: supply-demand coordination activities and activities for managing disruption risks.

Here, we decided to apply a different approach and to distinguish (1) *cause-oriented practices* and (2) *effect-oriented practices* of supply chain risk management.

17.2.4.1 Cause-Oriented Supply Chain Risk Management Practices

"If anything can go wrong, it will" says Murphy's Law. If this holds true, a good risk management approach is to *avoid* activities that are risky and "can go wrong."

Cause-oriented risk management practices attempt to do this, e.g., to reduce the probability of the occurrence of a disruption by aiming at its causes. Risk avoidance is possible for many types of disruptions. For instance, switching from a financially instable supplier to more stable one reduces the risk of a sudden supplier default. Or the relocation of manufacturing operations from geographic regions with a high exposure to natural hazards to safer regions reduces the probability to be directly affected by such events.

Another set of activities and measures in this context are *preventive* in nature such as preparative safety and security initiatives. Risk prevention can be used to get a grip on issues such as vandalism, sabotage, fire, and some sorts of industrial accidents. Rice and Caniato (2003) distinguish physical security (e.g., access controls), information security (e.g., education and training of employees for IS security), and freight security (e.g., C-TPAT). Zsidisin and Ellram (2003) emphasize that companies can reduce the probability of occurrence of various risks by influencing the risk awareness of their suppliers and by driving a risk culture into the supply base. Sheffi (2005) even argues that competitors should collaborate to control common risks. He names the example of TAPA (Technology Asset Protection Association) which was founded in 1997 by Intel and other high-technology firms with the objective to set standards for freight security.

Approaches that result in improved supply chain transparency and information exchange also support the effort of reducing the probability of occurrence of disruptions.

17.2.4.2 Effect-Oriented Supply Chain Risk Management Practices

In case of effect-oriented risk management practices, a firm decides to bear certain risks but at the same time makes attempts to limit or mitigate the negative consequences of a disruption. Thus, these measures aim at minimizing the level of damage in case of a risk event occurrence. In general, this can be achieved by seeking redundancy for activities or facilities which are particularly exposed to risk. Many of the risk handling activities proposed in the literature are rather effect-oriented than cause-oriented. In particular, buffering strategies are very prominent which aim to increase a company's tolerance to external resource shortage over a limited period of time. A usual approach in practice is to anticipate risk scenarios and to build slack (inventory, flexibility, or time buffers) into the supply chain in a way that the damage to the supply chain and the involved firms is limited in case of a materializing disruption. In this context, one often encounters the terms "resilience" (Sheffi 2005) as well as "robustness" (Christopher and Peck 2004, p. 2).

In the area of supply management and purchasing, the design of the supplier portfolio is a major target for effect-oriented measures – in particular, the decision of single sourcing versus multiple sourcing. The common ex-ante strategy to safeguard against the consequences of a sudden shortfall in supply – such as a supplier default – is the diversification of the supply base (Anupindi and Akella 1993; Treleven and Schweikhart 1988). The rationale is to install redundancy by developing contingency supply sources in order to decrease the vulnerability to supply-side disruptions (Sheffi and Rice 2005). The buying firm can diversify order quantities and hedge against the sudden demise of a single supplier by having multiple competing suppliers (Tomlin 2006).

Apart from improved forecasting of customer demands, a lot of risk mitigation potential resides in the design of products as well as in the layout of the manufacturing processes. Products can be modularized and components standardized so that the firm becomes more tolerant against the uncertainties on both supply and customer markets. In production and manufacturing, capacity buffers, stockpiling and flexibility are common measures.

Another important aspect is the creation of financial risk reserves which has to be considered to be an effect-oriented risk management measure. The risk bearing firm can build up financial reserves individually or it can try to transfer the risk to an insurance company that builds a collective reserve. Insurance companies offer many products pertaining to supply chain risks such as transportation insurances, inventory-related insurances (e.g., fire), or insurances against natural hazards (e.g., flooding). A rather new technique is the so-called Alternative Risk Transfer (ART) which offers a possibility to provide coverage for very specific risks or for risks where there is no insurance product available, catastrophe risks, for instance. Such catastrophe risks (e.g., gulf coast hurricanes) can be placed with investors by issuing corresponding catastrophe bonds or structuring derivative products (Lewis 2007). Contrary to traditional insurance the risk is transferred to the capital markets (Lane 2003). Finally, "business continuity plans" (Gilbert and Gips 2000) or "recovery plans" have to be cited as important tools to ex-ante optimize the "firefighting" after a disruption.

17.3 Questionnaire Development and Data Collection

The questionnaire for this study consisted of two sections: One referring to supply chain risks and one referring to supply risk management practices. We conducted several qualitative interviews as well as a thorough literature review to determine the initial item pool. After reviewing this pool with several researchers and supply chain management executives, some items were dropped or reworded. The remaining items were incorporated into a questionnaire and pretested. In the first section of the final questionnaire, the respondents were presented a list of relevant types of supply chain disruptions. Among these were aspects like supplier quality problems, supplier defaults, or terror attacks. For each disruption type, the respondents were asked to score on a five-point Likert-scale how their business unit had been negatively affected during the last 3 years in total by each of this specific issue. The scale ranged from "not at all" to "to a very large extent." The second part of the questionnaire consisted of a list of supply chain risk management measures. The respondents were asked to score the level of implementation of each risk management practices in their business unit on a five-point Likert-scale ranging again from "not at all" to "to a very large extent."

Based on this questionnaire, data were collected through a cross-sectional survey administered in Germany in 2005 to a sample of 4,946 top-level executives in logistics and supply chain management. The mailing and two follow-ups generated 760 usable responses, yielding a relatively high response rate of about 15.4%, considering the time constraints of top-level executives (Tomaskovic-Devey et al. 1994). Non-response bias was assessed on the notion that later respondents would be more like non-respondents (Armstrong and Overton 1977). For all questionnaire items, the responses of later respondents were compared to those of earlier. This comparison indicated absence of non-response bias.

Table 17.1 Sample characteristics

	Percent of total sample
1. Sector and industry	71.7
Industry sector	11.2
Automotive	10.1
Electro/Electronics	9.5
Machinery	8.4
Chemicals and pharmaceutical	6.6
Information technology	6.2
Materials and metal production	5.5
Food	4.2
Paper, pulp, and printing	3.0
Construction	2.5
Consumer goods	2.1
Aerospace and defence	1.3
Medical devices	0.9
Other industry	11.2
Service sector	19.5
Logistics services	17.1
Other services	2.4
Trade sector	8.8
2. Sales (in US$)	
Less than 10 million	14.9
10 million–under 50 million	23.9
50 million–under 100 million	16.3
100 million–under 250 million	14.7
250 million–under 500 million	8.7
500 million–under 1 billion	6.7
1 billion–under 10 billion	7.2
10 billion and more	5.0
n.a.	2.5
3. Number of employees	
Less than 100	21.4
100–499	42.2
500–999	11.6
1,000–4,999	15.3
5,000–9,999	2.8
10,000 and more	3.7
n.a.	3.0

We did not focus on a specific industry sector because we attempted to obtain a more general idea of supply chain risks in Germany. In particular, we wanted to include manufacturing and process industries.

The sample covered a broad range of sectors and firm sizes, e.g., industrial (71.7% of the sample), service (19.5%) and trade (8.8%) firms. The firms' annual sales ranged from less than US$ 10 million to US$ 90 billion (mean US$ 60.3 million), and the number of employees from less than 100 to 430,000 (mean 2,913), thus yielding a heterogeneous sample. Given the range and size of the firms studied and the diversity of industries, there was no prima facie reason to expect any systematic bias in the results. Most of the respondents held management positions in logistics and supply chain management (37.5%), or were in higher-level senior management positions (e.g., Executive VP, Senior VP) or owners of the business (23.8%). On average, the respondents have worked in this position for 7.0 years and have been with the firm for 10.9 years. A more detailed breakdown of the sample can be found in Table 17.1.

Additionally, we collected data pertaining to the types of supply chains the respondents were involved in (domestic vs. global and simple vs. complex). In particular, the respondents had to indicate (1) if their firms rely on global supplier networks, and (2) if they consider their supply chains to have a high degree of complexity. The means for these two items were on or close to the scale mean of three (3.00 and 3.18) and the standard deviations were around one (1.37 and 1.01). This shows that on average the respondents reported on a fairly homogeneous set of supply chains.

17.4 Results and Discussion

Based on the obtained sample of 760 top-level executives in logistics and supply chain management, the results of this study (1) present a detailed overview on the importance of specific supply chain risks and (2) shed light on the use and implementation of supply chain risk management practices in Germany.

17.4.1 Supply Chain Risks

Table 17.2 presents the investigated supply chain risks, their mean values, and standard deviations (SD).

The numbers reveal that demand side and supply side risk sources represent the dominant and most prevalent supply chain risks. In particular, the issues of volatile customer demand, information distortion in the supply chain, price fluctuation on the supply markets, as well as quality problems with sourced material have significantly affected the surveyed firms during the last 3 years.

Table 17.2 Supply chain risks and their prevalence

	Mean	SD
1. Demand side	3.03	
Unanticipated or very volatile customer demand	3.43	1.10
Insufficient or distorted information from customers about orders	3.08	1.14
Bad payment behaviour or payment defaults of customers	2.57	1.10
2. Supply side	2.55	
Price fluctuations on the supply markets	2.94	1.14
Supplier quality problems	2.80	1.03
Capacity fluctuations or shortages on the supply markets	2.68	1.08
Poor logistics performance of suppliers (e.g., delivery dependability)	2.68	1.09
Poor logistics performance of logistics service providers	2.16	0.93
Sudden supplier defaults	2.03	1.03
3. Regulatory, legal and bureaucratic	2.29	
Introduction of road pricing schemes	2.47	1.23
Changes in the political environment (e.g., new environmental laws)	2.31	1.12
Administrative barriers to the setup and operation of supply chains	2.08	1.03
4. Infrastructure	1.73	
Downtime of own production capacity due to technical reasons	1.83	0.93
Perturbation or breakdown of internal IT systems	1.80	0.87
Perturbation or breakdown of external IT systems	1.71	0.88
Downtime of own production due to local disruptions (e.g., fire, strike)	1.56	0.84
5. Catastrophic	1.55	
Terror attacks (e.g., London 2005)	1.61	0.92
Political instability, war, civil unrest or other socio-political crises	1.59	0.89
Natural disasters (e.g., earthquake, flooding)	1.51	0.84
Diseases or epidemics (e.g., SARS)	1.47	0.87

Scale: "Please indicate how your business unit has been negatively affected during the last 3 years by each of the following supply chain risks" (1: not at all – 5: to a very large extent).

Interestingly, bureaucratic risks, legal and regulatory risks and particularly catastrophic risks hardly affect firms operating in Germany. Not only are the mean values of these risks low but also the standard deviations indicate very little variation around the means. Of course, Germany has been a very "calm" place for these types of risks during the last years. Similar to most other Western European countries, it has a very low exposure to natural hazards and can be considered a very stable business environment. However, although all respondents were based in Germany and worked for firms sustaining operations in Germany, the supply chains reported in this survey were not dominantly domestic. The sample also included supply chains that extend the national borders. Due to such global supply chains, events can occur in other regions of the world and still have an effect on the surveyed firms. For this reason, the result that catastrophic events did hardly affect the surveyed firms is quite astonishing – and seems somewhat conflicting with the intensive recent interest concerning catastrophic risks. However, there is a well-accepted psychological reasoning for the misjudgement of the impact of supply chain disruptions. Research conducted by psychologists shows that people, instead of using statistics, rely on a limited number of heuristics to predict the impact of risks. These heuristics sometimes result in reasonable judgments and sometimes in serious errors (Kahneman and Tversky 1973). One such heuristic is called the "availability heuristic" (Slovic et al. 1982). Human beings make judgments based on what they can remember, rather than on complete data. These individuals can easily remember the pictures of terror attacks or natural disasters. The attention and awareness that these events receive is much higher than they should according to their probability (Stauffer 2003).

17.4.2 Supply Chain Risk Management

Table 17.3 shows the investigated supply chain risk management practices, the corresponding mean values, and standard deviations.

Table 17.3 Supply chain risk management practices

	Mean	SD
1. Cause-oriented	3.38	
We use only materials and products to which we know exactly their origin.	4.04	1.12
If possible, we do not sustain own operations in risky geographic regions.	3.44	1.54
We do not source from suppliers that produce in risky geographic regions.	3.36	1.41
We distribute our products only to markets that we know very well.	3.31	1.25
In collaboration with our customers and suppliers we are working on transparent supply chains and open information exchange.	3.14	1.20
We monitor regularly our suppliers with regard to potential supply chain risks.	3.01	1.15
2. Effect-oriented	2.82	
In our contracts with suppliers we usually try to transfer as much risk as possible to the suppliers.	3.18	1.01
Often, we use flexible contracts or options with our suppliers.	2.99	1.18
Our firm has elaborated business continuity or contingency plans addressing relevant supply chain risks.	2.73	1.33
We use late product differentiation to mitigate demand side risks.	2.71	1.20
We hold additional inventory and capacity buffers to mitigate the consequences of potential supply chain disruptions.	2.68	1.25
If possible, we insure against supply chain related risks.	2.61	1.21

Scale: "Please indicate to what extent your business unit has implemented the following supply chain risk management practices" (1: not at all – 5: to a very large extent).

The results reveal that the surveyed firms dominantly pursue cause-oriented activities of risk prevention and avoidance. It seems that most companies are rather risk averse and try to avoid problems wherever possible instead of waiting for a disruption and mitigating its consequences. In particular, the results highlight that there is an emphasis on avoiding risky geographic regions both from a purchasing as well as distribution perspective.

With one single exception, all means are below 3. This is an indication that supply chain risk management practices have not yet fully arrived in business practice. This finding is in line with results from other empirical studies in other countries that indicate that there is a lack of diffusion and implementation of supply chain risk management ideas and measures (Jüttner 2005).

Interestingly, the option to mitigate the consequences of disruptions by the effect-oriented approach "insurance" plays only a very minor role in practice.

17.5 Managerial Implications and Conclusions

The objective of this research was to examine the relevance of various supply chain risks and to provide a current picture of the implementation of supply chain risk management ideas in practice. Building on a thorough examination of the various supply chain risk taxonomies and risk management practices proposed in the pertinent literature as well as on interviews with practitioners, we developed a questionnaire and empirically investigated both aspects on a large-scale basis.

The data indicates that demand side and supply side issues are the most dominant risks for the surveyed firms. Primarily, this bolsters the notion that supply and demand coordination is the central issue in supply chain management (Kleindorfer and van Wassenhove 2004).

However, the results qualify the current interest on the subject, in particular with regard to infrastructure risks and catastrophic risk. Contrary to the general public perception, the prominent catastrophic risks are not a dominant factor for firms operating in Germany. Although the world witnessed a series of large-scale catastrophic events during the time of investigation, the surveyed firms did not experience significant losses from these disruptions. In contrast, the every-day issues of supply side and demand side risks – which are arguably the "bread-and-butter" issues of supply chain management – are the most prevalent supply chain risks.

Certainly, these results do not question the concept of supply chain risk management as a whole. However, they advocate managers to primarily turn their attention on supply side and demand side risk sources and on excelling in the "classic" supply chain management activities such as demand forecasting, supplier relationship management, cooperative information sharing with customers and suppliers, as well as quality management. As supply chain risk management is not for free, managers are compelled to seek an efficient allocation of risk management resources and a reasonable cost-benefit trade-off. Companies in Germany should be cautious about spending significant resources on the management of catastrophic risks.

With regard to supply chain risk management practices, our findings support the results of previous empirical studies which revealed a lack of implementation. Jüttner (2005) conducted a similar study in the UK (2005) and found that "practitioners have little guidance on their supply chain risk management approaches" (p. 139). Despite the numerous helpful articles on supply chain risk management in practitioner journals (e.g., Chopra and Sodhi 2004; Swaminathan and Tomlin 2007; Zsidisin et al. 2005a), there is obviously a need to put more emphasis on supply chain risk management education and to include these aspects into general supply chain management courses.

Finally, a remark has to be made on the generalizability of the result. Since the data was collected in Germany, the results can – if at all – only be generalized to firms based in countries with very similar geographic, political, and economic characteristics as Germany. Therefore, a replication of this survey in other countries with presumably different risk profiles (e.g., Japan or the United States) would be an important next step towards a better understanding of the dominant risks and risk management practices in supply chains.

References

Anupindi R, Akella R (1993) Diversification under supply uncertainty. Management Science 39(8):944–963.
Armstrong JS, Overton TS (1977) Estimating nonresponse bias in mail surveys. Journal of Marketing Research 14(3):396–402.
Barnes P, Oloruntoba R (2005) Assurance of security in maritime supply chains: Conceptual issues of vulnerability and crisis management. Journal of International Management 11(4):519–540.
Blaikie P, Cannon T, Davis I, Wisner B (1994) At Risk: Natural hazards, People's Vulnerability, and Disasters. Routledge, London.
Choi TY, Krause DR (2006) The supply base and its complexity: Implications for transaction costs, risks, responsiveness, and innovation. Journal of Operations Management, 24(5):637–652.
Chopra S, Sodhi MS (2004) Managing risk to avoid supply-chain breakdown. Sloan Management Review 46(1):53–61.
Christopher M, Lee HL (2004) Mitigating supply chain risk through improved confidence. International Journal of Physical Distribution & Logistics Management 34(5):388–396.
Christopher M, Peck H (2004) Building the resilient supply chain. International Journal of Logistics Management 15(2):1–13.
Coleman L (2006) Frequency of man-made disasters in the 20th century. Journal of Contingencies and Crisis Management 14(1):3–11.
Craighead CW, Blackhurst J, Rungtusanatham MJ, Handfield RB (2007) The severity of supply chain disruptions: Design characteristics and mitigation capabilities. Decision Sciences 38(1):131–156.
Elkins D, Handfield RB, Blackhurst J, Craighead CW (2005) 18 ways to guard against disruption. Supply Chain Management Review 9(1):46–53.
Fisher ML (1997): What is the right supply chain for your product?. Harvard Business Review 75(2):105–116.
Gilbert GA, Gips MA (2000) Supply-side contingency planning. Security Management 44(3):70–74.
Giunipero LC, Eltantawy RA (2004) Securing the upstream supply chain: A risk management approach. International Journal of Physical Distribution & Logistics Management 34(9):698–713.

Hallikas J, Karvonen I, Pulkkinen U, Virolainen V-M, Tuominen M (2004) Risk management processes in supplier networks. International Journal of Production Economics 90(1):47–58.
Harland C, Brenchley R, Walker H (2003) Risk in supply networks. Journal of Purchasing & Supply Management 9(2):51–62.
Hendricks KB, Singhal VR (2003) The effect of supply chain glitches on shareholder wealth. Journal of Operations Management 21(5):501–522.
Hendricks KB, Singhal VR (2005a) An empirical analysis of the effects of supply chain disruptions on long-run stock price performance and equity risk of the firm. Production and Operations Management 14(1):35–52.
Hendricks KB, Singhal VR (2005b) Association between supply chain glitches and operating performance. Management Science 51(5):695–711.
Hermann CF (1963) Some consequences of crisis which limit the viability of organizations. Administrative Science Quarterly 8(1):61–82.
Hult GTM, Ketchen Jr. DJ, Slater SF (2004) Information processing, knowledge development, and strategic supply chain performance. Academy of Management Journal 47(2):241–253.
Johnson ME (2001) Learning from toys: Lessons in managing supply chain risk from the toy industry. California Management Review 43(3):106–124.
Jüttner U (2005) Supply chain risk management – Understanding the business requirements from a practitioner perspective. International Journal of Logistics Management 16(1):120–141.
Jüttner U, Peck H, Christopher M (2003) Supply chain risk management: Outlining an agenda for future research. International Journal of Logistics: Research and Applications 6(4):197–210.
Kahneman D, Tversky A (1973) On the psychology of prediction. Psychological Review 80(4):237–251.
Kleindorfer PR, Saad GH (2005) Managing disruption risks in supply chains. Production and Operations Management 14(1):53–68.
Kleindorfer PR, van Wassenhove LN (2004) Managing risk in global supply chains. In: Gatignon H, Kimberly JR, Gunther RE (eds.) The INSEAD-Wharton Alliance on Globalizing. Cambridge University Press, Cambridge, pp. 288–305.
Knight FH (1921) Risk, Uncertainty and Profit. Hart, Schaffner & Marx, Boston.
Kraljic P (1983) Purchasing must become supply management. Harvard Business Review 61(5):109–117.
Lane M (ed.) (2003) Alternative Risk Strategies. Risk Books, London.
Lee HL, Billington C (1993) Material management in decentralized supply chains. Operations Research 41(5):835–847.
Lee HL, Wolfe M (2003) Supply chain security without tears. Supply Chain Management Review 7(1):12–20.
Lee HL, Padmanabhan V, Whang S (1997) Information distortion in a supply chain: The bullwhip effect. Management Science 43(4):546–558.
Lewis M (2007) In nature's casino. The New York Times (August 26, 2007):26.
Lin Z, Zhao X, Ismail KM, Carley KM (2006) Organizational design and restructuring in response to crises: Lessons from computational modelling and real-world cases. Organization Science 17(5):598–618.
March JG, Shapira Z (1987) Managerial perspectives on risk and risk taking. Management Science 33(11):1404–1418.
Martha J, Subbakrishna S (2002) Targeting a just-in-case supply chain for the inevitable next disaster. Supply Chain Management Review 6(5):18–23.

McKinnon A (2006) Life without trucks: The impact of a temporary disruption of road freight transport on a national economy. Journal of Business Logistics 27(2):227–250.

Munich Re (2007) Natural Catastrophes 2006: Analyses, Assessments, Positions. Munich Re Publications, Munich.

Nagurney A, Cruz J, Dong J, Zhang D (2005) Supply chain networks, electronic commerce, and supply side and demand side risk. European Journal of Operational Research 164(1):120–142.

Pearson CM, Clair JA (1998) Reframing crisis management. Academy of Management Review 23(1):59–76.

Peck H (2006) Reconciling supply chain vulnerability, risk and supply chain management. International Journal of Logistics: Research and Applications 9(2):127–142.

Peck H, Jüttner U (2002) Risk management in the supply chain. Logistics & Transport Focus 4(19):17–21.

Perrow C (1984) Normal Accidents. Basic Books, New York.

Perrow C (1999) Organizing to reduce the vulnerabilities of complexity. Journal of Contingencies and Crisis Management 7(3):150–155.

Rice Jr. JB, Caniato F (2003) Building a resilient and secure supply chain. Supply Chain Management Review 7(5):22–30.

Ritchie B, Brindley C (2007) Supply chain risk management and performance – A guiding framework for future development. International Journal of Operations & Production Management 27(3):303–322.

Sarathy R (2006) Security and the global supply chain. Transportation Journal 45(4):28–51.

Sheffi Y (2001) Supply chain management under the threat of international terrorism. International Journal of Logistics Management 12(2):1–11.

Sheffi Y (2005) The Resilient Enterprise. MIT Press, Cambridge.

Sheffi Y, Rice Jr. JB (2005) A supply chain view of the resilient enterprise. Sloan Management Review 47(1):41–48.

Shrivastava P, Mitroff II, Miller D, Miglani A (1988) Understanding industrial crisis. Journal of Management Studies 25(4):285–303.

Slovic P, Fischhoff B, Lichtenstein S (1982) Facts versus fears: Understanding perceived risk. In: Kahneman D, Slovic P, Tversky A (eds.) Judgment Under Uncertainty: Heuristics and Biases. Cambridge University Press, Cambridge, pp. 463–489.

Spekman RE, Davis EW (2004) Risky business: Expanding the discussion on risk and the extended enterprise. International Journal of Physical Distribution & Logistics Management 34(5):414–433.

Stauffer D (2003) Risk: The weak link in your supply chain. Harvard Management Update March:3–5.

Svensson G (2000) A conceptual framework for the analysis of vulnerability in supply chains. International Journal of Physical Distribution & Logistics Management 30(9):731–750.

Svensson G (2002) A conceptual framework of vulnerability in firms' inbound and outbound logistics flows. International Journal of Physical Distribution & Logistics Management 32(2):110–134.

Swaminathan JM, Tomlin B (2007) Risk management pitfalls. Supply Chain Management Review 11(5):34–42.

Tang CS (2006) Perspectives in supply chain risk management. International Journal of Production Economics 103(2):451–488.

Tomaskovic-Devey D, Leiter J, Thompson S (1994) Organizational survey nonresponse. Administrative Science Quarterly 39(3):439–457.

Tomlin B (2006) On the value of mitigation and contingency strategies for managing supply chain disruption risks. Management Science 52(5):639–657.

Treleven M, Schweikhart SB (1988) A risk/benefit analysis of sourcing strategies: Single vs. multiple sourcing. Journal of Operations Management 7(4):93–114.

Wagner SM, Bode C (2006a) An empirical investigation into supply chain vulnerability. Journal of Purchasing & Supply Management 12(6):301–312.

Wagner SM, Bode C (2006b) An empirical investigation into supply chain vulnerability experienced by German firms. In: Kersten W, Blecker T (eds.) Managing Risks in Supply Chains: How to Build Reliable Collaboration in Logistics. Erich Schmidt Verlag, Berlin, pp. 79–96.

Wagner SM, Johnson JL (2004) Configuring and managing strategic supplier portfolios. Industrial Marketing Management 33(8):717–730.

Wagner SM, Hoegl M (2007) On the challenges of buyer-supplier collaboration in product development projects. In: Gibbert M, Durand T (eds.) Strategic Networks: Learning to Compete. Blackwell Publishing, Malden, pp. 58–71.

Waring AE, Glendon IA (1998) Managing Risk. International Thomson Business Press, London.

Warren M, Hutchinson W (2000) Cyber attacks against supply chain management systems: A short note. International Journal of Physical Distribution & Logistics Management 30(7/8):710–716.

Webster N (1983) Webster's New Twentieth Century Dictionary, 2nd edn., Simon & Schuster, New York.

Zsidisin GA, Ellram LM (2003) An agency theory investigation of supply risk management. Journal of Supply Chain Management 39(3):15–27.

Zsidisin GA, Panelli A, Upton R (2000) Purchasing organization involvement in risk assessments, contingency plans, and risk management: An exploratory study. Supply Chain Management: An International Journal 5(4):187–197.

Zsidisin GA, Ragatz GL, Melnyk SA (2005a) The dark side of supply chain management. Supply Chain Management Review 9(2):46–52.

Zsidisin GA, Ragatz GL, Melnyk SA (2005b) An institutional theory perspective of business continuity planning for purchasing and supply management. International Journal of Production Research 43(16):3401–3420.

SECTION FOUR - SUPPLY CHAIN SECURITY

Chapter 18: Food Supply Chain Security: Issues and Implications

Douglas Voss* and Judith Whipple

*Corresponding author: College of Business, University of Central Arkansas, Conway, AR, USA

18.1 Introduction

Prior to September 11th, 2001 the private sector was well aware of the threat of terrorism. However, the terrorist threats they perceived were different than those perceived today. In the early 1990s, firms were mostly concerned about overseas employee kidnapping (Harvey 1993). Ports were more concerned with theft and smuggling (Thibault et al. 2006). Beyond airline hijackings, the use of supply chain assets as a method to inflict damage was far from the minds of corporate America.

This mindset changed drastically in the aftermath of September 11th. Following this event, the need to secure supply chains against terrorist-induced disruptions became more evident. As such, firms have begun to rethink "security" within the confines of their four walls as well as across the supply chain. Security is no longer just about theft or product damage, but now must incorporate an assessment of possible disruptions (intended as well as unintended) in an effort to prevent, detect, and potentially recover from such disruptions. This chapter will examine the issues and implications of supply chain security with particular focus on the food industry.

To begin, it is important to understand what a supply chain is, what supply chain management entails, and how supply chain security management is defined. A *supply chain* is defined as, "The combination of organizations and service providers that manage the raw material sourcing, manufacturing, and delivery of goods from the source of the commodities to the ultimate users" (Closs and McGarrell 2004 p. 8). In this sense, the supply chain represents a cradle to grave concept encompassing the flow of materials, information, and financial resources from production of raw materials to final consumption.

Managing these flows became recognized as a source of competitive advantage and led to the concept of supply chain management. *Supply chain management* is defined as "the inter- and intra-organizational coordination of the sourcing, production, inventory management, transportation, and storage functions with the objective of meeting the service requirements of consumers or users at the minimum cost" (Closs and McGarrell 2004 p. 8). Supply chain management seeks to leverage core competencies of supply chain partners in order to effectively and efficiently coordinate the flow of product, information, and money through the supply chain.

Terrorist threats complicated these efforts. Costs to secure the supply chain are estimated to reach $151 billion USD annually (Russell and Saldanha 2003). Motor and air carriers are expected to incur an extra $2 billion in costs (Wolfe 2001). Warehouses are expected to incur an extra cost of $1 to $2 per square foot (Warehousing Education and Research Council 2004). It has been estimated that ports will incur $1.1 billion in initial security expenses and an extra $656 million annually. Additional considerations include holding excess inventory to buffer against supply chain disruption (Lee and Whang 2003), a decreased ability to deliver goods on time due to new security processes and measures (Dobie 2005), and an increase in cycle times and lead-times that result (Lee and Whang 2003).

Supply chain managers do more than coordinate flows efficiently and effectively, but must also simultaneously protect the supply chain and its stakeholders from harm. The study of supply chain security management emerged from this dilemma. *Supply chain security management* is defined as, "the application of policies, procedures, and technology to protect supply chain assets (product, facilities, equipment, information, and personnel) from theft, damage, or terrorism, and to prevent the introduction of unauthorized contraband, people, or weapons of mass destruction into the supply chain" (Closs and McGarrell 2004 p. 8).

A number of observations can be drawn from this definition. First, supply chain assets are defined as not only the equipment and facilities used to carry out supply chain processes, but also the products, information, services, and human resources needed to conduct supply chain operations. Therefore, supply chain protection does not stop with securing a facility through gates and locks, but also encompasses maintaining the safety surrounding the product and people involved in supply chain activities. Second, a secure supply chain requires preventing the following actions: (1) biological, chemical or unauthorized agents from becoming incorporated in the product; (2) any illegal commodity to be intermingled with legal shipments; (3) transportation assets or a shipment's contents to be used as a weapon; (4) unauthorized access to the product and/or supply chain network; and (5) disruptions of the supply chain network and/or its infrastructure. Third, supply chain security incorporates traditional focal points (theft and damage) as well as new concerns surrounding terrorist activity.

The remainder of this chapter will focus on a particularly important and vulnerable subset of the broader supply chain concept: food supply chains. The first section will explore the importance and challenges associated with food supply chain security. Section 2 will detail best in class security practices used by firms in the food industry. Section 3 discusses the role of security in the supplier selection decision. Finally, the chapter concludes with managerial implications.

18.2 Importance and Challenges of Food Supply Chain Security

While security is important in any supply chain, the food supply chain is an important subset to protect. A safe food supply is a primary foundation upon which society is built. Not only is food of importance to the general population, but it is also important to the economic well-being of a country. Agricultural products, and their related industries, significantly contribute to the economic welfare of the United States through the creation of jobs and exporting opportunities (Rand Corporation 2003).

In addition, food supply chain security is important because the food supply chain is vulnerable to both unintended as well as intended supply disruption. The food supply chain is vulnerable for many reasons. First and foremost, food is susceptible to unintentional disruptions that could occur via disease (e.g., Asian bird flu), blight, infestation, improper handling, and perishability. Second, the food supply chain is vulnerable to intentional harm. Harl (2002) identified seven general areas of security susceptibility, and five of those areas relate to the agri-food supply chain. Further, on average food travels 1,300 miles from farm to fork (Harl 2002) which creates a ready-made distribution channel for rapid and widespread disruption (Bruemmer 2003).

There are a number of supply chain nodes where a disruption (intentional or unintentional) could occur within the food supply chain. The supply chain for food is complex due to both the various hand-offs throughout the system and its global nature. In 2006, for example, it was estimated that US agricultural exports would equal $68.7 billion, while agricultural imports would approach $64 billion (Economic Research Service 2006). This globalization will continue as worldwide economies rely on each other for agricultural and food production.

Rice and Caniato (2003) discussed the emerging expectations with respect to supply chain security when they described the need for creating both secure supply chains (e.g., supply chains that maintain advanced security processes/procedures) and resilient supply chains (e.g., supply chains that are able to react quickly and restore operations when unexpected disruptions occur). Given the widespread nature of the food supply chain, creating both a secure supply chain as well as a resilient supply chain is of utmost importance.

18.3 Why Firms Focus on Supply Chain Security

While the threat of potential terrorist acts is a primary driver of recent security initiatives, secondary drivers that encourage firms to develop security competencies also exist. The Aberdeen Research Group (2004) reported four key drivers of security initiatives: brand equity protection; prevention of brand piracy, gray market activity, and product counterfeiting; customer and trading partner requirements; and increased product safety and traceability concerns due to greater outsourcing activity.

Brand equity in the food industry is important not just to manufacturer's establishing national/international brand images, but also to retailers who are creating their own brand images via store branded/private label products. Related to brand equity is the issue of brand piracy, gray markets, and product counterfeiting. Firms secure supply chain assets in order to reduce theft and provide assurance to their customers of product origin. These issues have come to the forefront in recent discussions over purchasing pharmaceuticals on-line. The Food and Drug Administration (FDA) and other government agencies worry that unregulated pharmaceutical imports (in addition to blatant drug and nutraceutical counterfeiting) stand a greater chance of being ineffective or harmful. In the food industry, piracy has impacted food aid deliveries (O'Rourke 2005) and a new term "biopiracy" has emerged concerning intellectual property rights associated with biotechnology used to develop advancements in agriculture and medical fields (Powledge 2001).

Customer and trade partner security requirements are becoming more critical, and relate to increased product safety and traceability concerns. In order to protect the product, customers are beginning to require their suppliers to increase security measures, in general, but also as a means to provide greater product safety and product traceability. For example, product attributes that are often hard to distinguish (e.g., organic, dolphin-safe, fair trade, etc...) generally require more traceability in order to verify the existence of the sought after attributes (Golan et al. 2004).

EyeforTransport (2004) found customer requirements and government pressure were among the reasons firms implement security initiatives. Working with suppliers and customers to ensure their supply chains are secure is a primary tenet of the Customs-Trade Partnership Against Terrorism (C-TPAT)[1]. C-TPAT is a voluntary, public-private initiative encouraging firms to secure their supply chains, and those of their trading partners. The goal of C-TPAT is to certify enough firms to create a critical mass of supply chain protection.

While C-TPAT is voluntary, in other instances, the government has also taken a regulatory stance in the food industry. The Bioterrorism Act of 2002 requires that firms engaged in food processing be able to trace raw materials and output one step up, and one step down, the supply chain (USFDA 2003).

These drivers are prompting many firms to be proactive in establishing their security programs. Other firms are less prepared. As reported by Rogers et al. (2004), many managers feel their firm will not be the target of a terrorist attack because they only sell certain commodities; other managers feel their security obligations end when they transfer goods to a carrier. On the other hand, security may be a priority for firms, but they may lack guidance as to the most effective means to achieve a secure supply chain (Unisys 2005). Complicating the situation is the fact that individual companies and industries are vulnerable to different forms of disruptions, creating the need for firms to understand their own vulnerabilities (Sheffi and Rice 2005). The next section details the results of an extensive literature review and interviews conducted with appropriate food industry representatives to determine the best-in-class security competencies utilized by firms in the food industry.

---

[1] For further information on C-TPAT, see http://www.cbp.gov/xp/cgov/import/commercial_enforcement/ctpat/.

18.4 Security Best Practices in the Food Industry

Beginning in 2005, a research team at Michigan State University embarked on an initiative sponsored by the U.S. Department of Homeland Security, in conjunction with the National Center for Food Protection and Defense (http://www.ncfpd.umn.edu/), to examine best practices in security management within the food industry. Given the emerging nature of supply chain security and the lack of existing security frameworks, exploratory interviews were conducted in order to better understand the characteristics and current status of supply chain security in the food industry.

A standard, open-ended interview guide was developed and pre-tested with academic reviewers (from supply chain management, information technology, and criminal justice disciplines) and with industry practitioners familiar with supply chain security issues. The interview guide was structured, but allowed for the researchers to explore new issues raised during the interview process.

Interviews were conducted with 15 different entities focused on either manufacturing or distributing food products or involved in food safety and security. These interviews included over 25 managers. Participant responsibilities ranged from supply chain, quality control, and security functions, and their positions ranged from manager to executive officer. Small, medium, and large firms were included as well as firms focusing on different food commodities. Firms that are considered leaders in their industry segment were also included. Interviews were conducted over the telephone and in-person.

Information gathered from the literature and interviews was used to develop a proposed model of security competencies necessary to protect the food supply chain. The proposed model is shown in Fig. 18.1. Security competencies represent the broad set of skills, knowledge and capacity to develop and maintain a secure supply chain. In all, ten competencies were proposed including: (1) process strategy; (2) process management; (3) infrastructure management; (4) communications management; (5) management technology; (6) process technology; (7) metrics/measurement; and (8) relationship management; (9) service provider management; and (10) public interface management. Each of these competencies and their definitions are provided in Table 18.1, and are examined and explained below.

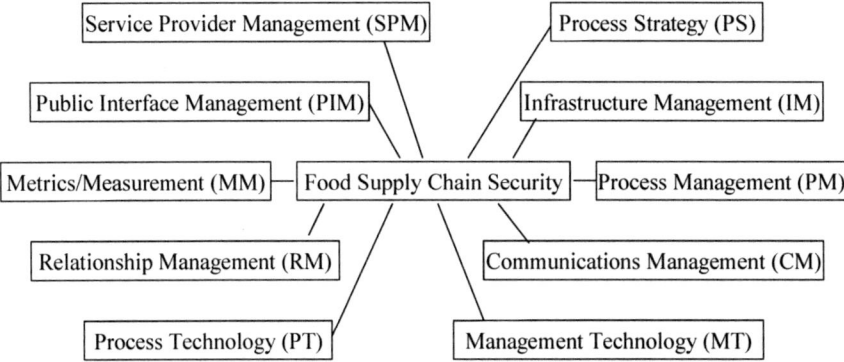

Fig. 18.1 Supply chain security competencies

Process Strategy involves high-level management support which is crucial to any successful security initiative. In order to foster a security culture, management should encourage open and honest discourse regarding security plans, incidents, and the role of security in protecting firm stakeholders as well as brand equity. Management's security commitment should be visible. In best-in-class firms, this visibility often occurred through the creation of a "chief security officer" to promote the security mission. This security culture helps empower front-line employees to act as the eyes and ears of the organization and report any suspicious activity.

Process Management develops as a means to properly secure a supply chain process wherein the firm must be intimately aware of the cross-functional activities involved in moving from "farm to fork." Activities should be documented and formalized via cross-functional flow charts. These flow charts should document each activity and each step of the process as well as track both product and information flows. Additionally, firms should engage in mock disruptions and table-top exercises in order to simulate the firm's reaction to a security event.

Table 18.1 Research competencies and definitions

Competency	Definition
Process strategy	The executive commitment to fostering the importance of security and instituting a culture of security within the enterprise
Process management	The degree to which specific security provisions have been integrated into processes managing the flow of materials and products into and out of the firm
Infrastructure management	Security provisions that have been implemented to secure the physical infrastructure (buildings, transportation vehicles)
Communication management	The internal information exchange between employees, managers, and contractors to increase security
Management technology	The effectiveness of existing information systems for identifying and responding to a potential security breach
Process technology	Specific technology (e.g., electronic seals, RFID) implemented to limit access and trace the movement of goods
Metrics	The availability and use of measurement to better identify and manage security threats
Relationship management	The information sharing and collaboration between supply chain partners
Service provider management	The information sharing and collaboration between the firm and its logistics service providers (transportation firms, warehousing providers, 3PLs, etc. ...)
Public interface management	The security related relationships and exchanges of information with the government and the public

Infrastructure Management involves the physical security measures that represent the basic building blocks of security best-practice. Infrastructure management activities help to form a secure perimeter around an entire facility and to protect sensitive areas within a firm's facilities. It is important to limit access to vulnerable processes via locks, gates, and ID cards and to monitor these areas with closed circuit television cameras (CCTV). Additionally, background checks should be performed on all employees as well as third party contractors with access to sensitive areas.

The physical security measures taken by firms to secure their operations are often cited as an effective method of not only reducing the threat of terrorist incidents, but also reducing theft. Theft is a major concern for firms for many reasons. First, theft obviously represents an opportunity cost in that the firm is "robbed" of the ability to sell the stolen product to customers. However, firms should also be concerned about the brand equity implications of product theft. Stolen product can be tampered with and then resold. This is a significant problem with food items that are subsequently ingested by consumers. Should this tampered product be resold and consumed, the producing firm is likely to be held liable. Further, stolen product can be resold for below market value. This may lead the consumers of that product to perceive it as "cheap," thus damaging a firm's ability to market its product as a premium good.

Communications Management is the proactive creation of a communication network that would come into play if and when an incident occurs. Should an intentional or unintentional disruption occur, efficient and effective communication between departments and across the supply chain is crucial. In order to achieve the ability to communicate effectively, firms should facilitate cross-functional communication via safety meetings and training as well as establish communication protocols. Communication management represents the tools used to create a security culture in the firm. It is important for firms to break down functional silos in order to improve the efficiency and effectiveness of communications in times of crisis. However, breaking down internal silos has the dual effect of promoting inter-departmental communication pertaining to issues other than security. If security related inter-departmental collaboration induces collaboration on other topics, then the effort to promote internal security measures has secondary benefits. For example, the purchasing and manufacturing departments may be encouraged to communicate on security related issues in order to ensure that purchased food product is sourced in such a manner that it is not spoiled by the time it reaches production. This interaction may lead to these two departments sharing inventory information and it may be revealed that a disconnect existed with respect to the amount of product purchased from suppliers and the amount of product required by production. In this case, excess inventory could be removed from the system and/or stockouts of raw materials may be prevented. Thus, promoting internal communication on security related issues may pay dividends to the firm in non-security related areas.

Management Technology represents the tools used to detect potential threats/ incidents and to share timely and accurate information both internally and externally. This includes historical databases of product movements, emergency contact information, and information technology resources used to share information inside the firm and between supply chain partners. Such technology includes e-mail, electronic data interchange (EDI), telephones, and faxes. It is crucial that such tools be in place in order to facilitate information exchange before, during, and after an incident. Management technology also plays a crucial role in facilitating inter-departmental communications in that it allows for the efficient and effective exchange of information. This information may be security related to begin with but then translate into the efficient and effective exchange of non-security related information that subsequently improves overall firm and supply chain performance.

Process Technology enables a firm to determine if an incident has occurred by creating a check and balance system for detecting anomalies within the system. Process technology represents the presence, use, and ability of information systems to track the movement of products and monitor processes both internally and across the supply chain. These tools may include Radio Frequency Identification (RFID) and smart-seals attached to transportation assets that not only track and trace these assets, but also determine if unauthorized entry has occurred. It should be noted that many tracking technologies have not reached a level of reliability necessary to derive a sufficient return on investment (ROI). The potential security and firm performance benefits derived from new technologies, such as RFID, are promising but mostly unrealized at this stage.

Metrics/Measurement is used as a means to track security performance. As the old adage goes, "you can't improve what you do not measure." Security metrics include continuously developing, using, testing, and redefining security guidelines that measure procedures, plans, and capabilities. Metrics are perhaps one of the more challenging security competencies to implement. It is often hard to determine the effectiveness of security efforts in the absence of a security incident. However, firms can employ metrics that indicate a procedural vulnerability. For example, a firm may choose to measure the number of broken trailer seals or the number of times unauthorized persons attempted to gain access to facilities.

Relationship Management and Service Provider Management acknowledges that, as adapted from Sheffi (2001), a supply chain security initiative is only as strong as its weakest link. In order to maintain security throughout the supply chain a firm must share security related information with supply chain partners and service providers in an accurate and timely manner. Further, it is important for firms to specify security expectations in supplier contracts and monitor suppliers to ensure they are complying with expectations. In addition, companies may consider auditing performance along the supply chain to ensure security compliance and/or include audits as a condition of prequalifying suppliers. Similar to communications management, if supply chain partners collaborate to improve security, this collaboration could subsequently lead to collaboration on non-security related issues. This collaboration stands to strengthen relationships with suppliers, customers, and third-party logistics providers (3PLs) that might lead to cost reductions and improved clarity of service needs.

Public Interface Management encourages firms to exchange security related information with the government and other public entities as needed. The government not only serves as an excellent source of security information, but also serves as a source of information regarding methods to secure operations. In the food industry, firms must have the capability to efficiently and effectively notify the public of product recalls. This is often accomplished through media contacts and subsequent dissemination of information to those potentially affected.

These security competencies culminate in improved overall supply chain security as presented in Fig. 18.1. Based on the literature review conducted as part of this research as well as the results of the in-depth interviews, it is clear that some firms are more advanced in their security efforts than others. There is a continuum of security initiatives and preparedness. As discussed by Rice and Caniato (2003), security competencies exist at four levels: (1) basic – where fundamental security and preparedness activities exist; (2) reactive – where a greater awareness exists with respect to security vulnerabilities; (3) proactive – where security and resilience practices beyond the norm are adopted; and (4) advanced – where a firm exists as a leader in progressive security initiatives and creates a highly resilient supply chain. Given that security is an emerging field, it is not surprising that Rice and Caniato (2003) report few companies have reached the advanced level.

18.5 Encouraging Security Enhanced Supply Chains

As discussed earlier in this chapter, business costs are likely to increase due to higher security requirements. These costs are derived, for example, from compliance with new regulations and/or voluntary programs, higher inventories to buffer against disruptions, increased insurance premiums, investment in new technologies, additional security measures and audits, and as well as reconfiguring the supply chain. For many companies, accepting that additional costs will be incurred is a challenge – particularly, in the food industry, where competition is fierce and margins are generally thin. Managers have to build a business case to encourage their firms and supply chain partners to increase security initiatives.

Two schools of thought exist with respect to the effect that security has on a firm's cost and service performance. One school of thought maintains that security is a net detractor from firm performance since security measures are expensive. The cost of gates, locks, guards, technology, and relationship management can seem prohibitive. Additionally, in order to improve firm and supply chain security, product may be inspected at multiple points throughout processes. These inspections add another layer of activities to a process. With any additional activity comes additional process variability. By definition, additional process variability increases the amount of uncertainty with respect to product availability. This uncertainty can translate into decreased delivery reliability performance unless the firm holds extra safety stock in anticipation of inventory shortages (Lee and Whang 2003).

Another school of thought maintains that, while security may have a negative effect on cost performance in the short-run, eventually security will pay for itself through, among other means, reductions in theft and decreased inventory levels gained through more detailed inventory monitoring. Further, security may also synergistically improve delivery reliability performance through improved process knowledge, leading to reductions in dwell times and overall process variability. The use of tracking technology for enhancing supply chain security can also lead to a greater ability to ascertain product location and inventory levels. This can lead to a decrease in the occurrence of stock outs and increase the ability to deliver product when promised.

A recent study highlights long-term payoffs and collateral benefits from engaging in security related investments (Peleg-Gillai et al. 2006). This study advocates that firms move beyond calculating direct expenses related to security and, instead, develop an understanding of secondary benefits derived from a secure and resilient supply chain, including the following:

1. Greater supply chain visibility;
2. Enhance supply chain efficiencies;
3. Greater customer satisfaction;
4. Improved inventory management;
5. Reduce cycle time and shipping time; and
6. Associated cost reductions achieved through the secondary benefits.

While advanced security initiatives are assumed to synergistically improve security and firm performance, most firms have not progressed beyond physical security measures (e.g., Infrastructure Management) and therefore have not derived the service benefits from "higher level" security measures (EyeforTransport 2005). This creates a dilemma for the firm that faces greater customer and/or government requirements. How can a firm improve security in order to not be placed at a competitive disadvantage, but incur the up-front costs necessary in the short-run to pay for security improvements or deal with lower performance levels due to added security steps? In particular, price and delivery reliability are likely to be challenged (at least initially) in the face of growing security concerns.

The consequence of this dilemma is even more pronounced when one considers that price and delivery reliability are two of the most important criteria used by firms to select suppliers (Lehmann and O'Shaughnessy 1974; Evans 1982; Wilson 1994). However, if customers value security to the extent that they are willing to pay a higher price and/or accept lower level delivery in return for supply security, then the supplier has a greater incentive to implement security measures.

18.6 The Impact of Security of Supplier Selection in the Food Industry

Research was undertaken at Michigan State University to determine the extent to which firms in the food industry value security in relation to quality, price, delivery reliability, and supplier location. A sample of 107 food industry purchasing professionals employed by food manufacturers completed a survey that asked them to choose between four hypothetical suppliers characterized by different performance levels across the above mentioned criteria. Utilizing conjoint analysis, a method of determining the value respondents place on the various characteristics of a product or service, respondent utility values for each of these factors/supplier characteristics were ascertained. The greater the amount of utility respondents attach to a given factor, the more important that factor is to the purchasing professional when choosing between suppliers and the greater the factor importance score attached to a given supplier characteristic. The results are presented in Table 18.2. As indicated in Table 18.2, respondents placed the most emphasis on delivery reliability followed by price, location, product quality, and security. These results indicate that food industry purchasing professionals are not likely to sacrifice price and delivery reliability for their suppliers in order to operate in a secure manner.

Table 18.2 Ranking of supplier selection criteria

Rank	Criteria	Factor importance score
1.	Delivery reliability	27.092
2.	Price	22.275
3.	Location*	20.904
4.	Quality	19.905
5.	Security	10.645

*Location = choice between domestic or international

However, certain factors may exist that exacerbate the importance of security to food industry purchasing professionals. For example, if a supplier is located in a region perceived to be at greater risk for terrorism, customers may be more concerned about supplier security. Second, if the purchasing firm has experienced a security incident, this experience would create more awareness of the importance of a secure supply chain. Third, if a purchasing firm manufacturers or distributes name-brand products, brand-equity is an issue. This firm might be more likely to utilize a secure supplier and protect their name-brand. Finally, if a purchasing firm considers the supplier's product to be of strategic importance, they might be more likely to employ advanced security measures.

18.7 Conclusion

This chapter has detailed the results of two separate research initiatives examining food supply chain security best practice and the willingness of food industry purchasing managers to sacrifice price and delivery reliability for security. The reader should pay particular attention to two key takeaways: (1) the competencies associated with supply chain security; and (2) the effect of security on firm performance. Given the emerging nature of supply chain security, a framework was presented in this chapter that is intended to guide readers in their efforts to implement and/or research supply chain security phenomena.

With regard to the effect of security on firm performance, it appears that suppliers who employ best-in class security measures may not immediately derive a competitive advantage from their programs unless they (1) are able to derive synergistic benefits from security, (2) are located in a higher-risk, international region, (3) sell goods to customers characterized by concern over security incidents or brand-equity, and/or (4) provide a strategically important product. As such, companies must understand there are secondary benefits to engaging in a security initiative. Those benefits are likely to be realized over the long run, and must be included in cost/benefit evaluations of security investments. These benefits are often not readily linked to security efforts (e.g., a reduction in the number of incidences), but rather may be more closely associated with supply chain productivity measures such as improved inventory management. While this may seem counterintuitive, the ten competencies proposed in this framework, when developed, can provide overall supply chain improvements. For example, improvements made through better relationship management and service provider management not only assist in providing better security, but also promote non-security improvements (e.g., inventory reductions, information sharing, collaborative relationships).

Given these secondary benefits, firms should not shy away from engaging in a security program. It simply means that firms must engage in a thorough business case analysis that goes beyond examining direct investments in security-related expenses. In addition, firms can analyze key vulnerabilities to prioritise where investments are most critical and offer the greatest return. Further, it is our position that security should be viewed as a matter of social responsibility, especially in the food industry. While it may be hard to quantify the benefits of security in the near term, the experience of others should teach us that it only takes a single incident to destroy a firm's brand image. Whether a firm sells a generic product or a nationally recognized name, it is not far fetched to state that severe repercussions result in the event of an incident.

If firms do not engage in security programs and significant breaches in security result, the public will demand government regulation to increase security and require compliance. In that vein, the security movement is analogous to the quality movement of the 1980s – it may not offer a potential competitive advantage in the short run, but makes good business sense in the long run.

References

Aberdeen Group (2004) How Supply Chain Leaders Protect Their Brands: A Benchmark Report on Regulatory Compliance and Product Safety Mandates for Food, Pharmaceuticals, Consumer Products, and Medical Devices. September.

Bruemmer B (2003) Food Biosecurity. Journal of the American Dietetic Association, 103:6, 687–691.

Closs DJ, McGarrell EF (2004) Enhancing Security Throughout the Supply Chain. IBM Center for the Business of Government.

Dobie K (2005) The Core Shipper Concept: A Proactive Strategy for Motor Freight Carriers. Transportation Journal, 44:2, 37–53.
Evans RH (1982) Product Involvement and Industrial Buying. Journal of Purchasing and Materials Management, 18:Summer, 23–28.
EyeforTransport (2005) Cargo and Supply Chain Security Trends 2005.
EyeforTransport Global Research (2004) North American Supply Chain Security: An Analysis of Eyefortransport's Recent Survey. July.
Golan E, Krissoff B, Kuchler F (2004) Food Traceability: One Ingredient in a Safe and Efficient Food Supply. Amber Waves, April, 14–21.
Harl NE (2002) U.S. Agriculture, Food Production is Threatened by Bioterrorism Attacks. Ag Lender, April.
Harvey MG (1993) A Survey of Corporate Programs for Managing Terrorist Threats. Journal of International Business Studies, 24:3, 465–478.
Lee HL, Whang S (2003) Higher Supply Chain Security with Lower Cost: Lessons from Total Quality Management. Stanford Graduate School of Business Research Paper Series, October, 1–28.
Lehmann DR, O'Shaughnessy J (1974) Difference in Attribute Importance for Different Industrial Products. Journal of Marketing, 38:April, 36–42.
O'Rourke M (2005) Piracy on the High Seas. Risk Management, 52, 8.
Peleg-Gillai B, Bhat G, Sept L (2006) Innovators in Supply Chain Security: Better Security Drives Business Value. The Manufacturing Innovation Series of the Manufacturing Institute, July.
Powledge F (2001) Patenting, Piracy, and the Global Commons. Bioscience, 51:4, 273–277.
Rand Corporation (2003) Agro Terrorism: What is the Threat and What Can Be Done About It. Rand Corporation Research Brief Series.
Rice JB, Caniato F (2003) Building a Secure and Resilient Supply Network. Supply Chain Management Review, 7:5, 22–30.
Rogers D, Lockman D, Schwerdt G, O'Donnell B, Huff R (2004) Supply Chain Security. Material Handling Management, 59:2, 15–17.
Russell DM, Saldanha JP (2003) Five Tenets of Security-Aware Logistics and Supply Chain Operation. Transportation Journal, 42:4, 44–54.
Sheffi Y (2001) Supply Chain Management Under the Threat of International Terrorism. International Journal of Logistics Management, 12:2, 1–11.
Sheffi Y, Rice Jr. JB (2005) A Supply Chain View of the Resilient Enterprise. MIT Sloan Management Review, 47:1, 41–48.
Thibault M, Brooks MR, Button KJ (2006) The Response of the U.S. Maritime Industry to the New Container Security Initiatives. Transportation Journal, 45:1, 5–15.
Unisys (2005) Secure Commerce Roadmap: The Industry's View for Securing Commerce. Unisys Corporation White Paper.
USDA Economic Research Service (2006) Outlook for U.S. Agricultural Trade. AES-52, November 22.
USFDA (2003) Risk Assessment for Food Terrorism and Other Food Safety Concerns. October 7.
Warehousing Education and Research Council (2004) The Cost of Security. Warehousing Education and Research Council Newsletter, September, 1–3.
Wilson EJ (1994) The Relative Importance of Supplier Selection Criteria: A Review and Update. International Journal of Purchasing and Materials Management, 30:3, 35–41.
Wolfe M (2001) Fifth EU/US Forum on Intermodal Freight Transport.

Chapter 19: Supply Chain Security: A Dynamic Capabilities Approach

Chad Autry* and Nada Sanders

*Corresponding author: Neeley School of Business, Texas Christian University, Fort Worth, TX

19.1 Introduction

Throughout history, human civilizations have been afflicted by and have responded to large-scale destructive events, impacting both the individuals living within them and the overall social environment. Earlier occurrences such as the repeated pandemics of the bubonic plague, the eruption of Mount Vesuvius, the Irish potato famine, and the 1777 Lisbon Earthquake each dramatically impacted the way civilizations formed, grew, lived, and interacted. Despite the temporally increasing efforts of humans to prepare for such disasters, and although many of the most prominent and well known disasters affecting human societies occurred long ago, large scale catastrophes have remained to some extent a constant, impacting cultures around the world throughout history and up to the present day (Quinn 2003).

Even in modern times there are several examples where widespread or extenuating circumstances have yielded significant human social and economic damage and/or disorder (Quinn 2003). For instance, in 1999, a major earthquake in Taiwan wreaked havoc on the Asian economic and physical infrastructure, disrupting production in the semiconductor and telecommunications manufacturing industries for over 18 months, displacing thousands of low-wage workers, and damaging or destroying billions of dollars worth of corporate and personal assets. Just over 2 years later, in September of 2001, terrorists piloted airplanes into major economic and political centres in the United States, leading to massive material losses and thousands of human casualties. In a third devastating episode within a 5-year span, a 2004 tsunami stemming from a tectonic shift off the coast of Sumatra wiped out coastal communities in five different nations and inflicted widespread destruction to property and infrastructure in several others.

Though these more recent incidents are all similarly destructive and socially disruptive, they appear at first consideration to be independent events, separated by time, cause, and geography. However, major events such as these, as well as other less deadly but similarly damaging incidents such as power grid failures, airport shutdowns, the spread of computer viruses, and large scale political demonstrations have had some common detrimental effects. One such effect, which represents the focus of this chapter, is the impact of disaster and associated responses on commerce. Large scale disastrous events throughout history have inhibited the business interactions that are unquestionably necessary for human social and economic need fulfilment. As a result, social scientists in the business disciplines have become increasingly interested in human reactions to large scale disorder emanating from disasters, and subsequently, several streams of research have emerged from scholars studying both the sociological and economic outcomes of disastrous world and/or national events.

The impact of disaster on modern business, business culture, and operations is fairly well documented by the popular press, albeit on an anecdotal basis. Adversities resulting from natural and human initiated hazards have significantly impeded many firms' efforts to serve customers, perform efficiently, and conduct normal exchange. These impacts have been catalogued economically as frequently costing in the millions or billions of dollars, notwithstanding other equally extensive social costs. However, unexpectedly, very little academic work has been written regarding thematic firm-level impacts of disastrous events affecting the business environment. A question commonly asked by concerned businesses addresses the continuity or discontinuity of business and their supply chain operations in the event of a catastrophic event. Managers wonder how they will continue to operate, and/or resume operations, in the after effects of a large scale disaster (Helferich and Cook 2002). Thus, business continuity refers to the minimization of disruption to the supply of products, services and information throughout the supply chain, over time, following a disastrous occurrence. Business continuity is important because it relates to how the firm performs from a financial and market perspective, whether customers continue to be satisfied, and whether the firm can continue to achieve a competitive advantage in the aftermath of an event that yields destruction to the local and global business infrastructure (Helferich and Cook 2002; Closs and McGarrell 2004). Business continuity is a primary goal of security and risk management initiatives at both the firm and supply chain level of analysis, and depends on resource bases and capabilities housed both internal to the firm and in the external supply chain environment (Helferich and Cook 2002; Zsidisin, Ragatz, and Melnyk 2005; Zsidisin, Melnyk, and Ragatz 2005).

This chapter presents a framework documenting business continuity planning capabilities that are thought to engender firm-level benefits. The context is that of supply chain management. Thus, in this chapter, we seek to address the following question: *What categories or types of capability development should businesses focus on in seeking to secure their supply chains, and thereby ensure continuity of operations?* In order to answer this question, we first introduce the concepts surrounding the focal context – those of supply chain security and supply chain security management, and the development of firm-level capabilities. Then, a conceptual framework is introduced that integrates the technology, processes, and human resources needed to develop supply chain security management capabilities, and thereby support business continuity planning in future business operations.

19.2 Supply Chain Security Concepts

Given the recent interest in the topic of supply chain security, it could be expected that firm-level security issues would be the focal topic of an emerging body of theoretical and confirmatory research. However, an examination of the supply chain security literature to date reveals relatively few investigative studies, ambiguous definitions and terminology, some inconsistency in theoretical development, and very little confirmatory research. We begin by defining these focal concepts and their interrelationships.

19.2.1 Supply Chain Security and Security Management

Supply chain security and supply chain risk are two related, although discrete, topics addressed by the dearth of publications to date focused on security-related issues. Terminologies pertaining to either security or risk have been used somewhat interchangeably. However, recent research is emerging that identifies these constructs as being distinctly different. *Supply chain security* is defined as the prevention of contamination, damage, or destruction of any supply chain assets or products. By contrast, *supply chain risk* is defined as the likelihood or chance of supply chain outcomes being susceptible to disruption, which would have detrimental effects on the firm (Closs and McGarrell 2004).

The processes for managing both supply chain security and supply chain risk are interrelated, as they are both involved in the promotion of the overarching concept of business continuity. The processes for managing supply chain security is known as *supply chain security management*, whereas the process for managing supply chain risk is known as *supply chain risk management*. Closs and McGarrell (2004, p. 8) define supply chain security management as the "application of policies, procedures, and technology to protect supply chain assets…from theft, damage, or terrorism, and to prevent the unauthorized introduction of contraband, people, or weapons of mass destruction into the supply chain". By contrast, the definition of supply chain risk management has evolved from Zsidisin's (2003) notion of supply risk management. It includes the processes involved in the reduction of the probability of occurrence and/or impact the damaging supply chain events have on the firm.

19.2.2 Supply Chain Security Orientation

More recent research has begun to identify common business philosophies that effectively promote supply chain security as a salient and necessary element for ensuring business continuity. Autry and Bobbitt (2007) introduce the idea of Supply Chain Security Orientation (SCSO), which represents an enterprise-wide attitude reflecting focus on supply chain security issues. Integrating both the security and risk management perspectives, the authors conceptualize SCSO as a broadly accepted company orientation reflecting the firm's collective attention to both supply chain security management and supply chain risk management principles. Employees of supply chain security oriented firms are posited as being concerned with both the potential for product/asset contamination, damage, or loss, as well as the potential for mitigating the likelihood and impact of disruptions between the firm and its supply chain partners. This group level attitude is manifested as four core premises that are consistently mentioned as vital for firms seeking to maintain effective levels of security and in minimizing and/or managing supply chain risk. The key premises include: (1) preparation and planning initiatives; (2) security-related partnerships; (3) organizational adaptation; and, (4) security-dedicated communications and technology.

Autry and Bobbitt (2007) suggest several drivers, outcomes, and moderating conditions affecting a firm's SCSO. Drivers of SCSO – conditions which cause a firm to become supply chain security oriented – include risk perceptions, past experiences with security failures, and mandates stemming from supply chain partners policies (e.g., supplier or customer concerns.) Top management support and employee factors are the primary conditions that affect the pervasiveness and/or acceptance of SCSO within firms. However, top management's awareness of security and risk management implications for breaches of security is not sufficient. Top management must also be committed to the process and be the driving force in the implementation of SCSO. Without commitment of top management to both sets of initiatives, other organizational employees will likely either not understand or appreciate the importance of the measures. In the judgment of the authors the critical role of implementing supply chain security and risk management initiatives rests on the middle and lower level employees of the firm. The authors suggested that when employees are motivated and empowered to cooperate with security/risk management initiatives, when they possess positive attitudes about security/risk management, and/or possess integrity and loyalty to the firm while dealing with sensitive information, they can better facilitate SCSO. Just as employees can be critical factors in securing supply chain assets and minimizing risks, they can also hinder SCSO initiatives if their attitudes are not supportive of security and risk management. Supply chain managers responding to questions within a structured interview setting were aware that the critical role in educating and conveying the importance of security measures to employees rests on the top management of the firm. To underscore this importance, it was even suggested that employees be measured and rewarded based on their efforts toward supply chain security and their contribution toward reductions in supply chain risk.

In addition to the primary factors discussed, other factors emerged that also have the potential to either facilitate or inhibit SCSO. Although these factors are less critical, they are still important to consider and include the following: (1) employee possession and use of security-related technology; examples include anti-tamper devices, x-ray/gamma-ray scanners, and radiation detection equipment; (2) employee-driven control processes for the continuous monitoring of security and risk; (3) integration of security measures into business processes; examples include inventory management and transportation; and (4) the allocation of human and financial resources to security and risk-related implementation efforts. It must be noted that these additional factors are driven by the strategic direction of the firm, which is ultimately under the direction of top management. As such, this suggests that security efforts need to be addressed at the strategic level of the firm and included in firm-level planning.

Along with the internal factors that influence the extent to which firms adopt SCSO, there were external factors uncovered that also influence adoption. Political/legal factors and supply chain partner factors were the two primary factors external to the firm found to be both facilitators and inhibitors of firm SCSO adoption. Government support was one important external facilitator cited. This government support could be either in the form of enforced regulations regarding security or merely as public support for a firm's security efforts manifested in informal statements or formal debate. The issue is further complicated by the fact that many firms today have global operations that are dependent upon the assistance of local authorities in protecting firm assets, and the movement of products and information through the supply chain. Unfortunately, the study revealed that government entities can also serve to stifle security management and risk management initiatives, such as by creating obstacles in the form of "government red tape" or "bureaucracy". Other respondents mentioned the lack of government regulations in foreign countries that results in the lack of security and in turn creates opportunities for threats. This latter issue may be difficult to counter if local governments do not cooperate.

The authors also identify the importance of the support and cooperation of supply chain partners, including suppliers and customers, and all entities directly or indirectly involved in the supply chain. Supply chain management has created an environment where processes and systems cross firm boundaries. As such, firms are affected by decisions made by their trading partner, including security and risk-related policies and capabilities. Poor risk-related policies and capabilities of supply chain partners can also impede SCSO adoption and implementation. In fact, one respondent to the SCSO study indicated, "there needs to be a common understanding between all parties" on supply chain security and risk management. It was suggested that supply chain partners need to be made aware of the long-term benefits that can be realized by all firms in the supply chain in order to cooperate on SCSO. Partnering with firms that have not made a commitment to their own security and do not have risk policies in place may be viewed as risky, as it potentially exposes the entire supply chain to threats.

Potential outcomes of SCSO are thought to be widely variable. The major themes identified by Autry and Bobbitt (2007) relate to firm performance, operational performance, market performance, customer satisfaction/dissatisfaction, and continuity of supply chain operations. Though enhanced performance and satisfaction stemming from security management initiatives are obviously desirable objectives, they are best viewed as "stretch goals" in most business contexts; supply chain security management is largely focused on the prevention of negative impacts on business processes rather than generating positive firm outcomes. Business continuity, alternatively, is the primary focus and driving force for security management, and thus represents the central focus of this chapter.

In order for firms to leverage SCSO to strengthen business continuity, they must begin to develop capabilities directed toward the detection and eradication of security threats. The literature from strategic management related to firm-level resources and the development of dynamic capabilities (e.g., the resource-based view of the firm) provides a sound and well-respected theoretical basis for the study of supply chain security capabilities (e.g., Penrose 1959; Helfat and Peteraf 2003). In the following section, we review this literature base, and extend its premises to the security management context. Then, in the final section of the chapter, a capability-based framework for security management is delineated for use by firms seeking to convert their SCSO into operational security protection mechanisms.

19.3 Securing Supply Chains Through Capability Development

Increasingly, firms are devoting resources to the protection of themselves and their supply chains. As noted in the previous section, to some extent, the increase in security-related investment in recent years may be attributable to well-known contemporary events that draw attention to the need for protecting company and supply chain assets (e.g., the 9/11 terror attacks or Hurricane Katrina), and thus could be considered a key theme or pillar of supply chain security management. The resource-based view of the firm (RBV) provides a basis for understanding the effectiveness and advisability of security-related investments designed to protect firm and supply chain assets. The RBV is a well-known strategic framework that speaks to competitive heterogeneity and associated advantages that stem from differences in internal organization across firms rather than industry characteristics and structure (Penrose 1959). The RBV is adaptable to the supply chain security context via a manipulation of the focal outcome variable; rather than seeking supernormal profits, Firms practicing supply chain security management organize themselves such that they mitigate losses. In other words, RBV principles are suggested herein as applying in the security management context, but the role of security-providing resources is to minimize losses rather than to maximize profits, though the overall focus on generating enduring industry advantage still applies (e.g., Porter 1979; Barney 1991).

As the RBV has evolved into a widely-accepted tool for explaining differential firm performance outcomes, it has become evident that simply acquiring and collecting resources is not a sufficient condition for firms to succeed – firm resources must be bundled intelligently with a given end purpose in mind in order to allow the firm to do things that its competitors cannot. The unique ability of a firm to perform a coordinated set of tasks utilizing organizational resources for the purpose of a specific end result is known as an organizational capability. In the context of supply chain security management, the development of organizational security capabilities may be what separates firms whose supply chains are breached from those that remain safe in the face of an impending risk. Drawing again from the strategic management literature, the development of supply chain security management capabilities may be properly identified as a *dynamic capability*. Dynamic capabilities are defined as the firm's ability to reconfigure resource stocks (Teece et al. 1997; Eisenhardt and Martin 2000). These sets of organizational capabilities do not directly involve the production of the focal good or provision of a marketable service, but rather, indirectly contribute to the success of the firm through their impact on regular operational capabilities (Helfat and Peteraf 2003). Supply chain security management capabilities are those dynamic capabilities that allow the firm to continue to perform operationally through protection of other firm resources such as plant, equipment, and human factors, and via this protection represent a differentiating factor for firms in the event of a disaster.

The invocation of the dynamic capabilities perspective in the supply chain security management context also requires consideration of several observations such as those catalogued by Eisenhardt and Martin (2000). Specifically, there are three sets of stipulations that must be considered when adopting a dynamic capabilities viewpoint on supply chain security management. First, dynamic capabilities are suggested to consist of very specific strategic and organizational processes and resource investments that create value through their manipulation. In the current supply chain security management context, this perspective is adjusted such as prescribed above; firms are thought to benefit most from an asset protection standpoint when they are able to optimally mobilize resources and processes to react to impending risks, and the realized benefit is presented as a loss prevention rather than profit maximization calculus. Second, dynamic capabilities that are thought to yield maximum expected benefits tend to be replicated across firms within and across industries, and therefore they have greater equifinality, homogeneity, and substitutability across firms than would operational capabilities. Thus, a body of "best practices" with respect to dynamic capabilities tends to emerge over time. This is verifiably true in the case of supply chain security management, where competing and non-competing firms alike tend to share and imitate maximally effective security practices (e.g., RFID or biometrics) when resource endowments are similar. Finally, in competitive markets, configurations of organizational resources and capabilities deemed to be effective for profit generation are contingent on the stability/volatility of the marketplace in question. The same is true for security management, where the effectiveness of a specific security-related dynamic capability depends on the risk characteristics of the given firm, supply chain, and industry.

Based on the above literature review, the dynamic capabilities approach is adopted for examination of the supply chain security management capabilities of modern firms. Drawing on additional knowledge from operations management and organizational safety literatures, the remainder of the chapter describes in detail the specific categories of dynamic firm capabilities that best facilitate security management in the supply chain organizational environment.

19.4 Building Supply Chain Continuity Through Security Management Capabilities

Continuity planning represents the process of managing security performance throughout the entire supply chain. Building continuity into the future involves ongoing management of numerous aspects of the supply chains which working together can serve to mitigate many supply chain risks and improve supply chain security, including the development of dynamic security management capabilities. Dynamic security management capabilities are resource bases that can be activated or deployed toward the goal of securing the firm's supply chain. They can be broadly divided into three categories: *processes, technology*, and *human resource*. These types of security management capabilities are depicted in Fig. 19.1.

Processes are the protocols and procedures followed by the organization and its supply chain partners. The fundamental aspect of building continuity into the future rests on well constructed processes which ensure integrity throughout the supply chain by providing monitoring, prevention, and responsiveness. It requires policies for everything from supplier quality checks to inventory management to forecasting disruptions.

Technology involves the use of systems, tools, and machines that gather, process, and communicate data and information within and between firms. As such, they enable real-time monitoring and visibility of the supply chain. Finally, people are the human entities that implement procedures and use the technology in order to build continuity.

Human Resources capabilities include organizational leadership, all of the organization's employees, suppliers, as well as government agencies responsible for putting into place programs that require security compliance. Building continuity into the future requires a comprehensive and holistic solution that is

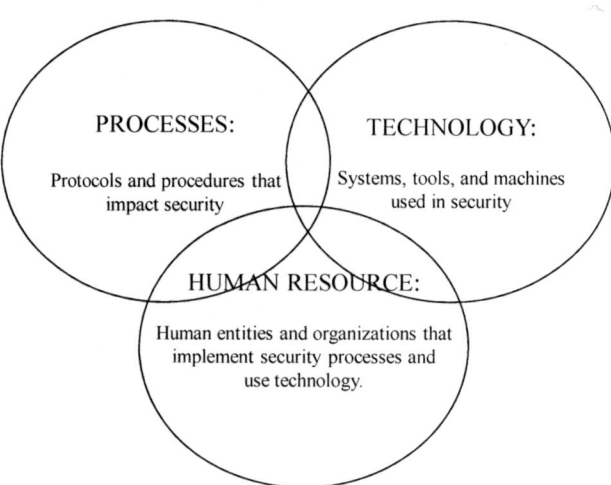

Fig. 19.1 Categories of capabilities used to build business continuity

integrated across the enterprise and between entities of the supply chain. As such, the three dynamic security management capability types – processes, technology, and human resource – cannot be managed independently of one another but must be designed to work together in unison. For example, technology capabilities must be developed and leveraged to support organizational and supply chain security procedures, and organizational members and suppliers must be trained in technology-related procedures and technology use with respect to security. In the remaining sections, each of these security management capability categories is described in detail.

19.4.1 Processes

One of the most important dynamic capabilities for the purpose of building business continuity is the capability of implementing processes, or protocols, for the management of the supply chain that offer effective security without compromising performance. This type of capability includes all processes within the organization, as well as processes throughout the supply chain. Security must be built into each process, such that processes provide for supply chain visibility, an effective sourcing strategy, strategically managed inventory, the capability for redesigning products and processes with security in mind, and effective forecasting procedures.

19.4.1.1 Build Security into the Process

Strategies for developing effective system wide processes that are all encompassing in nature can be learned from the total quality management (TQM) movement of the 1980s (Lee and Wolfe 2003). Prior to TQM, companies used inspection and statistical sampling to check for quality defects after the product was produced. The defects missed by inspection could result in product failure at the customer site and cause numerous costs, such as customer down time, product recalls, field repairs, liability, and loss of goodwill. Then in the 1980s, with the help of quality gurus such as W. Edwards Deming and Phil Crosby, there was recognition that quality defects can be costly to the company. There was a recognized understanding that it was important to build quality into the process – the hallmark of the TQM movement – rather than relying on inspection after the product was produced. In fact, according to TQM, inspection should be the last resort as inspection by itself does not improve quality. Similar to the TQM analogy, security should be built into the process, rather than managed through inspection.

Another lesson from TQM is that quality processes should be implemented at every stage of the supply chain, from the supplier site through the production process to final customer delivery. Applied to security, this means that processes to monitor security should be put in place at every stage of the supply chain, rather than just relying on inspection at delivery points and border crossings. In fact, the concept of acceptance sampling – meaning randomly sampling the finished product for defects – is considered out of date in TQM. The same should be true with security. Further, TQM requires monitoring of all processes to ensure that they are functioning in a pre-determined state of control, designed to produce items to specifications. The idea is that a process that is out-of-control will produce more items that do not conform to specifications. TQM teaches that processes should be regularly monitored and if it is detected that they are out-of-control, corrections should be immediately made. The same is true of security. Standards should be set for all processes that monitor security and any abnormalities noted immediately corrected.

For supply chain security, building security into the processes means developing processes that prevent tampering with containers before, during and after the loading process. The first step may be to thoroughly inspect credentials and backgrounds of all personnel, especially those responsible for cargo handling and with access to plants and warehouses. All processes should be closely monitored, including the flow of inbound materials, the pick-and-pack process, the staging of outbound loads, as well as the loading process. Documentation should be made of all goods being shipped and the processes of preparing and shipping the load. For example, this documentation can include the identities of pickers, packers, loaders, checkers, and all other individuals involved. The provided documentation can serve as proof of the level of security detail involved in the process.

19.4.1.2 Supply Chain Visibility

An important aspect of building continuity into the future is to have processes in place that enable a company to have complete supply chain visibility. Visibility of the supply network can help a company anticipate a disruption and mitigate their effects when they do occur. Part of the visibility is having a clear picture of the location and form of inventories in the supply network, whether they are in raw material form, work in process, stored at a warehouse location, or are in transit. Then when a disruption does occur, a company can respond more effectively as it knows which resources to marshal and how to manage them. With this information a company would immediately know which goods to reroute, which production resources to redeploy, how to adjust capacities, and how to revise production plans. Also, having a process in place that provides clear visibility gives reassurance to customers that the company has the foresight to devise an appropriate response to a disruption.

System-wide visibility is enhanced using near real time databases that can collect daily or hourly snapshots of demand, inventory, and capacity levels at key nodes in the supply chain, including ports and shipping locations. Systems can be put in place that automatically track breaking news and monitor media Web sites for information regarding problems at high risk locations. This information flow provides the needed visibility and more flexibility in planning a response to a supply chain disruption. In turn, the company needs to plan for adjusting its response as operating conditions change.

One example of network links that provide all entities of the supply chain with needed visibility is Cisco's eHub, a private exchange (Sheffi 2005). Multiple tiers of suppliers are linked together and instantly provided with a complete picture of the supply chain, including potential disruptions, capacity problems, and potential delays or shortages. The eHub also provides problem resolution options. In this manner, participants have resolution options to logistics problems once the problem is identified.

19.4.1.3 Sourcing Strategy

Since the 1980s and the focus on lean manufacturing there has been a trend toward streamlining the supply base. Many companies have continued to pare down their supply base and have adopted a single sourcing strategy. The objective of this strategy has been for companies to build stronger collaborative relationships with their suppliers, improving quality and long-term sustainability. Further, the argument for a single sourcing strategy is that committed supply chain partners can work together to reduce costs typically associated with multiple supplier relationships and these committed relationships would help companies during troubled economic times. Unfortunately, given the problems of supply disruption, as a result of either natural or man-made disasters, the wisdom of the single sourcing strategy is being revisited, as relying on a single supply source may be risky (Sheffi 2005).

There are numerous flexible sourcing options that are being used by companies in order to mitigate the risks of supply chain disruption and improve security. One strategy is to develop multiple supply sources for the same component in a way that is cost effective while providing flexibility. For example, Hewlett-Packard (HP) has a sourcing strategy that achieves this (Feitzinger and Lee 1997). For each component the company offers a fixed supply contract with guaranteed quantities to one supply source that specializes in efficiency. Then, another supply source is selected which specializes in flexibility, and is given a flexible contract with upper and lower volume limits. The idea is that the first source addresses issues of cost efficiency while the second provides flexibility. Then, if the demand exceeds the sum of the fixed and flexible sourcing amounts, HP relies on the spot market to make up the difference.

A second sourcing strategy is to use local supply sources as part of the sourcing strategy. Even for companies that rely heavily on global sources of supply, it is wise to at least in part develop local supply sources to supplement their supply base. A local supply source may be more expensive, but it also enables a company to respond more quickly as market needs change and can serve as a "backup". By contrast, relying exclusively on an offshore supply base can be risky. This dual sourcing strategy can be very effective as many companies found out when supply at the US border crossing was cut-off after the 9/11 terrorist attacks.

A third strategy is to rely on suppliers with multiple manufacturing sites at different locations, rather than just one site. The reason is that the risks associated with disruptions are minimized as supply can be shifted from one site to another.

19.4.1.4 Inventory Management

A key aspect of providing resiliency and building continuity into the future relies on the strategic management of inventory (Lee 2002). Strategic locations of inventory can provide resiliency and flexibility in times of disruption. One of the most basic inventory strategies is to improve visibility of inventory buffers in distribution channels. This includes visibility of inventory buffers at warehouse locations, manufacturing locations, and all distribution centres. Having visibility is a simple yet effective strategy as it can assist with real-time contingency planning and risk mitigation.

One of the key elements of contingency planning is having a process in place for rationing inventory when disruptions at one location result in shortages. Just-in-time inventory has promoted the notion of lean systems and the elimination of waste, including inventory. However, security problems and disruptions can make lean supply chains vulnerable to sudden stockouts. In fact, many companies with lean systems prior to the attacks of 9/11 found their inventories tied up at border crossings and their production processes coming to a halt. Due to this some companies have revisited the soundness of just-in-time inventory control principles (Sheffi 2001; Martha and Subbakrishna 2002). It would certainly be easy to simply put inventory back in the supply chain to provide added protection. However, the cost of doing so would be high and would slip organizations back into an era of inefficiencies and high warehousing costs, seen some decades ago. Rather than putting inventory back everywhere in the system, strategically placed buffer stocks of inventory are a necessity in today's security conscious environment.

One option in computing inventories is to recalculate the amount of safety stock needed in order to account for delivery unreliability. Traditionally, safety stock is computed as a function of both the average delivery lead time and the variance of the delivery lead time. In fact, of the two components the variance of the delivery lead time impacts safety stock levels more than the length of the lead time itself. It is because of this that lead time unreliability can have a greater impact on companies with typically stable demand patterns, as the amount of safety stock carried to buffer against uncertainty is lower. Therefore, delivery unreliability can be very costly in terms of shortages. A good option for companies is to assess the tradeoffs between the risks of stockouts and the cost of holding inventory. In most inventory systems the risks of stockouts are related only to *demand* uncertainties. Rather, stockouts should be computed considering uncertainties of both *supply* and *demand*.

19.4.1.5 Product and Process Redesign

The concept of designing products and processes with the intent of addressing risks and managing security problems has gained increased attention. One such strategy is to design products with increased commonality of parts. This strategy enables companies to increase the number of potential suppliers and the pool of available inventory the company can draw from. In case of a disruption or security problems at one location, the company can easily begin sourcing inventory from another location. Another strategy is the standardizing of the manufacturing process between manufacturing locations. This strategy enables companies to easily shift production from one facility to another in case of a disruption or security breach. For example, Intel uses a strategy called "*Copy Exactly*" which means that each manufacturing site is an exact copy of the other so that if a security breach or disruption occurs at one location, it can seamlessly switch to another (Sheffi 2005).

Maintaining as much flexibility in the production process can provide companies with the ability to rapidly respond in cases of security disruptions. The more standardized a product the greater the company's flexibility, as the company can more easily shift sources of supply (Sheffi 2001). An especially popular product design strategy that can help achieve flexibility is a technique called postponement, or delayed product differentiation. This strategy permits the product to remain in its generic form as long as possible before it is differentiated for customers and locations. Postponement enables companies to configure specific product characteristics at the last minute, allowing them to meet variable demands of different products at different locations. By keeping products in generic form for a longer period of time, postponement also provides greater flexibility in responding to security issues. Postponement is a strategy that has been particularly successfully in high tech industries, such as computers and printers (Sheffi 2005).

19.4.1.6 Demand Based Management

Traditionally companies have had competing goals between the marketing and operations functions. However, successful companies have discovered that having the two functions work together to make sure that the products offered match demand also provides them with some flexibility. This strategy is called demand based management and basically means offering the right product at the right price to match supply with demand. This is especially effective when security problems occur in one area of the business as companies can entice customers to buy what is available and avoid shortages in other areas of the business. Demand-based management requires marketing and operations to work together, as one provides an understanding of customer preferences and how customers might respond to price changes, while the other understands dynamics associated with product offerings.

A company that is especially successful in using demand based management is the Dell Computer Corporation (Sheffi 2005). Dell uses the direct sales model, rather than going through a retailer, and can make changes to product configurations and prices relatively easily. This flexibility permits Dell to steer customers to product configurations that consume surplus inventory. Dell is also able to change configurations easily that use up excess inventory. This flexibility permits Dell to guide customers to product configurations that use up ample supplies of inventory. This type of flexibility also has significant implications on the company's ability to respond to security breaches. A security problem that disrupts any source of supply will not be catastrophic for Dell, as the company will simply shift product designs to available sources. For example, when the 1999 Taiwan earthquake disrupted the supply of computer chips from a main supplier, Dell immediately switched designs to utilize another available component (Sheffi 2005).

19.4.1.7 Forecasting

Forecasting has always been a business challenge. However, it has become especially difficult over the past few years due to globalization, shortened product life cycles, and unforeseen disruptions (Sanders 2006). Similar to other organizational and supply chain processes, forecasting needs to be flexible in order to be most effective in times of change.

Statistical forecasting alone cannot be successful during rapidly changing times due to the fact that these forecasts are based on historical data that is often irrelevant. Managerial forecasts, in combination with statistical forecasts, have been shown to improve forecast accuracy (Sanders 2006). The reason is that managers often have information on last minute changes in the environment that is not available to the statistical model. One option is to rely on statistical forecasts that allow managers to make adjustments when there is specific information available about the environment. These adjustments can either be made manually, or done with software packages that provide for "overrides" of the computer system in order to account for unforeseen events.

Forecasting a range of potential outcomes, rather than forecasting a single demand figure, can be another good option in addressing uncertainty. This forecasting range can then serve as a guide for developing flexible contractual terms with suppliers. The range can also serve in the development of contingency plans for supply chain disruptions and security breaches. For example, it can help develop a plan for what should be done if demand is on the very high end versus on the very low end, outcomes that could be a result of security failures. The process can help the company think in terms of a range of possible outcomes, some of which may be unexpected under normal circumstances. This is important as it forces the company to be able to think in terms of flexibility rather than in fixed performance terms. This forecasting range can further be broken down into confidence intervals that specify high and low ranges of uncertainty. This permits companies to plan with certainty for one range, but be prepared for the possibility of responding at another end of the range.

A successful forecasting option is to rely on collaborative forecasting efforts with supply chain members rather than developing the forecast alone. One such effort is Collaborative Planning Forecasting and Replenishment (CPFR), which is a joint process between two trading partners that establishes formal guidelines for joint forecasting and planning. In addition to joint forecasting, or collaboratively attempting to predict future events, trading partners develop contingency plans to deal with exceptional events, which can include security breaches and failures. The premise behind CPFR is that companies can be more successful if they join forces to bring value to their customers, share risks of the marketplace, and improve their performance. Trading partners can also improve their performance relative to security by developing joint forecasts of the future and contingency plans for dealing with extreme outcomes.

19.4.2 Technology

Technology capabilities encompass tools and machines that can be used to solve problems. State of the art advances in technology have led to its use as a dynamic capability in security management to provide communication and information flow throughout the supply chain, as well as global inventory and schedule visibility. A wide array of technologies is available that range from active telecommunications to traditional visual bar codes and labels, to electronic sensors (Lee and Wolfe 2003). These technologies permit shippers to track their shipments at all times, monitor their status, and provide network visibility that can prevent tampering and mishandling, as well as provide overall shipment integrity. In addition to these obvious security benefits, the use of technologies can provide other benefits, which include increased efficiency and productivity, reliability, and improved service quality. The most popular complementary technologies providing dynamic capabilities for security include the following:

- Biometric systems – These are technologies used to measure and analyze human physical and behavioural characteristics for the purpose of authentication. This includes identification of physical characteristics such as fingerprints, eye retinas, facial patterns and hand measurements. For example, Walt Disney World uses biometric measurements taken from the fingers of guests to ensure that the person's ticket is used by the same person from day to day. These biometric systems can be valuable tools for the automated identification of personnel, including truck drivers.
- Wireless and mobile technologies – These are technologies that permit in-transit long and short distance communication using satellite or cellular devices, with global positioning system-like location determination.
- Sensors – These technologies include all types of fixed and portable devices to detect a range of violations, including the presence of explosives, human presence, and intrusion of any type.
- Electronic cargo seals – Electronic seals serve similar functions to physical seals by providing unauthorized access detection. These seals consist of digital sensors and use digital signatures for authorized access. Electronic seals are used to ensure shipment integrity and track cargo, and are especially useful for high-value loads and any shipments requiring enhanced security.

All of these technologies are currently in use and are continuously being improved. They are used by themselves or in concert together to add multiple layers of security. In addition to security, these technologies have the potential to improve efficiency and productivity.

19.4.2.1 Radio Frequency Identification

Of all the technologies currently used in security management, radio frequency identification (RFID) has received the most attention and needs to be discussed separately. RFID uses memory chips equipped with tiny radio antennas that can be attached to objects that transmit streams of data about the object. For example, RFID can be used to identify any product movement, reveal a missing product's location, or have a shipment of products "announce" their arrival. Empty store shelves can signal that it is time for replenishment using RFID, or low inventories can inform the suppliers that it is time to stock more products. In fact, RFID has the potential to become the backbone of an infrastructure that can identify and track billions of individual objects all over the world.

Consider the case of an incident where shelves in a Wal-Mart store in Broken Arrow, Oklahoma were equipped with RFID readers to track a Max Factor lipstick (US Dept of Transportation 2002). Procter & Gamble managers could tell when the lipsticks were removed from the shelves 750 miles away by using Webcam images placed on product shelves. Many of the large supply chain retailers have implemented RFID due to its enhanced capabilities and are requiring their suppliers to do so. Also, the United States Department of Defense (DOD) has successfully used RFID tags to reduce logistics costs and improve supply chain visibility for more than 15 years (US Dept of Transportation 2002).

RFID tags have numerous applications in security and their use has been broad and varied. RFID tags have been used in passports issued by many countries to improve security integrity and speed up throughput. These passports can record travel history, including time, date and place, as well as entry to and exit from a country. RFID is also used for electronic toll collection at toll booths on highways, such as Florida's SunPass and Georgia's Cruise Card. The tags are read remotely as vehicles pass through the booths. The information on the tag is then used to debit the toll from a prepaid account. In addition to speeding up traffic, the information can be used to keep track of shipments and other route status information. RFID tags are even used at some ski resorts to prevent unauthorized access to the resort and offer better service, such as providing skiers with hands-free access to ski lifts.

RFID has had tremendous use in monitoring inventory control. The visibility provided by RFID gives accurate knowledge of inventory levels and can eliminate any discrepancy between a physical count and inventory records, also reducing labour costs since a physical count isn't needed. This is one of the reasons many large organizations are mandating that their suppliers implement RFID tagging. For example, Wal-Mart has required its top 100 suppliers to apply RFID labels to all shipments. RFID tags have even begun to be used in library books to allow for book tracking and prevent theft.

RFID has particular use in product tracking in the commercial sector. For example, the Canadian Cattle Identification Agency began requiring the use of RFID tags to identify a bovine's herd of origin. This is important in being able to detect the origin of a carcass if it is found to harbour an illness. Also, RFID tags are commonly used to track cases, pallets, and shipping containers. RFID tags are also used to prevent building access control, airline baggage tracking, pharmaceutical item tracking to authenticate the pureness of the product, and in apparel tracking to ensure product originality.

Of course, RFID is not a panacea for security related problems and the technology itself raises certain security concerns. For example, there is an issue of illicit tracking of RFID tags. In addition, there is the issue that tags affixed to products that continue to remain functional even after the products have been purchased and taken home. The tags could potentially be used for surveillance, and world-readable tags pose a risk to personal location privacy. To address this issue some RFID tags have incorporated a defence system that involves cryptography to prevent tag cloning. For example, a rolling code can be used, where the tag identifier information changes after each scan. More sophisticated RFID technologies are being developed where a tag interacts with the reader, and information is not provided by the tag until the reader has enabled the circuit. Researchers are continuing to improve the value of the technology, to include those associated with security.

19.4.2.2 Specific Security Applications of Technology

State of the art technology such as RFID can provide a number of security functions, ranging from the monitoring of vehicle movement and their content, to the visibility of entire networks within which they move. Some specific security applications provided by technology are discussed next.

Asset Tracking. This function involves tracking and monitoring of vehicles, cargo, as well as adherence to route during movement. This can significantly reduce the risk of in transit tampering, intrusion, and any other security breach by providing visibility and control.

The most important devices for asset tracking are those that facilitate communication. This includes satellite and cellular systems for long-distance mobile communication. These devices provide significant benefits due to their ability to impart information at any time or point in the transportation cycle. For short-range communication, RFID is very effective. Although it is limited to reporting within 100 m or less from the fixed reader site or handheld, it can provide a rich range of information about the status of the product and cargo.

One important aspect of using these mobile communication systems to support security is to always note the current location when an event is recorded or a message sent. For satellite and cellular communication devices the most common method for doing this is to use on-board calculations of latitude and longitude from a global positioning system (GPS). For short-range communication, technologies such as RFID are an excellent choice.

Uses of these technologies for purposes of asset tracking in practice are numerous. For almost a decade the DOD has been using RFID tags to track containers being moved globally. The RFID tags, containing a wealth of information, are read by readers at terminals and gateway points around the world and are able to provide status information. In the commercial sector, the use of simple RFID tags by shippers for asset tracking at the container and pallet level is becoming increasingly common (US Dept of Transportation 2002).

Geofencing. Geofencing is a concept related to asset tracking that enables commercial dispatchers and law enforcement officials to monitor vehicle route adherence. The technology uses computer algorithms to analyze and then display location data. Exceptions to the preset pattern can be quickly noted, such as noting a vehicle that has deviated from its route, has entered restricted areas, or is developing some type of failure. Geofencing can be set up to work with most wireless and mobile communication devices and provide sophisticated asset tracking.

On-board monitoring systems for supply chain assets. An important aspect of security support that technology can provide is the monitoring of vehicle operating parameters, cargo and freight conditions, detection of intrusion and tampering, and remote locking and unlocking. Basic technology solutions operate by gathering sensor data, then transmitting it to a remote destination for evaluation and action. However, more sophisticated technology solutions gather the data, evaluate it, and can automatically trigger actions without having to gain authorization from a remote destination.

A simple example of this is the development of automated restart circuits on containers requiring refrigeration. A much more extreme example of this was developed in South Africa, where a trailer intrusion sets an automated series of internal pepper gas dispensers designed to discourage thieves. Other examples of onboard monitoring include temperature sensors and recorders in perishable shipments. Also, toxic sensors provide on-board monitoring of the shipments of hazardous materials (US Dept of Transportation 2002).

Other on-board security devices include emergency call buttons that enable drivers to call for help through a single click. These devices can be wireless remote communication devices that the individual can have with them at all times, captures the GPS location and then sends an emergency signal. More sophisticated on-board monitoring devices combine different types of technologies. For example, electronic seals can monitor the integrity of their closure for tampering and either use RFID to convey the results to a reader or some other type of mobile communication. Other types of seals use magnetic pressure-based door sensors that detect entry and either create an automated response or relay the information to a reader. Also, security can be elevated through the use of remote locking and unlocking systems which can include electronic contact "keys", programmable access codes, local RFID controls, long-distance command via cell or satellite communication. On-board security systems can even combine radio remote control locks with geofencing information to prevent unlocking of containers and specific coordinates.

Entry-point Passage Facilitation. There is an entire set of technologies that can improve processing at entry points, including terminals, inspection points, and border crossings. This includes technologies for automatic driver identification and verification, which can greatly reduce the risk of theft, terrorism and other security breaches, and can add efficiency to frequent drop-offs and pick-ups. An important technology for this purpose is biometric identification, which includes tools such as fingerprint and iris recognition. Identification information can be incorporated into smart identification cards (IDs) and to on-line access of support information which includes manifest data, vehicle information, and a driver database.

Entry-point passage is further facilitated through non-intrusive inspection technologies, which can create greater efficiency and lower cost over a manual method of inspection. For example, large gamma ray and x-ray machines that scan for anomalies in the density of cargo can be performed at entry points. Also, customs inspection officials can carry handheld devices that have sensors to detect radiation and other explosives. In addition, in-motion weight detection and electronic toll payment using RFID technology can be especially helpful in improving efficiency and productivity by increasing throughput and reducing delay time without compromising security. Sensor technology can perform a calculation of truck weights without stepping on fixed scales.

Shipment and Network Status Information. Numerous technology applications are designed to facilitate the exchange of information about shipments and provide status updates to various members of the supply chain. For example, both the commercial and public sectors use web-based freight portals. Carriers and third-party logistics providers provide web sites to their customers for shipment status and pick up information. Technology can aid this process by providing efficient shipments, making the best use of available transportation capacity and avoiding congestion. These technologies can collect information on network conditions and ultimately manage shipments and routes accordingly.

Most transportation technologies are capable of providing congestion alerts and avoidance routes. Information captured by cameras and road sensors can be fed into predictive models and made available through Web portals. This information is especially important when trying to get shipments through crowded gateways, such as ocean terminals and border crossings. It is invaluable as a tool that enables supply chain network visibility.

19.4.3 Human Resources

An important lesson from the Total Quality Management (TQM) movement is that overarching organizational effort cannot succeed without the participation and support of everyone in the organization (Lee and Wolfe 2003). With the rise of TQM, it became clear that quality was not merely the domain of the quality control department and quality inspectors, but that everyone was responsible for quality. The same analogy can be applied to security. Well thought out processes and sophisticated technology are only tools and procedures, but are not enough. Building continuity into the future cannot be accomplished without the engagement and involvement of dynamic capabilities emanating from organizational participants. This includes the capabilities of top management, employees, suppliers, and even government agencies. These are all discussed next.

19.4.3.1 Top Management Support

One of the most critical aspects of building security is to have top management support, direction, and have someone from the top echelons champion the effort (Martha and Subbakrishna 2002). Building continuity into the future requires large sweeping organizational changes in process, technology, and individual training. Security will not become a top organizational priority without top management leadership. Instead, a great deal of energy will be wasted on merely trying to piecemeal resources for ensuring supply chain security. A good way to make sure that security does become a priority is for the CEO to take direct oversight of the process. Another option is for the CEO to appoint a senior vice president to be the champion of this organization wide effort. When these types of mandates come from such a top organizational level they quickly permeate throughout the entire organization.

19.4.3.2 Employee

An effective security system requires the involvement of well-trained employees at all levels of the organization. Key personnel need to be knowledgeable on how to handle disruptions and oversee the response process. All employees need to be trained to monitor security in their jobs and the organization as a whole. These employees need to be the eyes and ears of the organization noting anything that is indicative that the process is performing outside of normal bounds. In fact, the entire organization needs to have a culture of disruption awareness. Like with TQM, a good strategy is to provide initial training in security procedures, followed up with periodic refresher workshops. Also, the organizational culture should reward efforts that improve security procedures and employees should be encouraged to offer suggestions. Companies can go as far as to mimic Quality Circles – groups of employees who meet regularly to discuss quality problems – and create Security Circles. These could be groups of employees who meet on a regular basis and come up with ideas and suggestions on improving organizational security aspects.

One aspect of security that needs particular attention is the area of employee screening and hiring practices. Hiring individuals that can compromise security is extremely dangerous and companies need to have systems in place to prevent this from happening. Effective screening should go beyond the hiring process and include ongoing tracking and assessment of each employee and third-party provider with security access. This process involves careful record keeping that automatically alerts the company of potential threats that can occur from someone with knowledge and intent.

19.4.3.3 Suppliers

Building continuity into the future cannot be accomplished without the involvement and participation of suppliers. First and foremost, it is important to screen and regularly monitor current and potential suppliers for the risks they may pose to security. Self-assessment by suppliers can be used as part of the screening process and it can be an internally developed risk-scoring approach. The risk assessment can include scores that relate to quality, financial viability, technology, location risk, shipping mode, and route risks. Using this approach, companies can identify high versus low risk suppliers and consider this in the request for quote (RFQ) process. The continuous monitoring of current and potential suppliers can be stored in a database of suppliers and the assessment results of risks monitored over time. In addition, some companies are even requiring their potential suppliers to show a business continuity plan of their own as part of the bid process (Elkin, Handfield, Blackhurst and Craighead 2005).

Once suppliers have been selected, it is important to get critical suppliers involved early in the security process and to develop an understanding of the supplier's organization. For all critical suppliers it is important to understand their operating units and how they will be interacting with the organization. Regular meetings with suppliers to deal with security plans are important. These meetings should address questions that pertain to system compliance, operational processes, financial viability, and supply chain response techniques. Together with the supplier it may be important to identify supply risks for strategic materials throughout the entire supply chain. Also, it is useful to identify potential vulnerabilities or events that create risks, and how the supplier will respond to them. These risks should include work stoppage, disruptions of raw material flow, shipment failures, and material unreliability. A joint plan should be developed for each risk identified. Critical suppliers should also be required to share a business continuity plan of their own, identifying capabilities that would be implemented in the event of a disruption. Strategic sourcing can continue to work with the chosen suppliers into the future to continue to manage their supply chain risks.

19.4.3.4 Governmental Security Initiatives

The final categories of participants in building continuity into the future are government agencies that develop and implement security regulation that affect business. The best known US government program is the Customs-Trade Partnership Against Terrorism (C-TPAT). The C-TPAT program is open to manufacturers, importers, carriers, and third parties and is similar to certifications in the area of quality. Applicants for certification are required to complete detailed questionnaires and self-appraisals of their supply chain security processes. Similar to ISO 9000 certification by the International Standards Organization in the area of quality, Customs reviews the self-appraisals, visits facilities, asks for modifications and reserves the right to perform unannounced verification visits. Once approved, C-TPAT participants get faster processing at ports and border crossings. The government also identifies participant firms, providing acknowledgement that these firms have met certain security standards. Similar to the move in many industries to hire only suppliers that are ISO 9000 certified, there has been a move with some retailers to use only C-TPAT compliance logistics providers. Some of these companies have even reported significant cost savings as a result of reduced theft.

19.5 Conclusion

Taking a dynamic capabilities perspective, this chapter suggests several ways that firms can leverage resources in order to improve and maintain security management initiatives. By blending organizational capabilities emanating from people, processes, and technology, the firm can embrace a supply chain security orientation and effectively protect itself and its supply chain assets from damage and/or destruction caused by internal and external threats. In modern times, many firms operate with constant concern that they are susceptible to human acts such as terrorism, sabotage, or theft, and often additionally worry that their differentiating assets are vulnerable to the effects of natural disasters such as hurricanes, floods, or other acts of God. Thus, it has become vital that firms take every prudent step toward securing themselves and their partners' future operations. As such, this chapter provides a framework for future disaster planning efforts, and guides investment in organizational resources that go beyond normal profit-driven motivations, and allow the firm to embrace supply chain security as a common goal of business.

References

Autry, C.W. and Bobbitt, L.M. (2007). Supply Chain Security Orientation: Conceptual Development and a Proposed Framework. Working paper.
Barney, J.B. (1991). Firm Resources and Sustained Competitive Advantage, *Journal of Management* 17: 99–120.
Closs, D.J. and McGarrell, E.F. (2004). Enhancing Security Throughout the Supply Chain, Special Report to the *IBM Center for the Business of Government*, Washington, DC.
Eisenhardt, K.M. and Martin, J.A. (2000). Dynamic Capabilities: What Are They?, *Strategic Management Journal* 21: 1105–1121.
Elkin, D., Handfield, R.B., Blackhurst, J. and Christopher, W.C. (2005). Ways to Guard Against Disruption, *Supply Chain Management Review* 9: 46–53.
Feitzinger, E. and Lee, H.L. (1997). Mass Customization at Hewlett-Packard: The Power of Postponement, *Harvard Business Review* 75: 116–121.
Helfat, C.E. and Peteraf, M.A. (2003). The Dynamic Resource-Based View, *Strategic Management Journal* 24: 997–1010.
Helferich, O.K. and Cook, R.L. (2002). *Securing the Supply Chain*, Council of Logistics Management: Oak Brook, IL.
Lee, H.L. (2002). Intelligent Demand-Based Management, International Commerce Review: *ECR Journal* 2: 61–73.
Lee, H.L. and Wolfe, M. (2003). Supply Chain Security Without Tears, *Supply Chain Management Review* 7: 12–20.
Martha, J. and Subbakrishna S. (2002). Targeting a Just-in-Case Supply Chain for the Inevitable Next Disaster, *Supply Chain Management Review* 6: 56–66.
Penrose, E.T. (1959). *The Theory of the Growth of the Firm*, Oxford University Press: UK.
Porter, M.A. (1979). *How Competitive Forces Shape Strategy*, Harvard Business School Press: Cambridge, MA.
Quinn, F.J. (2003). Security Matters, *Supply Chain Management Review* 7: 38–45.

Sanders, N.R. (2006). Getting the Best Forecast by Combining Judgmental and Statistical Methods. *APICS Magazine* 3/6: 29–32.

Sheffi, Y. (2005). *The Resilient Enterprise*, MIT Press: Cambridge, MA.

Sheffi, Y. (2001). Supply Chain Management Under the Threat of International Terrorism, *International Journal of Logistics Management* 12: 1–11.

Teece, D.J., Pisano, G. and Schuen, A. (1997). Dynamic Capabilities and Strategic Management, *Strategic Management Journal* 18: 509–533.

US Dept. of Transportation (2002). Technology to Enhance Freight Transportation Security and Productivity. U.S. DOT FHWA (http://www.ops.fhwa.dot.gov/freight/ transportation_security.htm).

Zsidisin, G.A. (2003). Managerial Perceptions of Supply Risk, *Journal of Supply Chain Management* 39: 14–25.

Zsidisin, G.A., Ragatz, G.L. and Melnyk, S.A. (2005). The Dark Side of Supply Chain Management, *Supply Chain Management Review* 9: 46–52.

Zsidisin, G.A., Melnyk, S.A. and Ragatz, G.L. (2005) An Institutional Theory of Business Continuity Planning for Purchasing and Supply Chain Management, *International Journal of Production Research* 43: 3401–3420.

Chapter 20: Securing Global Food Distribution Networks

Nicole Mau* and Markus Mau

*Tec de Monterrey, Mexico

20.1 Challenges in Providing Food Safety and Quality

The product quality of specific groceries can be viewed from at least three dimensions: *product oriented quality* (physical characteristics of the product), *process oriented quality* (process oriented characteristics of the product) and *utilization oriented quality* (subjective quality aspects of purchasers/consumers) (Grunert et al. 2002). Against the background of this multidimensionality, food safety in the European Community is defined in EU-Community law by the criteria "harmful to health" and "suitability for human consumption" (EC Art. 14).

Risk occurs from the possibility of contaminated products due to insecure processes along the supply chain. However, the existence of unsecure processes does not necessarily result in the contamination of products. Likelihood and result of the impact on the product need to be taken into consideration when discussing risk management requirements for specific supply and production chains.

The risk of contaminated groceries is related to the likelihood that food has a negative effect on the consumer when consuming the specific product. On the other hand the products are considered to be safe when they have an extremely low risk of damage – this does not necessarily mean that this risk must be equal to zero.

Due to the existing information asymmetry in contestable markets, risk reduction needs to be implemented by either governmental/multinational institutions (to set minimum requirements on food safety) or BtoB-trust based on an adequate supply chain security or both.

Problems may increase whenever products are traded internationally. Depending on different production standards in terms of allowed ingredients and their remaining quantity within the product used during production processes, no consistent picture of a quality map can be drawn.

Table 20.1 shows a comparison of United States versus mid-European (German and Austrian) limits of allowance-levels for the residue of selected pesticide ranges (see also Henson and Northern 1997) for *table grapes*. Differences exist even between neighbouring countries with matching cultural backgrounds and consumer behaviour. In the last column additional information is given from a variety of countries. Whereas hexagons show maximum allowances the circles in the column "others" highlight minimum allowances from the respective country. The differentiation in limiting values does not follow a logical explanation. Even this simple example shows a variation of a factor of 40 and even more. There are no common standards on quality, product safety and ethical practices that supports supply chain security globally. As long as the realization of a global standard is in the remote future, BtoB-solutions are required. Beside this quality definition another area of responsibility is safety: During the last several years there were a number of food scandals such as BSE and Avain Influenza (BMELV 2007). This is not only a problem for producers and retailers but also for consumers. The ensuing loss of confidence for the consumer on the one hand and the awareness of improving life through healthy eating on the other hand is increasing significantly. An additional point for all participants of the supply chain is the security that comes from a process view over the complete chain of activities. Therefore, to implement efficient risk management, it is necessary to identify risks.

Table 20.1 Measurable quality by comparison for *table grapes*

(mg/kg) Q-factors	USA	Germany	Austria	Others	
Benomyl	10	3.0	2.0	5.0	Canada
Cymoxanil	0.1	0.2	0.1	0.05	Netherlands
Imidacloprid	1.0	0.05	0.05	1.5	Canada
Glufosinate	0.05	0.1	0.1		
Carbofuran	0.4	0.1	0.1	0.01	Belgium
Captan	50.0	3.0	3.0	3.0	France

⬡ Maximum Allowances ⬯ Minimum Allowances

20.2 Problems in the Food Supply Chain

In order to identify the risks of food production, at least some of the multifarious problems in the food chain must be explained. These problems are financial incentives, moral hazard, and information asymmetry. First of all there are the different complex claims of each partner of the supply chain concerning the financial aspects. The producers as well as the retailers want to maximize their profit. This point shows the first potential security gap. In order to be able to produce at low costs, producers might reduce their own quality controls to provide their products at a lower price than their competitors. The second possible gap is the willingness to pay from the consumers' point of view. As governmental institutions guarantee an average quality level, the majority of consumers reduce buying decisions down to the price. In addition, there is the problem of high quality and good "ingredients". This shows another possible gap, the requirements and expectations of the consumers versus the ability of the producers to minimize their product costs.

The second problem area is moral hazard. In the EU there are many examples that show many food producers take risks – for example, risking the probability of being caught with a lax control system versus maximizing profit, as happened in the rotten meat scandal (Kinzinger 2006). Also there were several retailers who changed the labels on their products in order to "extend the shelf-life" of some products. As long as control frequencies and intensities are predictable and cause costs that are not passed on to the buyer, many food scandals or investigations in other industries show that without a valid monitoring and control system supply chain security cannot be taken for granted.

The existing information asymmetry is the third reason for risks in the food supply chain. With information asymmetry it is very difficult to develop a qualified risk management process. In addition, the diversity between the market power of the producers and the retailers plays another important role. Information that is readily available and accessible is required for downstream supply chain participants if they are to manage risks effectively.

Figure 20.1 shows the above-mentioned three problems in the food supply chain in a producer, retailer and consumer triangle. Additionally, market power is included as it has an increasing influence on the supply chain security installed in globalized networks. This is related to the international problem that governmental, public and private organizations still operate without having any agreed or consistent definitions of quality and appropriate security standards.

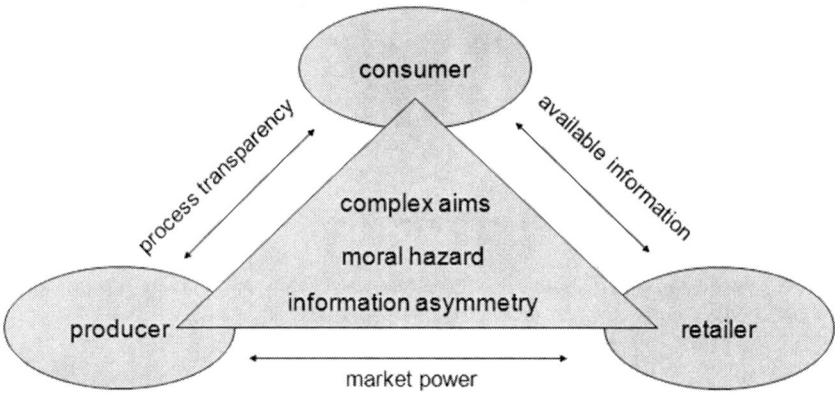

Fig. 20.1 Problems in the food supply chain

20.3 Requirement Shift Through Globalized Procurement and BtoB-Solutions

The inability of the average single company to secure Food Safety along the whole supply chain indicates the necessity for an interface-oriented exchange of risk management information. In the long run a data exchange that allows ongoing usage of information gathered in firms' supply chains could facilitate process transparency and risk management. This information exchange can be compared to using electronic data interchange (EDI) information for efficient replenishment. As a first step, process quality is required throughout all partners involved in the supply chain. If non-conforming behaviour is not traced due to missing information and documentation, it could be disastrous to the whole supply chain. For instance Fig. 20.2 shows the effect of additive usage of different pesticides without exceeding regulatory tolerance levels in any single residue. The contamination level of the final product is far beyond any acceptable point – the result is a so-called "pesticides cocktail". The example shown in Fig. 20.2 shows a sample with 17 different kinds of pesticides.

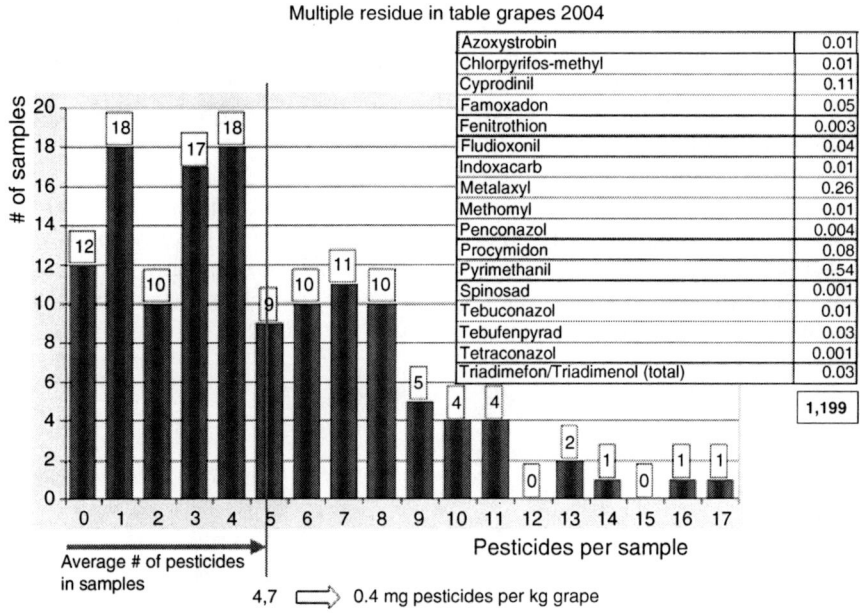

Fig. 20.2 Pesticide cocktails as a result of process ambiguity and missing documentation (CVUA 2004)

There might be a logical reason related to good agricultural practice (GAP), but several investigations concluded that some market players seem to take advantage of missing institutional (food) standards (CVUA 2006).

20.4 Risk Management

The risk problem is compounded as a result of multiple influences within the supply chain. Compared to former times the final goods pass through many more levels in the supply chain. And the more levels there are, the higher is the demand for efficient risk management. Figure 20.3 shows the process chain from the producer to the customer and the classical positions of the business partners as a risk driver, potentially creating greater vulnerabilities in the supply chains.

Although Fig. 20.3 is based on an e-business contract, it is adoptable for classical business relations. The missing in-house control of process performance in global procurement solutions can lead to a variety of issues that have negative effects on the enterprise. They are not restricted to problems directly related to products such as the quality of ingredients, long-term performance of the product and fade-resistance of the colours. Latest extensions in the supply chain perspective are labour force treatment, working conditions, child work and education as well as other ethical aspects in third world countries that can affect the whole supply chain design. But even by reducing the perspective to the classical process view (costs, performance, quality of process), an average reliability of processes of less than 1.00 will reduce the average received performance to a critical level (law of large numbers). A simple example with three activities along the chain that have a quality level of 0.995 deliver a result of an overall quality of 0.985 – much too low for many products.

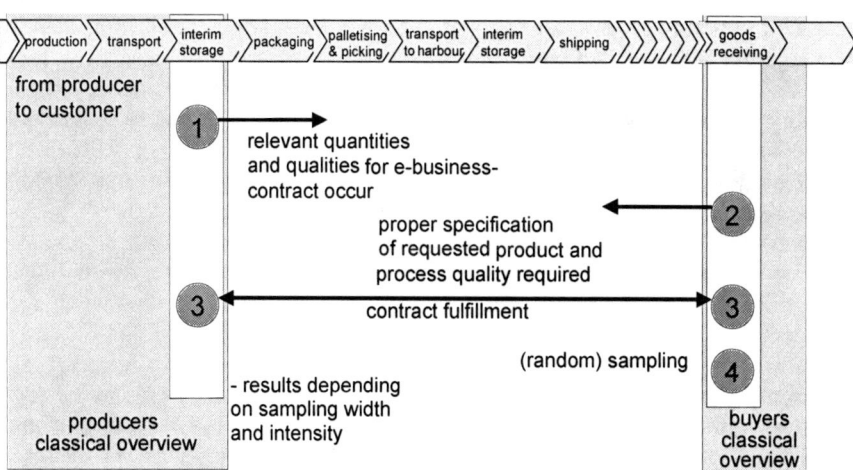

Fig. 20.3 Process chain from the producer to the customer

One problem with risk minimizing activities that don't start at the place of original production is that technical control instead of quality-oriented control allows hidden action of participants in the process chain. For example, if the temperature of deep frozen products rises above an acceptable level during transit but has the right temperature again at an arrival point, it might not be possible to trace and identify the point of temperature fluctuation.

There are legal requirements that exist which reduce the likelihood that moral hazard occurs. Importers are required to fulfill legal aspects of food safety and quality. The first steps toward a solution already exist as either companies' own inspection and monitoring systems, or by application of an industry-wide solution defined by global players. Whereas EurepGap (European Retailer Produce Group Good Agricultural Practices) and IFS (International Food Standards) represent industry wide solutions, a good example for a company solution is the British Retailer Tesco, known worldwide for its benchmark in supply chain security, and for setting standards above the prevailing level of their own industry (McEachern and Warnaby 2004).

20.5 Vertical Integration Versus Risk Management

As stated before, the internal solution by a vertically integrated supply chain may achieve the highest level of risk avoidance if all processes are run properly. It has the second advantage that IT-interfaces can be set up to the required interchange of traceable information. However, for most market players there is no possibility of bringing more value creating activities into their own enterprise. As long as transaction costs by external risk management solutions are competitive against internal solutions, there is no need for vertical integration. Thus, a system is required to handle all information that balances risk in a collaborative food network. Risk management is always balancing the cost of risk avoidance and the probability of an event resulting in direct damage, the costs related to it, and loss in goodwill (Schiller 2005; see also Caswell and Mojduszka 1996). Therefore a comprehensive overview on risk forcing/defending practices of (potential) trading partners available to the decision makers within each enterprise is required.

Figure 20.4 shows details of information and flow of goods in a real supplier-manufacturer-buyer relation for a fast moving consumer product. As the complexity of information requirements illustrate, there is need for combining risk management information that is detailed enough to avoid data loss on one side, and information overload that nobody can handle on the other side. Securing and documenting processes within their own production process (a mixture of processing of raw materials and probable contamination risks through packaging and machinery, for example) is as important as information on the suppliers' processes.

Therefore, relationship management and vertical cooperation becomes more and more important. As firms need to make sure that the quality of their products is continuously of high quality, there is a direct relation towards controlling the whole value chain for the products. However, this does not necessarily lead to the necessity for vertical integration. More important is an appropriate risk management scheme that combines trust through long-term relations with trading partners, as well as definitive process and product quality levels and control systems throughout the supply chain (Antle 1999).

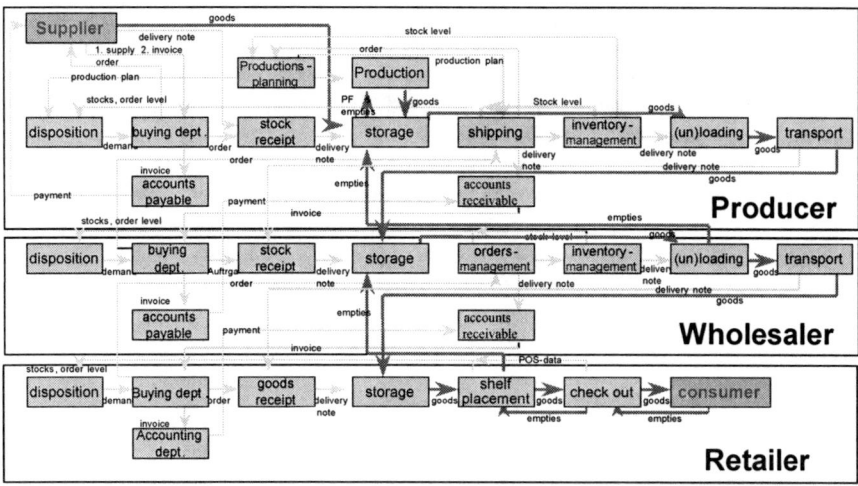

Fig. 20.4 Interfaces along the supply chain

In market solutions the average trade process has to take into consideration the multiple influences of risk within the specific supply chain. As mentioned before, the buyers cannot handle an overload of data. The decision maker requires an easy information system, where additional information may be available upon request in different layers.

The trust within a business network is mostly based on logistics performance and general quality aspects that end consumers identify with a particular food. However, consumer behaviour is a combination of factors that is not within the scope of this chapter. Essential for the BtoB risk and trust process are supply chain management solutions through logistics strategies that show how supply readiness, product conditions, as well as the ability to provide information and the level of process documentation are achieved.

20.6 Efforts to Solve Supply Chain Security Problems

In order to reduce the information asymmetry problem (Stiglitz 1987) as well as the issue of setting different standards for individual business relations, a group of 20 leading European grocery retailers established EurepGap. This organization communicates production, environmental, social and hygene standards for fruits and vegetables. EurepGap standards for fruit and vegetables is a normative document for certification, having been developed from a European group of representatives from all stages in the fruit and vegetable sector with the support from producer organizations outside the EU (EurepGap 2006). It is accredited by ISO 65 (EN 45011) and has worldwide applicability. Likewise, the International Food Standards (IFS) evolved from the Global Food Safety Initiative (GFSI) primarily to audit private label producers. These normative documents are the EurepGap General regulations for fruit and vegetables, the control points and compliance criteria, and the checklist. Since March 2004 some retailers already require an IFS-certification from their suppliers. Both standardization programs allow for a better control of risk relevant activities within the supply chain and support network efficiency through reduced fixed costs in maintaining such a system, as compared to one-to-one-relations.

As a result of long distance purchasing activities, the direct control of production, logistics processes and processing are out of the direct control of most of the persons in charge in downstream processes. As a result, vertical coordination through interorganizational agencies as EurepGAP and IFS has emerged and might displace the multinational developments of the Food and Agriculture Organization (FAO) of the United Nations. The contradiction of these democratic procedures versus the industrial solution may affect the sustainability of risk management.

20.7 A Potential Solution

The goals for single supply chains and the overall industry would be to secure worldwide and complete traceability of all involved products and inputs. This can only be achieved and safeguarded by a worldwide network of supply chain security providers. Monitoring food security with a central database may be necessary because participants are spread worldwide using different languages and business structures according to cultural background, size of enterprises, and information readiness. In order to offer the required service worldwide, the provider should work together with certificated and accredited inspection institutes and laboratories that supervise the flow of goods in place, and avoid the problems in actual supply chains as previously described.

Once the database-management-system is operative, the data access is secured by a permission and access routing scheme that allows selective data access according to job. Once access control and authorization is passed, all relevant data (upstream/downstream) are available.

Product flow data is enriched by information from inspection institutes, laboratories and additional evaluations and analyses from supply chain partners to ensure validity and consistency of the stored information as well as delivering secure additional data that usually is not accessible to the single partner of the supply chain. Supply chain partners add information for their input factors by combining data through the data management system (Fig. 20.5). As a result the benefits from participating in the network greatly exceed those that an in-house solution is able to deliver.

The control-system involves inspection levels at all the supply chain levels from feedstock supply to the food retail industry. The system allows batch tracing across all levels. Allowing self-monitoring as well as tracking and tracing across the supply chain to deliver goods according to EU legal standards can significantly reduce information asymmetries. Standardized online input at all levels allows continuous transaction data for inflows and outflows (Fig. 20.6).

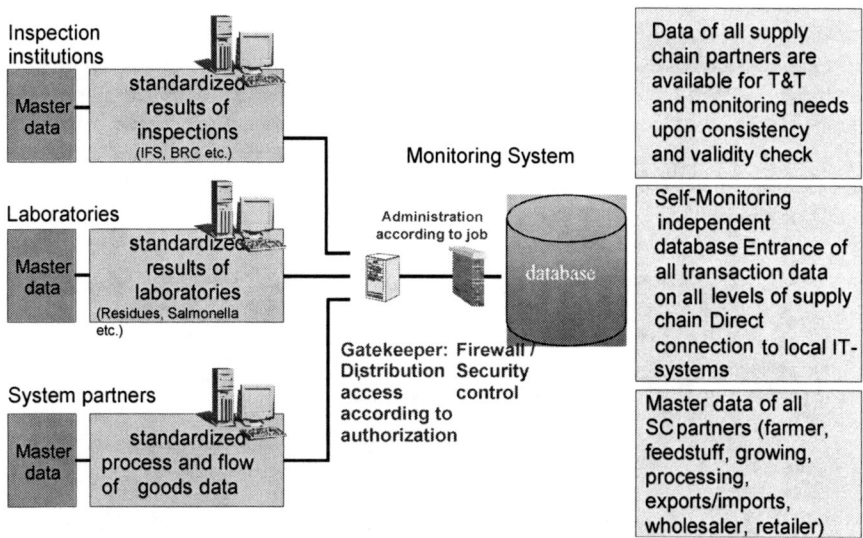

Fig. 20.5 Information management and supply chain solution for global procurement

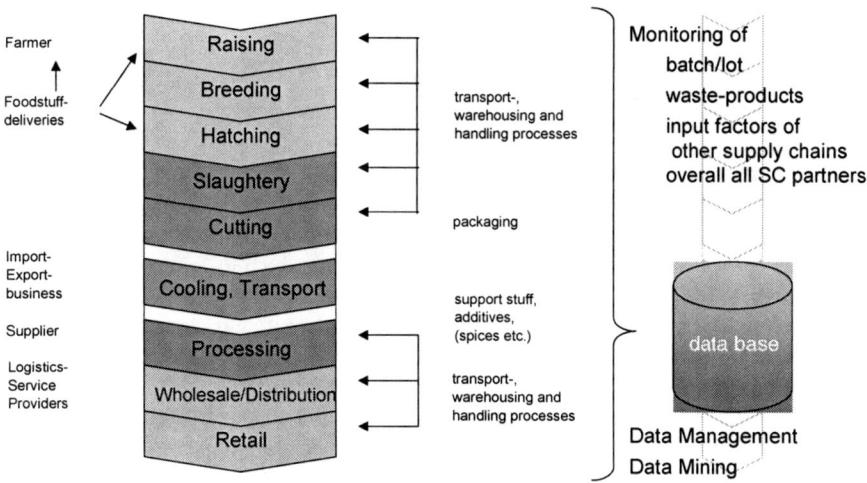

Fig. 20.6 Supply chain security by enriched information flow, shown for poultry production

Through an independent database a quality level can be reached that extends in-house solutions and offers better communication to customers, media and stakeholders. The effects on the products by third party inputs (e.g., packaging, storage, transport) that cannot be overseen by a single player in the chain are visible by having a centralized data management system. Monitoring covers batches/lots, waste products, and input factors from other supply partners throughout the process chain. All relevant information is stored in the database and in case of need is able to be retrieved by Data Mining according to specific needs of a request.

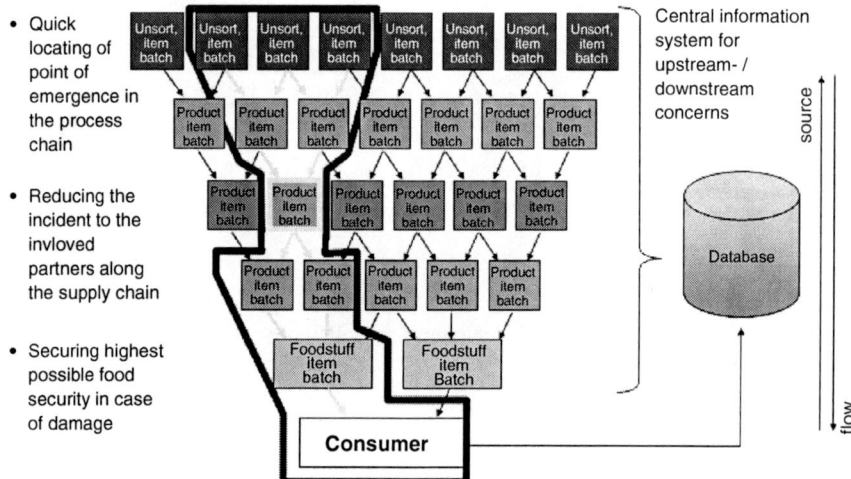

Fig. 20.7 Information management and supply chain solution

The database is the central information system that shows all worldwide data – safeguarding information readiness by decentralized data management. The concept allows fulfilling all needs of a global tracking and tracing strategy by securing process quality and documentation by diverse checking routines (participating laboratories, inspections institutions). The system gathers the right information at the right process point of the supply chain and controls the overall network at the global headquarters. It is there that all data are gathered and further connections can be made – this is essential for standardized, fast, qualified and valid information. As a result supply chain excellence occurs through quickly locating a point of contamination, thereby reducing the frequency of incidents along the supply chain and minimizing potential damage (Fig. 20.7). In day-to-day business access is limited to one level up and down.

Supply chain security can be extended to the monitoring of ethical practices of supply chain members. The leading European product testing organization *Stiftung Warentest* started its first ethical product test on ethically correct salmon production in 2005 (Stiftung Warentest 2005). Also, independently certified and accredited controls and reports about these performances can be added to the database that allows the same tracking and tracing, delivering valid transparency and security. A solution like this would allow better communication to consumers, media and stakeholders. Also, in most cases, it would reduce the cost of operating such a system compared to single enterprise solutions.

20.8 Setup of Supply Chain Security

Although supply chain security is continuously in focus after a (food) scandal, the request for intensified control levels (frequencies and range) is not necessarily the most efficient solution. The positive correlation between supply chain security and risk management for the participants is obvious, as long as the risk management level is low. However, when firms attain higher performance levels, further increases in controls (either process or product) will lead to significant cost increases with negligible benefits. This context is shown in Fig. 20.8. Furthermore the achievement of a 100% security is – if at all – associated with excessive costs in infrastructure and staff.

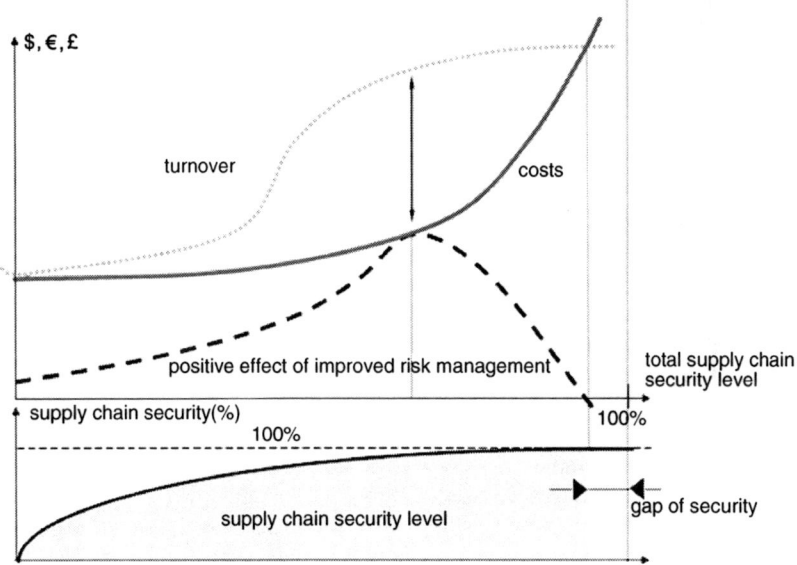

Fig. 20.8 Total risk management level

Therefore, a responsible and pragmatic solution needs to be established. Depending on where controls take place and where supply chain security instruments are implemented, there is a risk that smaller enterprises may be pushed out of the market. If they cannot cover the cost of control with their smaller financial resources, there is a potential risk for the industry that they may go out of business.

In order to establish an adequate Food Safety and traceability system, restrictions and typical issues of companies need to be taken into consideration. The basic idea behind this is to allow companies to evaluate the required level of food safety and supply chain control according to upstream, in-house, and downstream activities for specific products.

The appropriate level of detail in terms of available information and reliability of documented information would lead to product and process technical activities in the grocery supply chain. Examples of information that can be used include: lot numbers (as in 89/396/EWG), including best before date, lot and date of production); product labelling (labelling requirements as in 2000/13/EWG); number of pallets (barcode, transport labels with EAN 128 information); quality-control information; production planning information; audit reports; delivery notes (information accompanying the flow of goods); and accounting information (information following the goods flow).

According to the level of information required, details can be gathered by using all potential sources to close the remaining gap in information and transparency, assuming that all gathered information is reliable and trustworthy. Additional risk aspects through potential information manipulation need further study and action, such as those in the IFS and EurepGap procedures.

The higher the number of suppliers and customers the more technical efficiency in terms of electronic data interchange that is required (BLL-Online 2001). The level to be reached will depend on (re-)traceability efforts and the costs versus benefits from risk management solutions. Enterprises participating in the same value chain should negotiate which data source should be used for what kind of information to best meet requirements for delivery traceability, information readiness and acceptable security of the food supply chain.

The first step for a single enterprise is an evaluation of its own potential risk upstream, internally and downstream (Hammer 1988). With target orientated questions on the quality control and risk management system itself, as well as the relevant aspects regarding input and output factors, consensus can be reached to support risk management decisions. In order to make the questions operable for further evaluation of major areas such as calls for action for short-term activities, a percentage level of possible risk management performance is recommended. This can be delivered by converting qualitative answers into quantitative counts – weighing the results with risk relevant factors. As a result a differentiated position can be developed. By dividing this approach into process and product aspects for all three areas, a profile of the collaborative food network (from the single company point of view) is obtained.

Using the results of this evaluation in a condensed approach like this would deliver the accumulated information a buyer would need at the supplier's level in order to take risk management aspects into account. What this solution could look like is shown in Fig. 20.9.

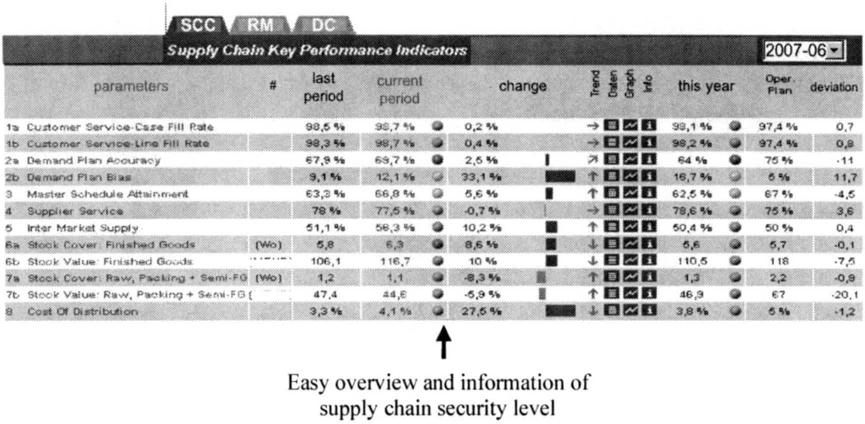

parameters	#	last period	current period	change	Trend / Daten / Graph / Info	this year	Oper. Plan	deviation
1a Customer Service-Case Fill Rate		98,5 %	98,7 %	0,2 %	→	98,1 %	97,4 %	0,7
1b Customer Service-Line Fill Rate		98,3 %	98,7 %	0,4 %	→	98,2 %	97,4 %	0,8
2a Demand Plan Accuracy		67,8 %	65,7 %	2,5 %	↗	64 %	75 %	-11
2b Demand Plan Bias		9,1 %	12,1 %	33,1 %	↑	16,7 %	5 %	11,7
3 Master Schedule Attainment		63,3 %	66,8 %	5,6 %	↑	62,5 %	67 %	-4,5
4 Supplier Service		78 %	77,5 %	-0,7 %	→	78,6 %	75 %	3,6
5 Inter Market Supply		51,1 %	56,3 %	10,2 %	↑	50,4 %	50 %	0,4
6a Stock Cover: Finished Goods (Wo)		5,8	6,3	8,6 %	↓	5,6	5,7	-0,1
6b Stock Value: Finished Goods		106,1	116,7	10 %	↓	110,5	118	-7,5
7a Stock Cover: Raw, Packing + Semi-FG (Wo)		1,2	1,1	-8,3 %	↑	1,3	2,2	-0,9
7b Stock Value: Raw, Packing + Semi-FG (		47,4	44,6	-5,9 %	↑	46,9	67	-20,1
8 Cost Of Distribution		3,3 %	4,1 %	27,5 %	↓	3,8 %	5 %	-1,2

↑

Easy overview and information of
supply chain security level

Fig. 20.9 Total risk management system – modified chart from food manufacturer

All additional information must be stored in a data warehouse or file system. Risk management tracking must become as normal as supply chain tracking is for leading companies nowadays. A standardized globally accepted approach such as the EurepGap and IFS approach has advantages of providing digitally available information on supply chain security for collaborative food networks. The positive effects on risk reduction, such as process transparency and continuous information readiness along the supply chain, may outweigh the disadvantage of forcing such a system onto suppliers by the market power of the buyers.

A business solution needs to be found to assure supply chain security on the one hand but that doesn't result in a process of concentration and reduced competition on the other hand. The heterogeneity of the industries and consideration that solutions that are too large and cost intensive are not affordable for many small and medium sized enterprises (SME) must also be taken into account.

A good example for an innovative business solution is the Future Store concept of the German retailer Metro Group allowing consumers to trace back the origin and call for additional information about the production process right at the point of sale. The solution was provided by cooperation with T-Systems. A user-friendly platform was built with access related to access authorization. This is possible by not only a complete logistical overview but also by including additional information in the data system, as shown in the possible business solution above. From the consumer's perspective another example is the complete monitoring system for eggs in Germany, where each egg has its own product code (KAT 2007). By typing this code into the related website, a picture of the farm of origin and additional production information is made available.

In order to achieve the potential benefits of an adequate supply chain security strategy a proper specification of required quality (product, process) is essential. Therefore all desired quality factors need to be quantified (e.g., maturity parameters, health parameters, social standards, operating resources, level of detail). The balance of wants and potential of operating such a system is represented in the reliability of the whole documentation and control system. The better the linkage of payment for process quality and supply chain security, the easier the implementation and acceptance.

References

Antle, John M. (2005) The new economics of agriculture, in: American Journal of Agricultural Economics, http://www2.montana.edu/jantle/trc/pdf/research/ paper/rdp33.pdf, (accessed: January 1, 2005).
BLL-Online (2001) BLL-Leitfaden Rückverfolgbarkeit, Information und Überwachung, 1. Auflage, Bonn.
BMELV (2007) German Federal Ministry of Food, Agriculture and Consumer Protection, reports on avian influenza (AI) and bovine spongiforme enzephalophathie (BSE).
Caswell, J. and Mojduska, E. (1996) Using information labelling to influence the market for quality in food products, in: American Journal of Agricultural Economics, 78, pp. 1248–1253.
CVUA (2004) Chemisches und Veterinäruntersuchungsamt Stuttgart, Ergebnisse der Rückstandsuntersuchungen von Pflanzenschutzmitteln in Trauben, Stuttgart.
CVUA (2006) Chemisches und Veterinäruntersuchungsamt Stuttgart, Annual report.
EC (2002) European Community, Regulation (EC) No 178/2002 of the European Parliament and of the Council of 28 January 2002.
Eurepgap (2005) Euro-Retailer Produce Working Group Good Agricultural Practices, The Global Partnership for Safe and Sustainable Agriculture, EUREPGAP Global Report.
Grunert, K., Fjord, T. and Brunss, K. (2002) Consumer's Food Choice and Quality Perception. MAPP working paper No. 77, Aarhus School of Business, Denmark.
Hammer, G. F. (1988) Der Aufbau eines qualifizierten Prüferpanels für sensorische Prüfungen, in: Sensorik und Lebensmittelqualität, Arbeiten der DLG, Band 192, Frankfurt.
Henson, S. and Northern, J. (1997) Public and private regulation of food safety: The case of the UK Fresh Meat Sector, in: Schiefer, G., Helbig, R., Quality Management and Process Improvement for Competitive Advantage in Agriculture and Food, Vol. 1, Proceedings of the 49th Seminar of the European Association of Agricultural Economists (EAAE), Bonn.
IFS (2007) International Food Standard, Standard for auditing retailer (and wholesaler) branded food products, Paris, Berlin, 2006.
KAT (2007) Verein fuer kontrollierte alternative Tierhaltungsformen e. V., Was steht auf dem Ei?.
Kinzinger, A. (2006) Gammelfleischskandal – Im Zweifel gegen den Wirt, Spiegel vom 04. Sept. 2006.
McEachern, M. and Warnaby, G. (2004) Retail quality assurance labels as a strategic marketing communication mechanism for fresh meat, The International Review of Retail, Distribution and Consumer Research, 14(14), pp. 255–271.
Schiller, W. et al. (2005) Risikomanagement für Marken, Weinheim.
Stiftung Warentest (2005) Vertraeglich für Mensch und Tier, Unternehmensverantwortung, test 01/2005, pp. 22–24.
Stiglitz, J. E. (1987) The causes and consequences of the dependence of quality on price, in: Journal of Economics Literature, 25(1), pp. 1–48.

INDEX

A
Actor specific risk analysis, 47
Adaptability, 156, 312
Agile systems, 2, 68, 137, 148, 150
Agility, 3, 68, 150, 151, 156
Alignment, 156
Analytic hierarchy process, 76
Asymmetric information, 236–239, 241–243

B
Balanced scorecard, 9, 254, 258, 263
Bayesian analysis, 229–230
Bayesian decision-making, 229
Bayesian framework, 229
Behavioural risks, 235–245
Bullwhip effect, 275, 276
Business continuity, management
    planning, 10, 308
    processes, 309, 314, 315
Business ecosystem, 37–39, 51
Business risk
    categories, 275
    definition, 54

C
Capability, 20, 35, 50, 51, 60, 117, 128, 129, 137, 151, 161, 168, 171, 208, 223, 301, 312–327
Capital asset pricing model (CAPM), 257
Catastrophic events, 228, 271, 284, 286, 308
Catastrophic risk, 275, 277–278, 284, 286
Causal pathway, 252, 253, 268

Classical statistical analysis, 114–116, 119
Clusters, 43, 60, 61, 183, 184
Comprehensive risk analysis and management network, 85
Computer simulation, 7, 110, 114
Conditional probability, 229
Conflict, 95, 139, 194, 205, 235, 284
Consequences, 2–8, 15, 17, 19–25, 28–32, 54, 84, 88, 90–93, 125, 130, 131, 134
Consequence analysis, 15, 21, 25, 28, 90
Contingency planning, 56, 58, 59, 95, 151, 172, 209, 285, 318, 320, 321
Continuity planning, 10, 308
Corporate performance and risk, 256
Crisis, 56, 57, 97, 156, 161, 274, 278, 299
Criticality ranking, 24, 25, 29
Customs-trade partnership against terrorism (C-TPAT), 277, 280, 296, 327

D
Data mining, 6, 59–62, 64, 339, 340
Decision maker, 8, 9, 11, 12, 16, 26, 68, 76, 77, 81, 86–90, 92, 94, 95, 97, 98, 220–221, 258, 259, 268, 336, 337
Decision support, 231
Dependency matrix, 131
Design led risk management, 7, 137
Discrete event simulation, 103, 104, 111, 114, 119
Downside risk, 73, 263

Downstream, 2–4, 9, 10, 68–70, 73, 76, 80, 105, 106, 110, 162, 208, 212, 264, 265, 267, 275, 276, 333, 338, 340, 342
Dynamic capabilities approach, 10, 307–328

E
E-business, 335
Effectiveness, 16, 17, 19–20, 38, 39, 42, 67, 94, 96, 98, 108, 113, 181, 203, 220, 225, 254, 260, 261, 266, 268, 298–300, 312, 313
Efficiency, 16, 17, 19–22, 74, 94, 96, 98, 112, 146, 189, 203, 209, 254, 260, 261, 265, 299, 317, 321, 324, 325, 338, 342
Electronic data interchange (EDI), 300, 334, 342
Enterprise risk management (ERM), 1, 8, 133–134, 177–184, 279
Event tree analysis (ETA), 59
Exposure, 7, 54, 60, 72, 74, 77, 90–92, 140, 164, 195, 219, 249, 250, 252, 256–258, 260–262, 264, 266, 277, 278, 280, 284

F
Failure mode effect analysis (FMEA), 6, 59, 61–62, 64, 208
Fault tree analysis (FTA), 62
Flexibility, 7, 8, 16, 72, 74, 77, 112, 113, 127, 131, 132, 139, 144, 147, 149, 150, 155–172
Food supply chain security, 293–304

G
Global food distribution networks, 331–344
Global sourcing, 105, 138–141, 143–145, 148, 149, 201, 278, 317
Global supply chain risk management, 7, 137–151

H
Hazard and operability analysis (HAZOP), 27, 28
Hazard identification, 90
Health and safety, 1, 49, 105
Heuristic, 9, 12, 220, 222–226, 284, 285

I
Influence map, 48, 49
Information and communication technologies (ICT), 6, 26, 36, 37, 43–46, 48, 151, 244
Information resource, 300
Infrastructure risk, 275, 277, 286
International supply chain risk management (ISCRIM), 11, 12

J
JIT, 130

K
Keystone strategy, 39
Knowledge transfer, 40

L
Lean, manufacturing
  supply chain management, 57, 118, 205, 317
  systems, 112, 318
Likelihood of occurrence, 4, 156, 252, 253
Linear supply chain, 3
Logistics, analysis, definition, 21

M
Marine accidents, 89
Metrics, 8, 9, 12, 54, 106, 112, 250, 254, 267, 268, 297, 298, 300
Mitigation-supply chain risks, 9, 200, 202, 210
Multiple sourcing, 3, 7, 104, 125–134, 280

Index

N
Natural hazards, 271, 277, 278, 280, 281, 284
Networks, 3, 6, 10–12, 16, 17, 32, 35–52
Nodes, 3, 6, 16–18, 26, 40, 109, 111, 250, 255, 278, 295, 316
Normal accident theory, 22, 23, 278

O
Off-shoring, 7, 128, 138, 139, 149, 150, 161, 272
Opportunistic behaviour, 235–239, 242–244, 276
Outsourcing, 37, 138–141, 148, 150, 151, 161, 203, 272, 276, 295

P
Packaged dangerous goods, 84, 88, 89
Partnering, 187, 188, 190, 191, 201–204, 209, 213, 262, 311
Performance driver, 260, 261, 266, 268
Performance measurement
 balanced scorecard, 9, 254, 263
 financial, 112, 256
 operational, 113, 263, 266
 strategic, 112–113
Performance metrics, 8, 9, 112, 164–166, 169, 256, 268
Performance profile, 252, 256–260
Portfolio of investments, 9, 252, 260, 263
Predictive–proactive methodology
 predictive mode, 62, 63
 proactive mode, 63–64
Prime contracting, 187–192, 194–197
Prime contracting areas of risk, 190, 197
Principal agent theory, 236, 237
Private financing initiatives (PFI), 8, 187–197
Product development management (PDM), 69
Project and process management, 71–72
Public private partnerships (PPP), 8, 188, 196

R
Radio frequency identification (RFID), 57, 298, 300, 313, 322–325
Relationship mapping, 26
Relationship marketing, 261
Residual risks, 95
Resilience engineering, 22, 23
Resource based view (RBV), 312
Return on assets (ROA), 257
Return on investment (ROI), 9, 155, 156, 166, 196, 257, 300
Risk
 categorisation
  behavioural, 235
  bureaucratic, 275, 276, 284
  catastrophic, 275, 277–278, 284, 286
  consequences, 3, 4, 6, 7, 15, 21, 90, 92, 93, 131, 191, 205, 208, 213, 220, 231, 250, 255, 257, 262
  contractual, 8, 128, 250
  demand side, 275–276, 283, 284, 286
  downside, 54, 73, 263, 273, 274
  infrastructure, 48, 49, 275, 277, 284, 286
  legal, 90, 93, 129, 275–277, 284
  regulatory, 49, 50, 93, 275, 276, 284
  supply side, 145, 275, 276, 283, 284, 286
  systematic, 128, 133, 257, 259, 260, 262, 279
  unsystematic, 257, 259–262, 265
  upside, 73, 263, 273
 definition, 3, 4, 70, 73, 84, 85, 88, 207, 252

driver, 250, 253, 259, 261, 263, 265, 335
perception, 9, 92, 97, 104, 251, 255, 310
portfolio, 209, 210, 213, 214
preparedness, 252
profile, 47–49, 51, 52, 64, 78, 143, 202, 205, 206, 213, 214, 257, 258, 260–263, 268, 287
sources, 6, 55, 64, 92, 132, 208, 210, 212, 253, 257, 260–262, 265, 266, 268, 274, 275, 283, 286
triggers, 4, 201, 208, 278
Risk aversion index, 92
Risk communication for chemical risk management, 97
Risk management
assessment, 6, 23, 26, 56–58, 69–71, 73–79, 81, 83–86, 94–98, 104, 133, 147, 148, 155, 172, 180, 194, 196, 197, 208, 209, 213, 220, 222, 223, 279
evaluation, 6, 76, 86, 87, 91–94, 208
identification, 4, 9, 47, 50, 57, 58, 62, 74, 172, 180, 191, 207, 209, 264, 268, 279,
measurement, 70, 268
modelling, 40
monitoring, 4, 279
Risk performance relationship, 251, 252
Risk tools
analysis, 5–7, 32, 64, 86, 89, 209
assessment, 5–7, 98, 104
communication, 85, 87, 96–97, 299
evaluation, 59, 86, 91–94, 209
exposure, 60, 72, 77, 195
identification, 86, 202

S

Safety theory, 21–23
Sanctions, 238, 239, 241–243
Security, 2, 5, 8, 10–12, 18, 21, 44, 46, 55, 63, 69, 83, 84, 89, 104,
Security competencies, 295–297, 300, 301
Sensitivity analysis, 91, 93, 192
Single sourcing, 7, 125–127, 129–134, 158, 278, 280, 317
Simulation
discrete event, 7, 103, 104, 111, 114, 119
framework, 110
Small and medium sized enterprises (SMEs), 8, 199–202, 204–210, 212, 213, 216
Sourcing
multiple, 3, 7, 104, 125–134, 280
single, 7, 125–127, 129–134, 158, 278, 280, 317
Stakeholder interests, 72
Stakeholder mapping, 26
Stakeholder theory, 262–263
Strategic risk management, 72
Supply chain
complexity, 278
definition, 2, 3, 252, 274
dependability, 2, 16
design, 5, 7–8, 58, 276, 335
disruption, 2, 7, 9, 103–107, 109, 111, 114, 115, 119, 213, 216, 272
disturbance, 255, 274–278
indeterminacy, 16
integrity, 310, 314
flexibility, 7, 156
multiple sourcing, 3, 126–128
regularity, 16
reliability, 16, 105, 302
resilience, 6, 22–23
risk, 1, 3–12, 53–64, 69, 74, 104, 137, 139, 140, 143, 148, 150, 151, 155
characteristics, 7, 129
security, 5, 10–12, 104, 293–295, 297, 300–303, 308, 309
single sourcing, 7, 132, 280, 317

vulnerability, 199, 212, 273, 278–279
Supply chain context, 4, 6, 15, 25, 55, 235, 250, 252, 278, 279
Supply chain disruptions-causes
 demand shifts, 105
 financial, 105
 human/organizational behaviour, 105
 information technology, 105
 legal/regulatory, 105
 natural, 105
 supplier problems, 105
Supply chain management (SCM), 1, 2, 5, 15–17, 25, 58, 69–71, 137, 139, 142, 150, 187, 190, 197, 203–205, 227, 231, 249, 250, 268
Supply chain mapping, 202, 210, 212
Supply chain models, 110
Supply chain nodes, 295
Supply chain operations reference (SCOR), 204
Supply chain processes–internal and external, 210, 213
Supply chain risk management (SCRM), 1, 5, 7–11, 53, 54, 56–58, 62, 63, 137, 139, 143
 strategies, 58
  alternate sourcing, 112
  buffer-related, 112
  component substitution, 112
  information-related, 11–112
Supply chain security management, 293, 294, 308–313
Supply chain security orientation (SCSO), 309–312
Supply chain stakeholders, 257–259
Symmetric information, 236–239, 241–243
Systematic risk, 133, 257, 260, 262, 265, 279

T
Third party logistics, 300
Time frame, 9, 184
Time series analysis, 114, 118, 119
Total quality management (TQM), 158, 190, 197, 315, 325
Transportation of dangerous goods, 4, 6, 83, 84
Triggering event, 106–109, 113, 116, 117, 226, 273, 274
Triple 'A' principles
 adaptability, 156
 agility, 156
 alignment, 156
TRIZ, 28

U
Uncertainty, 6, 35, 36, 38, 47, 53–55, 59, 64, 70, 73, 92, 95, 105, 141, 150
Unsystematic risk, 257, 260, 262, 265
Upside risks, 73
Upstream, 2–4, 10, 68–70, 73, 76, 80, 105, 106, 109–111, 117, 162, 200, 212, 275, 338, 340, 342

V
Value chain, 36, 37, 39, 40, 179, 203, 336, 342
Value engineering, 189–191
Value for money (VFM), 9, 147, 189
Value network analysis, 40, 41
Value networks, 6, 33, 35–42, 44, 50, 52
Vulnerability analysis, 16, 17, 20, 21, 23–32

Early Titles in the
INTERNATIONAL SERIES IN
OPERATIONS RESEARCH & MANAGEMENT SCIENCE
Frederick S. Hillier, Series Editor, *Stanford University*

Saigal / *A MODERN APPROACH TO LINEAR PROGRAMMING*
Nagurney / *PROJECTED DYNAMICAL SYSTEMS & VARIATIONAL INEQUALITIES WITH APPLICATIONS*
Padberg & Rijal / *LOCATION, SCHEDULING, DESIGN AND INTEGER PROGRAMMING*
Vanderbei / *LINEAR PROGRAMMING*
Jaiswal / *MILITARY OPERATIONS RESEARCH*
Gal & Greenberg / *ADVANCES IN SENSITIVITY ANALYSIS & PARAMETRIC PROGRAMMING*
Prabhu / *FOUNDATIONS OF QUEUEING THEORY*
Fang, Rajasekera & Tsao/ *ENTROPY OPTIMIZATION & MATHEMATICAL PROGRAMMING*
Yu / *OR IN THE AIRLINE INDUSTRY*
Ho & Tang / *PRODUCT VARIETY MANAGEMENT*
El-Taha & Stidham / *SAMPLE-PATH ANALYSIS OF QUEUEING SYSTEMS*
Miettinen / *NONLINEAR MULTIOBJECTIVE OPTIMIZATION*
Chao & Huntington / *DESIGNING COMPETITIVE ELECTRICITY MARKETS*
Weglarz / *PROJECT SCHEDULING: RECENT TRENDS & RESULTS*
Sahin & Polatoglu / *QUALITY, WARRANTY AND PREVENTIVE MAINTENANCE*
Tavares / *ADVANCES MODELS FOR PROJECT MANAGEMENT*
Tayur, Ganeshan & Magazine / *QUANTITATIVE MODELS FOR SUPPLY CHAIN MANAGEMENT*
Weyant, J./ *ENERGY AND ENVIRONMENTAL POLICY MODELING*
Shanthikumar, J.G. & Sumita, U./ *APPLIED PROBABILITY AND STOCHASTIC PROCESSES*
Liu, B. & Esogbue, A.O./ *DECISION CRITERIA AND OPTIMAL INVENTORY PROCESSES*
Gal, T., Stewart, T.J., Hanne, T./*MULTICRITERIA DECISION MAKING: Advances in MCDM Models, Algorithms, Theory, and Applications*
Fox, B.L. / *STRATEGIES FOR QUASI-MONTE CARLO*
Hall, R.W. / *HANDBOOK OF TRANSPORTATION SCIENCE*
Grassman, W.K. / *COMPUTATIONAL PROBABILITY*
Pomerol, J-C. & Barba-Romero, S. / *MULTICRITERION DECISION IN MANAGEMENT*
Axsäter, S. / *INVENTORY CONTROL*
Wolkowicz, H., Saigal, R., & Vandenberghe, L/*HANDBOOK OF SEMI-DEFINITE PROGRAMMING: Theory, Algorithms, and Applications*
Hobbs, B.F. & Meier, P. / *ENERGY DECISIONS AND THE ENVIRONMENT: A Guide to the Use of Multicriteria Methods*
Dar-El, E. / *HUMAN LEARNING: From Learning Curves to Learning Organizations*
Armstrong, J.S. / *PRINCIPLES OF FORECASTING: A Handbook for Researchers and Practitioners*
Balsamo, S., Personé, V., & Onvural, R./ *ANALYSIS OF QUEUEING NETWORKS WITH BLOCKING*
Bouyssou, D. et al. / *EVALUATION AND DECISION MODELS: A Critical Perspective*
Hanne, T./*INTELLIGENT STRATEGIES FOR META MULTIPLE CRITERIA DECISION MAKING*
Saaty, T. & Vargas, L./*MODELS, METHODS, CONCEPTS and APPLICATIONS OF THE ANALYTIC HIERARCHY PROCESS*
Chatterjee, K. & Samuelson, W. / *GAME THEORY AND BUSINESS APPLICATIONS*
Hobbs, B. et al. / *THE NEXT GENERATION OF ELECTRIC POWER UNIT COMMITMENT MODELS*
Vanderbei, R.J./*LINEAR PROGRAMMING: Foundations and Extensions, 2nd Ed.*
Kimms, A./*MATHEMATICAL PROGRAMMING AND FINANCIAL OBJECTIVES FOR SCHEDULING PROJECTS*
Baptiste, P., Le Pape, C. & Nuijten, W. / *CONSTRAINT-BASED SCHEDULING*
Feinberg, E. & Shwartz, A. / *HANDBOOK OF MARKOV DECISION PROCESSES: Methods and Applications*
Ramík, J. & Vlach, M. / *GENERALIZED CONCAVITY IN FUZZY OPTIMIZATION AND DECISION ANALYSIS*

Early Titles in the
INTERNATIONAL SERIES IN
OPERATIONS RESEARCH & MANAGEMENT SCIENCE
(Continued)

Song, J. & Yao, D. / *SUPPLY CHAIN STRUCTURES: Coordination, Information and Optimization*
Kozan, E. & Ohuchi, A. / *OPERATIONS RESEARCH/ MANAGEMENT SCIENCE AT WORK*
Bouyssou et al. / *AIDING DECISIONS WITH MULTIPLE CRITERIA: Essays in Honor of Bernard Roy*
Cox, Louis Anthony, Jr. / *RISK ANALYSIS: Foundations, Models and Methods*
Dror, M., L'Ecuyer, P. & Szidarovszky, F. / *MODELING UNCERTAINTY: An Examination of Stochastic Theory, Methods, and Applications*
Dokuchaev, N. / *DYNAMIC PORTFOLIO STRATEGIES: Quantitative Methods and Empirical Rules for Incomplete Information*
Sarker, R., Mohammadian, M. & Yao, X. / *EVOLUTIONARY OPTIMIZATION*
Demeulemeester, R. & Herroelen, W. / *PROJECT SCHEDULING: A Research Handbook*
Gazis, D.C. / *TRAFFIC THEORY*
Zhu/ *QUANTITATIVE MODELS FOR PERFORMANCE EVALUATION AND BENCHMARKING*
Ehrgott & Gandibleux/ *MULTIPLE CRITERIA OPTIMIZATION: State of the Art Annotated Bibliographical Surveys*
Bienstock/ *Potential Function Methods for Approx. Solving Linear Programming Problems*
Matsatsinis & Siskos/ *INTELLIGENT SUPPORT SYSTEMS FOR MARKETING DECISIONS*
Alpern & Gal/ *THE THEORY OF SEARCH GAMES AND RENDEZVOUS*
Hall/*HANDBOOK OF TRANSPORTATION SCIENCE - 2nd Ed.*
Glover & Kochenberger/ *HANDBOOK OF METAHEURISTICS*
Graves & Ringuest/ *MODELS AND METHODS FOR PROJECT SELECTION: Concepts from Management Science, Finance and Information Technology*
Hassin & Haviv/ *TO QUEUE OR NOT TO QUEUE: Equilibrium Behavior in Queueing Systems*
Gershwin et al/ *ANALYSIS & MODELING OF MANUFACTURING SYSTEMS*
Maros/ *COMPUTATIONAL TECHNIQUES OF THE SIMPLEX METHOD*
Harrison, Lee & Neale/ *THE PRACTICE OF SUPPLY CHAIN MANAGEMENT: Where Theory and Application Converge*
Shanthikumar, Yao & Zijm/ *STOCHASTIC MODELING AND OPTIMIZATION OF MANUFACTURING SYSTEMS AND SUPPLY CHAINS*
Nabrzyski, Schopf & Węglarz/ *GRID RESOURCE MANAGEMENT: State of the Art and Future Trends*
Thissen & Herder/ *CRITICAL INFRASTRUCTURES: State of the Art in Research and Application*
Carlsson, Fedrizzi, & Fullér/ *FUZZY LOGIC IN MANAGEMENT*
Soyer, Mazzuchi & Singpurwalla/ *MATHEMATICAL RELIABILITY: An Expository Perspective*
Chakravarty & Eliashberg/ *MANAGING BUSINESS INTERFACES: Marketing, Engineering, and Manufacturing Perspectives*
Talluri & van Ryzin/ *THE THEORY AND PRACTICE OF REVENUE MANAGEMENT*
Kavadias & Loch/*PROJECT SELECTION UNDER UNCERTAINTY: Dynamically Allocating Resources to Maximize Value*
Brandeau, Sainfort & Pierskalla/ *OPERATIONS RESEARCH AND HEALTH CARE: A Handbook of Methods and Applications*
Cooper, Seiford & Zhu/ *HANDBOOK OF DATA ENVELOPMENT ANALYSIS: Models and Methods*
Luenberger/ *LINEAR AND NONLINEAR PROGRAMMING, 2nd Ed.*
Sherbrooke/ *OPTIMAL INVENTORY MODELING OF SYSTEMS: Multi-Echelon Techniques, Second Edition*

Early Titles in the
INTERNATIONAL SERIES IN
OPERATIONS RESEARCH & MANAGEMENT SCIENCE
(Continued)

Chu, Leung, Hui & Cheung/ *4th PARTY CYBER LOGISTICS FOR AIR CARGO*

Simchi-Levi, Wu & Shen/ *HANDBOOK OF QUANTITATIVE SUPPLY CHAIN ANALYSIS: Modeling in the E-Business Era*

Gass & Assad/ *AN ANNOTATED TIMELINE OF OPERATIONS RESEARCH: An Informal History*

Greenberg/ *TUTORIALS ON EMERGING METHODOLOGIES AND APPLICATIONS IN OPERATIONS RESEARCH*

Weber/ *UNCERTAINTY IN THE ELECTRIC POWER INDUSTRY: Methods and Models for Decision Support*

Figueira, Greco & Ehrgott/ *MULTIPLE CRITERIA DECISION ANALYSIS: State of the Art Surveys*

Reveliotis/ *REAL-TIME MANAGEMENT OF RESOURCE ALLOCATIONS SYSTEMS: A Discrete Event Systems Approach*

Kall & Mayer/ *STOCHASTIC LINEAR PROGRAMMING: Models, Theory, and Computation*

Sethi, Yan & Zhang/ *INVENTORY AND SUPPLY CHAIN MANAGEMENT WITH FORECAST UPDATES*

Cox/ *QUANTITATIVE HEALTH RISK ANALYSIS METHODS: Modeling the Human Health Impacts of Antibiotics Used in Food Animals*

Ching & Ng/ *MARKOV CHAINS: Models, Algorithms and Applications*

Li & Sun/ *NONLINEAR INTEGER PROGRAMMING*

Kaliszewski/ *SOFT COMPUTING FOR COMPLEX MULTIPLE CRITERIA DECISION MAKING*

Bouyssou et al/ *EVALUATION AND DECISION MODELS WITH MULTIPLE CRITERIA: Stepping stones for the analyst*

Blecker & Friedrich/ *MASS CUSTOMIZATION: Challenges and Solutions*

Appa, Pitsoulis & Williams/ *HANDBOOK ON MODELLING FOR DISCRETE OPTIMIZATION*

Herrmann/ *HANDBOOK OF PRODUCTION SCHEDULING*

* A list of the more recent publications in the series is at the front of the book *

Printed in the United States
131967LV00005B/22-48/P